Scale-Up and Optimization in Preparative Chromatography

CHROMATOGRAPHIC SCIENCE SERIES

A Series of Textbooks and Reference Books

Editor: JACK CAZES

1. Dynamics of Chromatography: Principles and Theory, *J. Calvin Giddings*
2. Gas Chromatographic Analysis of Drugs and Pesticides, *Benjamin J. Gudzinowicz*
3. Principles of Adsorption Chromatography: The Separation of Nonionic Organic Compounds, *Lloyd R. Snyder*
4. Multicomponent Chromatography: Theory of Interference, *Friedrich Helfferich and Gerhard Klein*
5. Quantitative Analysis by Gas Chromatography, *Josef Novák*
6. High-Speed Liquid Chromatography, *Peter M. Rajcsanyi and Elisabeth Rajcsanyi*
7. Fundamentals of Integrated GC-MS (in three parts), *Benjamin J. Gudzinowicz, Michael J. Gudzinowicz, and Horace F. Martin*
8. Liquid Chromatography of Polymers and Related Materials, *Jack Cazes*
9. GLC and HPLC Determination of Therapeutic Agents (in three parts), *Part 1 edited by Kiyoshi Tsuji and Walter Morozowich, Parts 2 and 3 edited by Kiyoshi Tsuji*
10. Biological/Biomedical Applications of Liquid Chromatography, *edited by Gerald L. Hawk*
11. Chromatography in Petroleum Analysis, *edited by Klaus H. Altgelt and T. H. Gouw*
12. Biological/Biomedical Applications of Liquid Chromatography II, *edited by Gerald L. Hawk*
13. Liquid Chromatography of Polymers and Related Materials II, *edited by Jack Cazes and Xavier Delamare*
14. Introduction to Analytical Gas Chromatography: History, Principles, and Practice, *John A. Perry*
15. Applications of Glass Capillary Gas Chromatography, *edited by Walter G. Jennings*
16. Steroid Analysis by HPLC: Recent Applications, *edited by Marie P. Kautsky*
17. Thin-Layer Chromatography: Techniques and Applications, *Bernard Fried and Joseph Sherma*
18. Biological/Biomedical Applications of Liquid Chromatography III, *edited by Gerald L. Hawk*
19. Liquid Chromatography of Polymers and Related Materials III, *edited by Jack Cazes*
20. Biological/Biomedical Applications of Liquid Chromatography, *edited by Gerald L. Hawk*
21. Chromatographic Separation and Extraction with Foamed Plastics and Rubbers, *G. J. Moody and J. D. R. Thomas*
22. Analytical Pyrolysis: A Comprehensive Guide, *William J. Irwin*
23. Liquid Chromatography Detectors, *edited by Thomas M. Vickrey*
24. High-Performance Liquid Chromatography in Forensic Chemistry, *edited by Ira S. Lurie and John D. Wittwer, Jr.*
25. Steric Exclusion Liquid Chromatography of Polymers, *edited by Josef Janca*
26. HPLC Analysis of Biological Compounds: A Laboratory Guide, *William S. Hancock and James T. Sparrow*

27. Affinity Chromatography: Template Chromatography of Nucleic Acids and Proteins, *Herbert Schott*
28. HPLC in Nucleic Acid Research: Methods and Applications, *edited by Phyllis R. Brown*
29. Pyrolysis and GC in Polymer Analysis, *edited by S. A. Liebman and E. J. Levy*
30. Modern Chromatographic Analysis of the Vitamins, *edited by André P. De Leenheer, Willy E. Lambert, and Marcel G. M. De Ruyter*
31. Ion-Pair Chromatography, *edited by Milton T. W. Hearn*
32. Therapeutic Drug Monitoring and Toxicology by Liquid Chromatography, *edited by Steven H. Y. Wong*
33. Affinity Chromatography: Practical and Theoretical Aspects, *Peter Mohr and Klaus Pommerening*
34. Reaction Detection in Liquid Chromatography, *edited by Ira S. Krull*
35. Thin-Layer Chromatography: Techniques and Applications. Second Edition, Revised and Expanded, *Bernard Fried and Joseph Sherma*
36. Quantitative Thin-Layer Chromatography and Its Industrial Applications, *edited by Laszlo R. Treiber*
37. Ion Chromatography, *edited by James G. Tarter*
38. Chromatographic Theory and Basic Principles, *edited by Jan Åke Jönsson*
39. Field-Flow Fractionation: Analysis of Macromolecules and Particles, *Josef Janca*
40. Chromatographic Chiral Separations, *edited by Morris Zief and Laura J. Crane*
41. Quantitative Analysis by Gas Chromatography: Second Edition, Revised and Expanded, *Josef Novák*
42. Flow Perturbation Gas Chromatography, *N. A. Katsanos*
43. Ion-Exchange Chromatography of Proteins, *Shuichi Yamamoto, Kazuhiro Nakanishi, and Ryuichi Matsuno*
44. Countercurrent Chromatography: Theory and Practice, *edited by N. Bhushan Mandava and Yoichiro Ito*
45. Microbore Column Chromatography: A Unified Approach to Chromatography, *edited by Frank J. Yang*
46. Preparative-Scale Chromatography, *edited by Eli Grushka*
47. Packings and Stationary Phases in Chromatographic Techniques, *edited by Klaus K. Unger*
48. Detection-Oriented Derivatization Techniques in Liquid Chromatography, *edited by Henk Lingeman and Willy J. M. Underberg*
49. Chromatographic Analysis of Pharmaceuticals, *edited by John A. Adamovics*
50. Multidimensional Chromatography: Techniques and Applications, *edited by Hernan Cortes*
51. HPLC of Biological Macromolecules: Methods and Applications, *edited by Karen M. Gooding and Fred E. Regnier*
52. Modern Thin-Layer Chromatography, *edited by Nelu Grinberg*
53. Chromatographic Analysis of Alkaloids, *Milan Popl, Jan Fähnrich, and Vlastimil Tatar*
54. HPLC in Clinical Chemistry, *I. N. Papadoyannis*
55. Handbook of Thin-Layer Chromatography, *edited by Joseph Sherma and Bernard Fried*
56. Gas–Liquid–Solid Chromatography, *V. G. Berezkin*
57. Complexation Chromatography, *edited by D. Cagniant*
58. Liquid Chromatography–Mass Spectrometry, *W. M. A. Niessen and Jan van der Greef*
59. Trace Analysis with Microcolumn Liquid Chromatography, *Milos Krejcl*

60. Modern Chromatographic Analysis of Vitamins: Second Edition, *edited by André P. De Leenheer, Willy E. Lambert, and Hans J. Nelis*
61. Preparative and Production Scale Chromatography, *edited by G. Ganetsos and P. E. Barker*
62. Diode Array Detection in HPLC, *edited by Ludwig Huber and Stephan A. George*
63. Handbook of Affinity Chromatography, *edited by Toni Kline*
64. Capillary Electrophoresis Technology, *edited by Norberto A. Guzman*
65. Lipid Chromatographic Analysis, *edited by Takayuki Shibamoto*
66. Thin-Layer Chromatography: Techniques and Applications, Third Edition, Revised and Expanded, *Bernard Fried and Joseph Sherma*
67. Liquid Chromatography for the Analyst, *Raymond P. W. Scott*
68. Centrifugal Partition Chromatography, *edited by Alain P. Foucault*
69. Handbook of Size Exclusion Chromatography, *edited by Chi-san Wu*
70. Techniques and Practice of Chromatography, *Raymond P. W. Scott*
71. Handbook of Thin-Layer Chromatography: Second Edition, Revised and Expanded, *edited by Joseph Sherma and Bernard Fried*
72. Liquid Chromatography of Oligomers, *Constantin V. Uglea*
73. Chromatographic Detectors: Design, Function, and Operation, *Raymond P. W. Scott*
74. Chromatographic Analysis of Pharmaceuticals: Second Edition, Revised and Expanded, *edited by John A. Adamovics*
75. Supercritical Fluid Chromatography with Packed Columns: Techniques and Applications, *edited by Klaus Anton and Claire Berger*
76. Introduction to Analytical Gas Chromatography: Second Edition, Revised and Expanded, *Raymond P. W. Scott*
77. Chromatographic Analysis of Environmental and Food Toxicants, *edited by Takayuki Shibamoto*
78. Handbook of HPLC, *edited by Elena Katz, Roy Eksteen, Peter Schoenmakers, and Neil Miller*
79. Liquid Chromatography–Mass Spectrometry: Second Edition, Revised and Expanded, *W. M. A. Niessen*
80. Capillary Electrophoresis of Proteins, *Tim Wehr, Roberto Rodríguez-Díaz, and Mingde Zhu*
81. Thin-Layer Chromatography: Fourth Edition, Revised and Expanded, *Bernard Fried and Joseph Sherma*
82. Countercurrent Chromatography, *edited by Jean-Michel Menet and Didier Thiébaut*
83. Micellar Liquid Chromatography, *Alain Berthod and Celia García-Alvarez-Coque*
84. Modern Chromatographic Analysis of Vitamins: Third Edition, Revised and Expanded, *edited by André P. De Leenheer, Willy E. Lambert, and Jan F. Van Bocxlaer*
85. Quantitative Chromatographic Analysis, *Thomas E. Beesley, Benjamin Buglio, and Raymond P. W. Scott*
86. Current Practice of Gas Chromatography–Mass Spectrometry, *edited by W. M. A. Niessen*
87. HPLC of Biological Macromolecules: Second Edition, Revised and Expanded, *edited by Karen M. Gooding and Fred E. Regnier*
88. Scale-Up and Optimization in Preparative Chromatography: Principles and Biopharmaceutical Applications, *edited by Anurag S. Rathore and Ajoy Velayudhan*

ADDITIONAL VOLUMES IN PREPARATION

Handbook of Thin-Layer Chromatography: Third Edition, Revised and Expanded, *edited by Joseph Sherma and Bernard Fried*

Chiral Separations by Liquid Chromatography and Related Technologies, *Hassan Y. Aboul-Enein and Imran Ali*

Scale-Up and Optimization in Preparative Chromatography

Principles and Biopharmaceutical Applications

edited by
Anurag S. Rathore
Pharmacia Corporation
Chesterfield, Missouri, U.S.A.

Ajoy Velayudhan
Oregon State University
Corvallis, Oregon, U.S.A.

MARCEL DEKKER, INC.   NEW YORK • BASEL

ISBN: 0-8247-0826-1

This book is printed on acid-free paper.

Headquarters
Marcel Dekker, Inc.
270 Madison Avenue, New York, NY 10016
tel: 212-696-9000; fax: 212-685-4540

Eastern Hemisphere Distribution
Marcel Dekker AG
Hutgasse 4, Postfach 812, CH-4001 Basel, Switzerland
tel: 41-61-260-6300; fax: 41-61-260-6333

World Wide Web
http://www.dekker.com

The publisher offers discounts on this book when ordered in bulk quantities. For more information, write to Special Sales/Professional Marketing at the headquarters address above.

Copyright © 2003 by Marcel Dekker, Inc. All Rights Reserved.

Neither this book nor any part may be reproduced or transmitted in any form or by any means, electronic or mechanical, including photocopying, microfilming, and recording, or by any information storage and retrieval system, without permission in writing from the publisher.

Current printing (last digit):
10 9 8 7 6 5 4 3 2 1

PRINTED IN THE UNITED STATES OF AMERICA

To Csaba Horváth
Mentor and Friend

Preface

Preparative chromatography is arguably the most widely used purification technique in the pharmaceutical and biotechnological industries. Scaling up a preparative separation, however, continues to be seen as a difficult and time-consuming task. While this perception may have had some validity a generation ago, it is not accurate today. In fact, many industrial separations can be scaled up using one of the simple methods that have been described in the literature. The goals of this book are to bring wider recognition to these simple methods and to show that they are effective in many practical problems.

Another perception that persists is the view that results obtained in academia are either inapplicable or not applied to the day-to-day problems faced by industrial professionals. While there is no doubt that a greater degree of collaboration between academia and industry is desirable, it is nevertheless true that useful results obtained in academia have been, and are being, used in industry. Equally, the results and experience gained by industrial experts have informed and refined the academic approach to these problems. The first six chapters of this book describe the state of the art in methods and approaches to scale-up and optimization of preparative chromatography. These chapters are followed by a set of industrial case studies, which show how scale-up is carried out for a variety of important separations.

The editors contribute an overview that includes a simple quantitative approach as well as a discussion of the various practical aspects of scale-up. Signposts are provided to later chapters in which more detail is provided on specific topics that are discussed in earlier chapters.

Lightfoot et al. contribute an incisive overview to mass-transfer effects in the current chromatographic context. In addition to laying out problems that are currently being addressed, the chapter outlines fruitful lines of future study.

There has been a great deal of work on numerical optimization of nonlinear chromatographic separations, aimed at making it possible to obtain the production rates and yields for every operational mode for a given separation, thereby allowing a rational choice of the best operational mode for that separation. Felinger summarizes such work and indicates how these results can be used to advantage in guiding the scale-up process.

Simulated moving beds (SMBs) are becoming increasingly popular in preparative work, and may even have become the method of choice for some enantiomeric separations. Antia's chapter describes clearly and succinctly the issues involved in designing and controlling SMB units.

Watler et al. give a detailed analysis of the theoretical issues that must be addressed in successfully scaling up ion-exchange separations. The various theoretical issues—selectivity, bandspreading, optimization—are presented in a way that allows direct application to other separations.

Levison describes a variety of ways in which ion-exchange adsorbents are used at large scale, going beyond batch and column modes to suspended and fluidized beds. This detailed assessment provides a valuable complement to Watler et al.'s analysis of theory applicable to ion-exchange chromatography.

The second section of the book contains a set of industrial case studies that provide the practical approach taken in industry to scale-up of realistic separations. Fahrner et al. describe in detail the development of the crucial cation-exchange step in the kilogram-scale purification of a recombinant antibody fragment. Useful discussions of appropriate analytical methods and scouting for stationary phases are followed by the development of an optimized stepwise elution protocol. The purification of supramolecular assemblies is a topic of great current interest. Sagar et al. report their experiences with large-pore adsorbents in the purification of plasmid DNA. The entire purification sequence for producing monoclonal antibodies to tumor necrosis factor (TNF) is detailed by Ng. A variety of ion-exchange steps as well as a size exclusion step were used in this process. Miller and Murphy report on a normal-phase separation of a synthetic intermediate that was scaled up to the kilogram level. Guhan and Guinn describe the removal of a trace impurity from the parenteral drug alatrofloxacin. Rathore provides a detailed development of how a chromatographic process was implemented and optimized, based on the simple approach to scale-up described in Chapter 1.

Preface

The book as a whole is intended to provide a useful summary of the methods and approaches current in preparative chromatography, enabling the reader to transfer the insights gained here to her or his specific separation problem. In addition to serving as a reference for industrial practitioners of preparative chromatography, the material should be readily accessible to graduate students in many disciplines including chemistry, pharmacy, bioengineering, and chemical engineering.

As the editors of this book, we wish to express our indebtedness to Professor Csaba Horváth of Yale University, in whose research group we both received our doctorates, and to whom the book is dedicated. There is little doubt that science is one of the areas in which apprenticeship to an expert continues to be the premier way to gain, both explicitly and osmotically, insight into the field. Discussing scientific problems with Csaba—how to choose, approach, and solve them—was a vital part of our training in becoming independent researchers.

Anurag S. Rathore
Ajoy Velayundhan

Contents

Preface v
Contributors xi

Part I   Methods and Approaches

1. An Overview of Scale-Up in Preparative Chromatography 1
   Anurag S. Rathore and Ajoy Velayudhan

2. Interaction of Mass Transfer and Fluid Mechanics 33
   Edwin N. Lightfoot, John S. Moscariello, Mark A. Teeters, and Thatcher W. Root

3. Optimization of Preparative Separations 77
   Attila Felinger

4. Engineering Aspects of Ion-Exchange Chromatography 123
   Peter Watler, Shuichi Yamamoto, Oliver Kaltenbrunner, and Daphne N. Feng

5. A Simple Approach to Design and Control of Simulated Moving Bed Chromatographs 173
   Firoz D. Antia

6	Large-Scale Ion-Exchange Chromatography: A Comparison of Different Column Formats *Peter R. Levison*	203

Part II Case Studies

7	Development and Operation of a Cation-Exchange Chromatography Process for Large-Scale Purification of a Recombinant Antibody Fragment *Robert L. Fahrner, Stacey Y. Ma, Michael G. Mulkerrin, Nancy S. Bjork, and Gregory S. Blank*	231
8	Case Study: Capacity Challenges in Chromatography-Based Purification of Plasmid DNA *Sangeetha L. Sagar, Ying G. Chau, Matthew P. Watson, and Ann L. Lee*	251
9	Case Study: Purification of an IgG_1 Monoclonal Antibody *Paul K. Ng*	273
10	Case Study: Normal Phase Purification of Kilogram Quantities of a Synthetic Pharmaceutical Intermediate *Larry Miller and James Murphy*	289
11	Case Study: Development of Chromatographic Separation to Remove Hydrophobic Impurities in Altrafloxacin *Sam Guhan and Mark Guinn*	303
12	Case Study: Process Development of Chromatography Steps for Purification of a Recombinant *E. coli* Expressed Protein *Anurag S. Rathore*	317

Index *339*

Contributors

Firoz D. Antia, Ph.D. Department of Chemical Engineering Research and Development, Merck & Co., Inc., Rahway, New Jersey, U.S.A.

Nancy S. Bjork Department of Analytical Chemistry, Genentech, Inc., South San Francisco, California, U.S.A.

Gregory S. Blank, Ph.D. Department of Recovery Sciences, Genentech, Inc., South San Francisco, California, U.S.A.

Ying G. Chau Department of Chemical Engineering, Massachusetts Institute of Technology, Boston, Massachusetts, U.S.A.

Robert L. Fahrner Department of Recovery Sciences, Genentech, Inc., South San Francisco, California, U.S.A.

Attila Felinger, Ph.D. Department of Analytical Chemistry, University of Veszprém, Veszprém, Hungary

Daphne W. Feng, B.S. Department of Process Development, Amgen, Inc., Thousand Oaks, California, U.S.A.

Sam Guhan, Ph.D. Department of Bioprocess Research and Development, Pfizer Global Research and Development, Groton, Connecticut, U.S.A.

Mark Guinn, Ph.D. Department of Bioprocess Research and Development, Pfizer Global Research and Development, Groton, Connecticut, U.S.A.

Oliver Kaltenbrunner, Dipl. Ing. Dr. Department of Process Development, Amgen, Inc., Thousand Oaks, California, U.S.A.

Ann L. Lee, Ph.D. Merck Research Laboratories, Merck & Co., Inc., West Point, Pennsylvania, U.S.A.

Peter R. Levison, B.Sc., M.B.A., Ph.D. Department of Science and Technology, Whatman International Ltd., Maidstone, Kent, England

Edwin N. Lightfoot, Ph.D. Department of Chemical Engineering, University of Wisconsin–Madison, Madison, Wisconsin, U.S.A.

Stacey Y. Ma, Ph.D. Department of Analytical Chemistry, Genentech, Inc., South San Francisco, California, U.S.A.

Larry Miller Department of Global Supply Early Process Research and Development, Pharmacia Corporation, Skokie, Illinois, U.S.A.

John S. Moscariello Department of Chemical Engineering, University of Wisconsin–Madison, Madison, Wisconsin, U.S.A.

Michael G. Mulkerrin, Ph.D. Department of Analytical Chemistry, Genentech, Inc., South San Francisco, California, U.S.A.

James Murphy Pharmacia Corporation, Skokie, Illinois, U.S.A.

Paul K. Ng Purification Development, Department of Biotechnology, Pharmaceutical Division, Bayer Corporation, Berkeley, California, U.S.A.

Anurag S. Rathore, Ph.D. Department of Bioprocess Sciences, Pharmacia Corporation, Chesterfield, Missouri, U.S.A.

Thatcher W. Root, Ph.D. Department of Chemical Engineering, University of Wisconsin–Madison, Madison, Wisconsin, U.S.A.

Sangeetha L. Sagar, Ph.D. Merck Research Laboratories, Merck & Co., Inc., West Point, Pennsylvania, U.S.A.

Contributors

Mark A. Teeters, B.S. Department of Chemical Engineering, University of Wisconsin–Madison, Madison, Wisconsin, U.S.A.

Ajoy Velayudhan, Ph.D. Department of Bioresource Engineering, Oregon State University, Corvallis, Oregon, U.S.A.

Peter Watler, B.S., M.S., Ph.D. Department of Process Development, Amgen, Inc., Thousand Oaks, California, U.S.A.

Matthew P. Watson, B.S. Merck Research Laboratories, Merck & Co., Inc., West Point, Pennsylvania, U.S.A.

Shuichi Yamamoto, Ph.D. Department of Chemical Engineering, Yamaguchi University, Ube, Japan

Scale-Up and Optimization in Preparative Chromatography

1
An Overview of Scale-Up in Preparative Chromatography

Anurag S. Rathore
Pharmacia Corporation, Chesterfield, Missouri, U.S.A.

Ajoy Velayudhan
Oregon State University, Corvallis, Oregon, U.S.A.

I. INTRODUCTION

Preparative chromatography continues to be the dominant purification technique in the production of biological compounds, especially in the pharmaceutical and biotechnological industries. However, the conceptual complexity of a purely theoretical approach to preparative chromatography is formidable, because we are dealing with systems of highly coupled, nonlinear partial differential equations [1,2]. Although theoretical work is progressing, it can currently capture predictively only a few aspects of realistic biotechnological separations, especially given the extremely complex biochemical feedstocks often used in these applications. It is not entirely coincidental that the current approach to scale-up and optimization in industry is highly empirical. Although this is natural, especially given the constraints of process validation, the first few chapters of this book attempt to show that current theoretical understanding does give insight into the practical issues involved in scale-up and optimization. These chapters show that a careful combination of basic theory with experiments can reduce the time needed to achieve an effective scale-up of a realistic chromatographic separation.

It will be convenient to introduce some terminology [3] to clarify the ensuing discussion. The various kinds of physiochemical interactions that are used in chromatography to produce selectivity are called *modes of interaction*. Examples include electrostatic interactions in ion-exchange or ion chromatography, hydrophobic interactions in reversed-phase and hydrophobic interaction chromatography, and specific interactions in affinity chromatography. Once a mode of interaction has been chosen, the various ways in which a separation can be achieved (isocratic or gradient elution, stepwise elution, displacement, frontal analysis) are called *modes of operation*. For many separations, the best mode of interaction is easily specified, and scale-up or optimization focuses on the choice of mode of operation.

Finally, when the concentrations of all adsorbable components are low enough to lie within the linear or Henry's law region of their respective adsorption isotherms, the separation is called *linear*. Even if one component's concentration reaches the nonlinear region of its (multicomponent) adsorption isotherm for some fraction of the separation, the process is called *nonlinear*.

The basic ideas for scale-up and optimization given in the beginning chapters are applied to real separations in the subsequent chapters in which industrial case studies are presented. An issue of practical importance in a separation sequence is that of how to achieve the global optimum in the parameter of interest (typically maximum productivity or maximum recovery or minimum cost; mixed or combined optimization criteria are also possible). This issue is not discussed in detail in this chapter, because Chapter 3 deals with it comprehensively. Further, the case studies in subsequent chapters often allude to constraints from one separation step limiting or otherwise affecting the choice of conditions in other steps.

The structure of this chapter is as follows. An introductory section on method development places in perspective the various steps involved in arriving at an effective separation protocol at the bench scale. This is, of course, a necessary preliminary to scale-up, which by definition seeks to maintain upon scale-up the quality of a separation that has already been developed. Section III begins with heuristic rules for scale-up and then develops a simple quantitative model that clarifies when such heuristic rules can be used with reasonable accuracy. The issue of bed heterogeneity and its implications for scale-up are also discussed. In Section IV practical considerations characteristic of the various modes of interaction and operation are described briefly. Although considerations of space preclude the full discussion of all these issues, key points are brought out and important references in the literature are highlighted.

An Overview

II. METHOD DEVELOPMENT

Method development is a multistep process that precedes scale-up and operation at large scale. The general practice is to perform optimization at small scale due to relatively smaller requirements of material and resources as well as the ease of performing several runs in a parallel fashion.

A. Decoupling of Thermodynamics from Kinetics

A variety of parameters—choice of stationary and mobile phases, the particle size of the stationary phase, the column dimensions, the flow rate, the feed loading—affect the production rate and recovery obtained in a preparative separation. Trying to understand the interplay of all of these parameters simultaneously is a daunting task. In addition, testing all the various possibilities experimentally is likely to be extremely tedious and is impractical under typical industrial constraints. However, the following simplification is available to us at little cost. It is likely that equilibrium parameters (the choice of stationary and mobile phases, leading to selectivity) can be selected independent of "kinetic" parameters such as flow rate, feed loading, and particle size. Such a decoupling of thermodynamic and kinetic parameters is probably rigorously justifiable only in linear chromatography, but even in nonlinear chromatography it is likely that choosing the mobile and stationary phases first does not significantly decrease the attainable production rates and recoveries.

The first step is therefore to choose the most effective mode of interaction. This is often clear from the fundamental properties of the feedstock or the product. Other important factors include the objective and nature of the separation problem, literature precedents, and prior experience with the product. Inputs from the vendors of chromatographic media and instrumentation may also be useful at this stage. The strategy is depicted schematically in Fig. 1 [4].

B. Optimization of Thermodynamics at Bench Scale

We present here a simple and rapid approach to the thermodynamic component of method development. We take the view that for many separations the choice of stationary phase is far more important than the choice of mobile phase (this is particularly true of ion-exchange runs, where standard salts are used as mobile phase modulators). Of course, there are many cases where specific binding of various kinds can require the use of special additives for the mobile

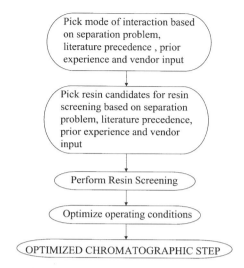

Figure 1. Strategy for optimization of a chromatographic separation.

phase, but we ignore these situations in order to make the general approach clear. We therefore intend to use a standard mobile phase, and we wish to screen a variety of stationary phases rapidly and equitably, i.e., we have reduced the problem to one of resin screening.

Once a list of resin candidates has been prepared, screening is performed to select the best resin to perform a particular separation. Selection of the resin for a chromatography step is perhaps the most important step in method optimization [4–9]. A resin screening protocol is illustrated in Fig. 2. In most cases the primary criterion for resin screening is selectivity. However, other screening criteria may also be identified and used depending on the particular separation problem.

The general approach is as follows (the specifics in what follows are for ion-exchange chromatography in the gradient mode of operation, but the arguments can easily be generalized to other contexts). The process takes place in two stages.

Stage 1. All stationary phases are packed into columns of identical size. If possible, all columns should be run at the same flow rate. This is not always practical (e.g., if the particle sizes available for different stationary phases are markedly different, then pressure drop constraints may limit the range of flow rates). Run a *test gradient* that spans a wide range of modulator levels, so that feed retention is facilitated. Make the gradient as shallow as

An Overview

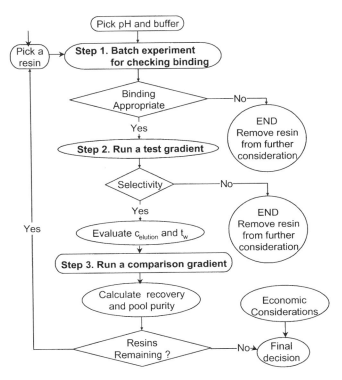

Figure 2. Resin screening protocol. (Reprinted courtesy LCGC North America, Advanstar Communications, Inc.)

practicable, in order to simultaneously get as much resolution as possible under these conditions. Stationary phases that exhibit little or no retention of the product are excluded at this stage. In addition, if almost no resolution is found between the product and the primary impurities, these stationary phases may be excluded. This latter decision should be made carefully, because the test gradient may not be a fair indicator of a sorbent's resolution. In other words, a sorbent may provide poor resolution of the product under the test gradient but high resolution under another gradient. Thus, the latter decision is to be made only if there is good reason to believe that this sorvent is unlikely to be effective.

Stage 2. For each of the stationary phases remaining, determine a tailored *comparison gradient* that is intended to show each sorbent under its most effective conditions for the given feed mixture. Parameters such as the feed loading and equilibration buffer should be kept the same for all stationary phases. If the flow rate was the same for all runs in stage 1, then it should be maintained in this stage. If different sorbents were run with different flow rates in stage 1, then use the same flow rate for each sorbent in this stage.

The comparison gradient is centered around the modulator concentration at which the product eluted in the test gradient in stage 1. Then, making the assumption that the band spreading of the peaks is inversely proportional to the gradient slope, all other parameters being constant, we have

$$\alpha w = \beta m \tag{1}$$

where α and w are respectively the gradient slope and product peak width in the test gradient, and β and m are the corresponding parameters in the comparison gradient. If we require that all comparison gradients have the same time (for standardization), then the starting and ending modulator concentrations (c_x and c_y, respectively) can be determined from the equations

$$c_x = c_{\text{elution}} - \frac{\alpha n t_w F}{2mV} \tag{2}$$

and

$$c_y = c_{\text{elution}} + \frac{\alpha n t_w F}{2mV} \tag{3}$$

where c_{elution} and t_W are respectively the concentration and time at which the center of the product peak eluted in the test gradient, n is the number of column volumes over which the comparison gradient is run, F is the flow rate, and V is the column volume. Note that if the beginning concentration c_x is found to be negative from Eq. (2), it is set to zero.

The comparison gradient provides an equitable way of comparing different stationary phases for the given feed mixture, because each stationary phase is provided with a gradient that is optimized for its particular retention behavior. Now the usual quantitative parameters of production rates and recovery and purity can be used to determine which stationary phase is best.

The simple approach described above provides a rapid way to choose the best resin. However, if the chromatography step is intended for operation at preparative scale, particulary for commercial manufacture, several other issues must be addressed before final resin selection. These include the cost of the resin, the physical and chemical stability of the resin at the bed height and the number of cycles to be used at the manufacturing plant, media availability with respect to the demand at commercial scale, resin lifetime, leaching

An Overview

of ligands, regulatory support files offered by the vendor, batch-to-batch variations in resin quality, etc.

It should be noted that the assumption that peak width is inversely proportional to gradient slope is an approximation and is not always expected to be valid (e.g., significant competition among the product and impurities for binding sites on the adsorbent could cause the assumption to fail). However, it is likely to be a reasonable approximation for many realistic separations. More detailed methods of this kind can be established (see, e.g., Quarry et al. [10]) but would usually require more data for each sorbent. Similarly, Jandera et al. [11] and Jandera [12] determined optimal gradients in normal and reversed-phase systems through numerical optimization of the governing equations; this is a significant advance in the field but is not yet at the level of accessibility where industrial practitioners would use it routinely. The method outlined here was chosen for its simplicity and ease of use in an industrial context.

This approach to resin screening is now demonstrated in detail for a practical separation problem. Rathore [4] showed that the stationary phase of choice for an anion-exchange separation was found rapidly using this approach.

1. Resin Screening for an Anion-Exchange Chromatography Column

This case study presents data obtained during the optimization of an anion-exchange chromatography column used in the process of purifying a protein molecule derived from microbial fermentation.

Nine anion-exchange resins—BioRad High Q, BioRad DEAE, Pharmacia DEAE FF, Pharmacia Q FF, Pharmacia Q HP, Whatman Q, Whatman QA52, Whatman DE53, and TosoHaas Q650M—were chosen for screening. All chromatography experiments were performed using an Äkta Explorer (Amersham Pharmacia Biotech). The buffer and other operating conditions were chosen on the basis of prior experience with the molecule. (Pre-equilibration buffer: 1 M Tris, pH 8.5. Equilibration buffer: 50 mM Tris, pH 8.5. Protein loading: 10 mg/mL resin.) Because the objective of this chapter is to lay out an efficient resin screening protocol and not to recommend a particular resin, the resins that were used will be referred to as resins 1–9 (not in the order in which they are named above). The optimum resin is expected to vary with the separation problem.

Columns were packed with 1 mL of resin and equilibrated for 30 min with the equilibration buffer. Equilibration was followed by loading of protein solution containing 1–2 times the intended protein loading for the respective column (mg protein/mL resin). After 30 min, a wash was performed with 5 mL equilibration buffer and the "flow-through" stream was collected. Protein

Table 1 Final Comparison of Resins Considered

Resin[a]	Binding[b]	Selectivity[b]	Recovery[b,c] (mAU/mL)	Pool purity[c] (%)	Final decision[b]
Resin 1	X				
Resin 2	√	X			
Resin 3	√	X			
Resin 4	√	X			
Resin 5	√	X			
Resin 6	√	√	X		
Resin 7	√	√	75.4	95.6	X
Resin 8	√	√	25.3	100	X
Resin 9	√	√	81.5	97.7	√

[a] Resins included BioRad DEAE, BioRad High Q, TosoHaas Q650M, Whatman Q, Whatman QA52, Whatman DE53, Pharmacia DEAE FF, Pharmacia Q FF, and Pharmacia Q HP, not numbered in this order.
[b] √ = Satisfactory column performance; X = unacceptable column performance.
[c] Based on measurements by anion exchange HPLC (AE-HPLC)
Source: Reprinted courtesy of LCGC North America, Advanstar Communications Inc.

was eluted with 10 mL of elution buffer (1 M NaCl in the equilibration buffer), and the eluant was collected separately. The flow-throughs and the eluants were analyzed for protein by UV absorbance at 280 nm and for its purity by anion-exchange high performance liquid chromatography (AE-HPLC). As shown in Table 1, it was found that most resins showed satisfactory binding characteristics with the product. Only resin 1 showed anomalous behavior in that the product was not retained under these conditions, so resin 1 was not considered further.

Next, columns were packed with 10 mL of the remaining eight resins, and separations were performed using an identical test gradient of 0–500 mM NaCl in 20 column volumes (CV) of equilibration buffer. Peak fractions were analyzed by AE-HPLC. Figure 3 illustrates the performance of resins 3, 4, and 5 under a test gradient of 0–500 mM NaCl in 20 CV. The Y axis in Figs. 3 and 4 denotes the peak area obtained upon analysis by AE-HPLC (mAU) per unit injection volume (µL). The flow velocity and the fraction sizes are given in the figure legends. It is evident that running identical gradients with different resins leads to very different elution profiles in terms of the peak

Figure 3. Column performance under test gradients. (Reprinted courtesy LCGC North America, Advanstar Communications, Inc.)

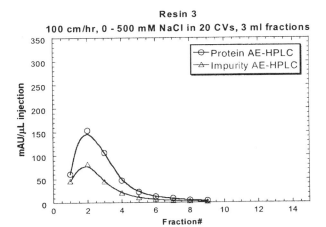
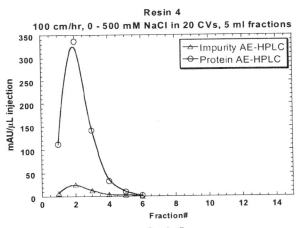
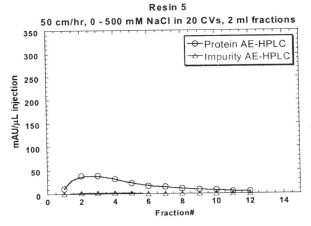

Table 2 Calculation of the Comparison Gradients

Resin[a]	Migration time (min)	Elution (mM)	Test gradient, start–end (mM)	Gradient volume (CV)	Column volume (mL)	Flow velocity (mL/min)	Comparison gradient, start–end mM
Resin 7	5.21	110	100–200	20	10	1.7	90–130
Resin 8	6.61	109	100–200	20	10	0.9	80–130
Resin 9	7.35	146	100–300	20	10	1.7	100–200

[a] Same as in Table 1.
Source: Reprinted courtesy of LCGC North America, Advanstar Communications Inc.

width and peak position in the overall gradient. The poor selectivity obtained with resins 3–5 led to their elimination from further consideration.

As listed in Table 1, it was found that only resins 6–9 showed satisfactory selectivity between the product and the impurity. Moreover, because resin 9 exhibited better resolution than resin 6 and they had identical matrix and ligand chemistry, the former was chosen over the latter for further consideration.

Comparison gradients were calculated according to the procedure described above for resins 7–9. Product recovery was defined as the sum of product peak areas (in mAU) in the pooled fractions (having >90% purity by AE-HPLC) per milliliter of injected sample. Pool purity was defined as the purity of the total pool formed by mixing the fractions that meet the pooling criteria. Table 2 shows the calculation of the comparison gradient for these three resins, and Fig. 4 illustrated the protein and impurity profiles that were obtained after fraction analysis by AE-HPLC.

Figure 4 reinforces the understanding that performing separations with the designed "comparison gradients" yields very similar elution profiles with different resins and leds to a fair comparison of resin performance. It also follows from Fig. 4 that resin 8 showed good purity but poor recovery. Resins 7 and 9 showed comparable recovery and pool purity. However, because of its better selectivity, resin 9 was chosen as the resin for this purification process and selected for further optimization of buffer pH, protein loading, feed flow rate, elution flow rate, gradient slope, and column length.

Figure 4. Column performance under comparison gradients. (Reprinted courtesy LCGC North America, Advanstar Communications, Inc.)

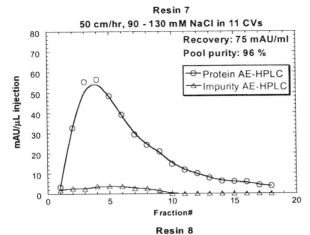
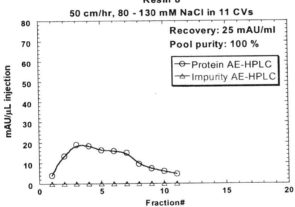
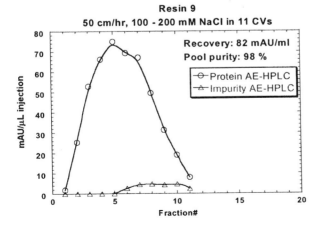

C. Optimization of Kinetics (Operating Conditions) at Bench Scale

Once the stationary and mobile phases have been chosen, we turn to the determination of optimal operating conditions, i.e., determination of the kinetic, as opposed to thermodynamic, contributions. Thus, the particle size and column dimensions are determined in these studies, along with the optimal gradient slope and feed loadings. The following general approach is suggested.

First, experiments are performed to evaluate the effect of various operating parameters that affect resin performance in terms of the selectivity and protein loading. These parameters may include the mobile phase conditions (pH, organic content, buffer composition, etc.) and the gradient slope and design. Optimum mobile phase conditions and the gradient design are chosen from the experimental data obtained.

Next, the effect of flow velocity and protein loading on the quality of separation is evaluated and, on the basis of resin performance, the bed height, protein loading, and flow velocity are chosen to obtain satisfactory resolution and cycle time. It is desirable that laboratory experiments be done at the bed height that will be used at pilot scale in order to obtain comparable column performance at large scale.

A detailed analysis of the interaction among these kinetic parameters is complicated and is not described here. Many of the underlying issues are brought out clearly by Felinger in his chapter on optimization (Chap. 3). In industrial practice, a heuristic approach similar to the one just described is often used. It is likely to produce effective, if not necessarily optimal, operating conditions in the hands of an experienced practitioner. More details of these practical approaches are given in several of the industrial case studies in this book.

This separation of very large molecules and particles such as viruses is an important industrial topic and is beginning to be addressed in the literature [13,14]. However, the field is still in its infancy and is likely to change rapidly. We therefore do not feel that it would be appropriate to attempt a summary here, and we refer the reader to the growing literature on this subject.

III. THEORETICAL CONSIDERATIONS IN SCALE-UP

A. Physical Overview

The performance of a chromatography column depends on a variety of design and operating factors. In order to have a successful scale-up it is desirable to

An Overview

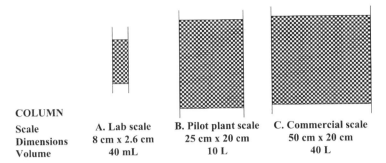

COLUMN			
Scale	A. Lab scale	B. Pilot plant scale	C. Commercial scale
Dimensions	8 cm x 2.6 cm	25 cm x 20 cm	50 cm x 20 cm
Volume	40 mL	10 L	40 L

Figure 5. Three different scales of columns used frequently during process development.

maintain kinetic (particle size, pore size, ligand chemistry, temperature, mobile phase) and dynamic (bed height, flow velocity, packing density) equivalence between the chromatography columns used in the laboratory and the pilot plant. This objective can be accomplished by using identical stationary and mobile phases in the two columns and operating them at identical bed height, linear flow velocity, protein loading (mg protein per mL of resin), feed conditions, gradient length, and gradient slope [6]. To handle the increased volume of load at pilot scale, the most common procedure used to increase column volume is to increase the column diameter so that the column volume increases proportionately [8,15]. This keeps the residence time of the product constant and avoids causing any product stability issues.

Figure 5 is a schematic illustration of the three sizes of columns that are often used at laboratory, pilot plant, and commerical scales. Scouting experiments in the laboratory are mostly done in small columns to conserve the materials and also because several experiments can be done in parallel simultaneously at lab scale. However, as discussed above, it is extremely important to maintain bed height constant while scaling up, so the best approach is to perform the final optimization steps at the bed height that will later be used at the pilot plant and commercial scale. This approach is illustrated in Fig. 6 and 7.

These general considerations are frequently used in industry as the basis for scale-up. In the next section, a quantitative analysis is given that shows when such simple "volumetric" scale-up can be used and describes alternatives that are appropriate when the column length must be changed on scale-up.

The van Deemter equation is widely used to characterize band broadening in a chromatography column and is expressed as

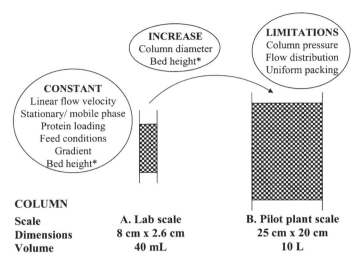

Figure 6. Scaling from laboratory (or bench) to pilot-plant scale.

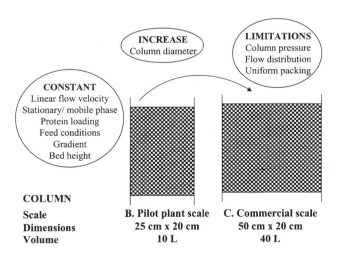

Figure 7. Scaling up from pilot-plant scale.

An Overview

$$H = A + B/u + Cu \tag{4}$$

where u is the linear flow velocity, H is the plate height of the column, and A, B, and C are constants. The plate height H is equal to the length of the column divided by the total number of plates N, so H is smaller for a more efficient column. A reflects the quality of the packing of the column and is independent of the linear flow velocity. A is small when the column is packed well and is homogeneous throughout its length. B is a measure of the band broadening due to longitudinal diffusion of the sample components along the edge of their respective bands as they travel across the column. It decreases with increasing linear flow velocity because the sample components spend less time undergoing diffusion inside the column. C includes contributions from the binding kinetics (adsorption/desorption) as well as the mass transfer of the sample components to and from the packing particles.

Preparative chromatography is usually carried out at high flow velocity in order to increase throughput. Then the C term usually dominates Eq. (4), leading to the simplified form

$$H = Cu, \quad \text{or, equivalently,} \quad N = L\,(1/Cu) \tag{5}$$

In an ideal case, when the column packing and operating conditions are kept the same while scaling up (C is a constant), the scale-up involves just a volumetric increase in column dimensions. For such a case, Eq. (5) can be rewritten as

$$L/u = CN \tag{6}$$

To preserve the efficacy of separation, the total number of plates is to be kept constant, so it follows from Eq. (6) that if the bed height needs to be increased or decreased for some reason (e.g., pressure drops too high), the linear flow velocity might also be altered appropriately so as to keep the ratio of L/u constant. This ensures a constant number of plates in the column (N), and the column performance is maintained. This simple analysis is expanded and generalized in the following section. In particular, if the particle size needs to be changed upon scale-up (for economic or other reasons), the more general treatment must be used.

This very simple physical introduction to scale-up sets the scene for a straightforward quantitative analysis of the problem in the next section.

B. Simple Scale-Up Calculation

The basic idea behind scale-up is to preserve the quality of the separation achieved at small scale [16,17]. Implicit in this approach is the admission that we are not yet able to determine optimal operating conditions *a priori* for different scales of operation. Thus, we settle for determining effective, near-optimal operating conditions at bench sale. Effective scale-up rules should then produce comparable results at larger scale. (The alternative approach of finding optimal operating conditions is currently practicable for some important classes of separation problems; this approach is discussed in Chapter 3).

A typical scale-up from laboratory to pilot plant is on the order of 50–100-fold. This is frequently followed by a 10–50-fold scale-up from pilot plant to final commercial manufacturing scale.

The usual approach is to hold the plate count constant upon scale-up and increase the feed volume and column volume proportionately. This approach was originally based on the assumption of linear adsorption. Later in this section we discuss how this assumption can be relaxed.

If the subscripts b and l are used to describe parameters at bench and large scale, respectively, we have

$$N_l = N_b \tag{7}$$

$$\frac{V_{\text{feed},l}}{V_{\text{column},l}} = \frac{V_{\text{feed},b}}{V_{\text{column},b}} \tag{8}$$

If band spreading is dominated by pore diffusion, as is often the case in realistic separations [18–20], then the plate count can be described by

$$N \sim \frac{L}{u d_p^2} \tag{9}$$

where L is the column length, u the mobile phase linear velocity, and d_p the particle diameter. The proportionality constant includes geometrical factors such as the phase ratio and thermodynamic factors such as the retention factor. This result can be derived from the van Deemter equation [16]. Combining Eqs. (7) and (9) we get

$$\frac{L_l}{u_l d_{p,l}^2} = \frac{L_b}{u_b d_{p,b}^2} \tag{10}$$

This represents one constraint on the three variables L_l, u_l, $d_{p,l}$. Recall that this approach is based on mimicking bench-scale results at large scale; the variables L_b, u_b, $d_{p,b}$ are therefore assumed to be known.

Quantifying the pressure drop across the columns give another result; we use Darcy's law in the form

An Overview

$$u = k \frac{d_p^2}{\mu} \frac{\Delta p}{L} \tag{11}$$

where k is the permeability *without* the dependence on particle diameter, which has been factored out, and μ is the mobile phase viscosity. In general, best results are obtained at the maximum allowable pressure drop [2]. If the maximum permissible pressure drop is different at bench and large scales, there follows

$$\frac{u_l L_l}{d_{p,l}^2} = P \frac{u_b L_b}{d_{p,b}^2} \tag{12}$$

where P is the ratio of maximum pressure drops at large and bench sale. In the common case where $P = 1$, the result becomes

$$\frac{u_l L_l}{d_{p,l}^2} = \frac{u_b L_b}{d_{p,b}^2} \tag{13}$$

Dividing Eq. (10) by Eq. (13) gives the simple expression:

$$u_l = u_b \tag{14}$$

Thus equality of plate counts and maximum pressure drops leads to equality of mobile phase velocity across scales. Substituting Eq. (14) into either Eq. (10) or Eq. (13) gives the familiar result

$$\frac{d_{p,l}^2}{L_l} = \frac{d_{p,b}^2}{L_b} \tag{15}$$

Typically, the choice of particle size at the large scale is limited by cost or availability. Once a particle size is chosen, equation (15) specifies the column length. An approximate theoretical calculation for the optimal d_p^2/L is given in Guiochon et al. [2]; this can also be used to give another estimate of the column length, given the particle size. The chapter by Felinger (Chap. 3) discusses such optimal calculations in detail.

Finally, in order to determine the column diameter at large scale, Eq. (8) can be rewritten as

$$\frac{V_{\text{feed},l}}{V_{\text{feed},b}} = \sigma = \frac{V_{\text{column},l}}{V_{\text{column},b}} = \frac{L_l}{L_b} \left(\frac{D_{c,l}^2}{D_{c,b}^2} \right) \tag{16}$$

Here, σ is the scale-up factor (which must be specific before scale-up can begin) and D_c is the column diameter. Because the column length at large scale has been determined from Eq. (15), the column diameter is obtained from the equation

$$\frac{D_{c,l}}{D_{c,b}} = \left(\sigma \frac{L_l}{L_b}\right)^{1/2} \tag{17}$$

When the particle diameter is kept constant on scale-up, we obtain the results for scale-up by "volume overloading": From Eq. (15), the column length remains constant (in addition to the particle diameter and the mobile phase velocity); thus, scale-up consists simply of increasing the column diameter by a factor of $\sqrt{\sigma}$.

These results have been obtained for pore diffusion as the process that dominates band spreading. Analogous results for the case where external mass transfer (film diffusion) is controlling are given by Pieri et al. [21], Grushka et al. [16], and Ladisch and Velayudhan [22]. The case where pore diffusion and film diffusion are comparable has been considered briefly by Ladisch and Velayudhan [22]. A more general treatment including the effects of axial dispersion (which is typically negligible for liquid chromatography) is found in Lee et al. [23].

Although the approach taken above was based on linear adsorption [which allowed the specialization of the van Deemter equation to obtain Eq. (9)], extensions to nonlinear adsorption are possible. Knox and Pyper [24] showed that Eq. (15) applied for scale-up of noninteracting compounds whose bands in isocratic elution can be approximated as right triangles (i.e., band spreading is dominated by isotherm nonlinearity). Wankat and Koo [25] showed that Eq. (6) holds for single compounds that have nonlinear single-component isotherms. Golshan-Shirazi and Guiochon [26,27] demonstrated that Eq. (12) also holds for two compounds with binary Langmuirian isotherms. Wantak [28] presents a general scale-up argument based on the constancy of plate count at both scales, which also results in Eq. (12) for pore-diffusion-controlled runs and applies even for arbitrary multicomponent isotherms. In fact, it is quite possible that similar results hold for multicomponent adsorption as long as the isotherms are locally concave downward (so that the leading edge of a band is self-sharpening and the trailing edge shows a "proportionate pattern"). The approach described above is therefore a good starting point for scale-up, although variations due to more complex adsorption behavior should be kept in mind.

There has been some discussion in the past about whether small particles are needed in scale-up, particularly when the column is highly overloaded. This is based on the view that under conditions of strong overloading the band spreading caused by isotherm nonlinearity dominates that caused by "kinetic" factors such as film and pore diffusion that depend on particle size. There is an important distinction to be drawn here. Under such conditions that the product band can be separated from its nearest impurities "thermodynamically," i.e., when a sequential stepwise elution schedule can be found such that no impurity elutes in the step in which the product elutes, then it is clear that plate count and its causal kinetic factors are unimportant. Such thermodynamic separations are sometimes possible when on–off binding of the feed components occurs, i.e., the feed components are either very strongly bound or almost completely unbound. This kind of "all-or-nothing" adsorption often occurs for macromolecules [29,30]. The same mechanism is often exploited in solid-phase extraction protocols. Under these conditions, it is clear that large particles can be used with impunity, at both bench scale and process scale. The only problem in scaling up such a separation is the possibility of overloading the column too heavily. Because the product is selectively displaced from binding sites by more retentive impurities, it is possible that the product may move faster than expected and thus emerge in more than one step of the stepwise elution schedule. This problem can be avoided by reducing the loading or increasing the column length.

However, there are many separations of practical importance for which such stepwise elution protocols *cannot* be found. Then the co-migration of the product peak with one or more impurities must be considered, and now kinetic factors play a vital role in determining the extent of mixing between adjacent peaks in the chromatogram [2,31] and thus recovery and production rates. In such cases, isocratic elution, gradient elution, and displacement chromatography are used and are reasonably well described by the general scale-up equations developed above. More detailed models, especially for gradient elution, are presented in Chapter 4 of this volume by Watler et al.

C. Constancy of Phase Ratio with Scale

An important assumption in the calculations above was that the phase ratio (and therefore the interstitial and intraparticulate porosities) remained constant upon scale-up. This is not always a good assumption when the overall scale-up factor is very large (above 100). A recent example of such variations in phase ratio is given in Heuer et al. [32]. In practice, it is worthwhile to estimate the phase ratio experimentally at each scale and use the results above with caution if the phase ratio changes appreciably with scale.

Experimental techniques to avoid such changes in packing structure as the column diameter increases include axial compression, radial compression, and annular expansion (mixed radial and axial compression). Useful reviews of these packing methods are found in Jones [33,34] and Colin [35]. In many cases, these approaches have results in improved column performance.

Fundamental and applied work on packing continues, and results from soil physics and other fields are being applied to the chromatographic problem (e.g., Bayer et al. [36] and Cherrak and Guiochon [37]). It is too early to say to what extent packing heterogeneities can be removed by further improvements in technology, but such observations raise the question of whether a lumped parameter like the phase ratio is sufficient to capture packing geometry. Close collaboration between manufacturers, industrial users, and academics is needed to arrive at highly effective designs for process columns. The next section includes some practical observations on packing properties as a function of scale under the heading of bed stability.

An assessment of these and related issues from a fundamental viewpoint is presented in Chapter 2 of this book by Lightfoot et al.

IV. PRACTICAL CONSIDERATIONS IN SCALE-UP
A. Practical Guidelines

In practice, several issues must be kept in mind when attempting to scale up a separation. In this subsection, these issues are dealt with briefly. They will recur constantly in the case studies discussed in subsequent chapters.

1. Bed Stability (Physical)

In a laboratory scale column the column wall offers support to the column bed and contributes to the stability of the column. However, when the column is scaled up and its diameter increases, the wall support contribution to bed stability starts to decrease. For column diameters greater than 25–30 cm, the lack of wall support may become an issue and could cause redistribution of packing particles and settling of the bed. The total drag force on the packing particles is a function of the liquid velocity, the liquid viscosity, and the bed height. The supporting force that keeps the particles in place decreases with increasing column diameter as a smaller fraction of the particles are supported by the column wall. Therefore, for large-scale columns with compressible packings, the maximum velocities are restricted and decrease with increasing column diameter under identical bed height and pressure drop, and the situation worsens with duration of column use [38–41]. This phenomenon is more prominent for nonrigid gel materials

An Overview

and is often reversible within limits but almost always with a marked hysteresis [39,42]. As illustrated in Fig. 6, the issue of physical stability of the column bed becomes particularly significant at large scale because the bed height is larger than in the lab and the column is put to more frequent reuse during commercial manufacturing. The bottom of the column is most vulnerable because it feels the pressure drop across the column as well as that due to the weight of the column itself, which can be appreciable for large columns. These effects must be taken into account for robust column design. For cases where bed compression is a problem and the maximum permissible bed height is smaller than the minimum required to obtain satisfactory separation, a suggested solution is to use stacked columns [42]. This reduces the pressure difference across any single section of the column and permits the use of smaller particles and nonrigid gels to obtain enhanced resolution.

2. Bed Stability (Chemical)

Chemical stability of the packing material includes any factors that may result in deterioration of the column performance over a period of use. It may be the leaching of ligands into the mobile phase as often in affinity chromatography destruction of the matrix in the mobile phases used for column operation, regeneration, or storage (e.g., silica packings at high pH) or irreversible binding at the packing surface [41]. The issue of chemical stability of the column bed becomes particularly significant when the column is reused many times during commercial manufacturing.

3. Product Loading

The product loading (milligrams of product loading per milliliter of resin) is generally held constant during scale-up. In most cases the resolution is found to decrease with increasing product loading after the loading has reached a certain level. Further, this behavior is more prominent when the paricle size is small. To ensure a successful scale-up and successful operation at large scale, studies must be conducted at lab scale to determine the maximum product loading with which satisfactory resolution can still be achieved. It is common to operate the column at 80–90% of this maximum loading.

4. Gradient Separations

Gradient elution is widely used owing to its ability to provide higher efficiency, reduced process times and solvent consumption, and a concentrated product stream. However, as process scale increases, buffer volumes increase also, and it becomes increasingly difficult to form accurate and reproducible gradients. Thus, either the possibility of performing step gradients should be explored or the ability of the chromatography skid to perform adequate buffer mixing and form controlled gradients must be evaluated [15].

5. Flow Distribution

For large diameter columns, uniform flow distribution at the column head may become difficult to achieve. This may result in deviations from the desired plug flow and lead to peak tailing. Use of a flow distributor at the column inset is generally found to be the most effective way of ensuring uniform flow distribution in the column [41]. The rational design of inlet and outlet headers to ensure uniform distribution is discussed at length in Chapter 2.

6. Packing Quality

Packing large columns such that the resulting packed column is homogeneous is very critical for obtaining uniform flow distribution. Channeling inside the column often leads to peak tailing and/or peak splitting. A variety of approaches have been developed by the different chromatographic equipment vendors to alleviate this problem. The most popular technique at preparative scale is axial compression and the use of self-packing columns [35]. Some degree of compression has been known to enhance resolution [41]. The connection between packing and the phase ratio used in quantitative scale-up was touched upon in the previous section.

7. System Design

The system dead volume arising from the piping and other support equipment for chromatography columns such as the valves, flow meters, air sensors, and tubings is much larger at pilot and particularly manufacturing scale than at lab scale. This leads to dilution effects and higher pressure drops as well as additional band broadening, so the impact these factors may have on the overall column performance must be evaluated. The general guidelines are to keep the system dead volume to the minimum, to have bypasses through devices such as air traps and filters for use during sample load, and to choose tubing diameter to achieve turbulent flow so as the reduce undesirable axial mixing [43]. Further, the chromatographic system should be designed such that all inlet sources are at or above the level of the column, whereas all outlet sinks

An Overview

are at or below the level of the column. This ensures that the column is not being operated against a hydrostatic pressure head.

8. Fraction Collection

The peak width and shape as seen in the chromatogram depends on several factors such as the column dimensions, extracolumn effects, operating conditions, and sample volume [42]. Thus, even if the scale-up is carried out following a well-thought-out methodology, it is still very likely that the peak width and shape may differ from that obtained at lab scale. Thus, the fraction collection strategy must be revisited at preparative scale based on column performance at that scale. The critical monitoring device, e.g., the UV detector, should be placed as close to the fraction collector as possible to ensure good representation of the process stream. Also, it is good to have a fraction collector that can collect fractions based on several process parameters, such as UV absorbance, conductivity, time, volume, first or second derivative of the signal, etc.

9. Media Availability

Although selectivity for a separation may be the primary criterion for selection of a resin for analytical separation, several other factors need to be considered before a resin is selected for preparative separation. These include the availability of large quantities of the resin, cost, continuity of supply, batch-to-batch consistency, column lifetime, and support documentation to aid in regulatory filing [44].

10. Costing

The cost of the feedstock is generally not given adequate consideration when the process optimization is carried out at bench scale. However, as the process is scaled up, the process should be modeled and raw material and facility costs must be examined. Resin costs usually account for the biggest contribution to the raw material costs. Thus, although a certain resin may offer the best selectivity at bench scale, it may be too expensive, under linear scale-up, at process scale [8].

11. Sample Pretreatment

Often cells or inclusion bodies undergo a rupture step by homogenization or some other mechanism prior to a chromatography step with the purpose of capturing the product from the cell culture or fermentation broth. In these situations the process stream is replete with lipids, nucleic acids, proteins, and other macromolecular contaminants. These impurities affect the passage of the product through the column and have a very significant effect on column performance. Because most cell-rupture operations are more effective at large scale than at bench scale, pretreatment of the process stream at large scale to remove all the interfering contaminants may become crucial to ensuring satisfactory and robust column performance [8].

12. Scale-Down

Although our focus is on scale-up, it is worth mentioning that some issues related to validation are often addressed by scaling down. Thus, using a small-scale study for viral clearance is often faster, safer, and cheaper while being at least as accurate as larger scale studies [45]. Many of the scale-up approaches discussed above apply, *mutatis mutandis*, to scale-down.

B. Other Modes of Operation

Although isocratic and gradient elution are the most common modes of operation and displacement chromatography has a small but significant role, there are other modes of interest at large scale. In this context, frontal chromatography has always played an important role under the guide of "adsorption steps" in a variety of applications, especially in the chemical industries. The process of feed introduction in isocratic, gradient, and displacement runs is nothing more than frontal chromatography, so it is clearly an important part of the run. Here, expanded bed chromatography and simulated moving bed (SMB) chromatography are discussed briefly. Though the basic scale-up rules mentioned above are applicable to both these techniques, there are some unique considerations that are worth highlighting.

Expanded Bed Chromatography

An expanded bed consists of specially designed packing particles that are fluidized in a column with controlled flow distribution so as to provide a large number of plates (high mass transfer) with minimal back-mixing. Thus, although expanded bed chromatography (EBC) offers more plates than batch adsorption, it also allows better utilization of binding capacity of the adsorbent [46].

Scale-up in EBC is a little more complicated than in other modes of chromatography because of the additional requirements of maintaining the stability of the expanded bed. The flow distribution, flow velocity, composi-

An Overview

tion, and other physical/chemical properties of the feed and the particle distribution and the physical/chemical stability of the adsorbent can all have a significant effect on the bed stability. Conditions optimized for high productivity may not be able to provide long-term stability, so a process optimized at small scale should be carefully evaluated with these two objectives in mind [47].

Simulated Moving Bed Chromatography

Simulated moving bed chromatography (SMB) is a continuous chromatographic process in which a mobile phase and the sample components are injected into and withdrawn from a ring of chromatography columns at points that are rotating between the columns during the process. SMB is slowly becoming the technique of choice for performing efficient enantiomeric separations [48].

The key to successful SMB operation is proper selection of the operating flow rates and the valve-switching times of the feed and eluant streams [49]. Pumps typically used in this technique provide control of flow rates of better than 1%. These are important both for stability of the different zones and for achieving satisfactory separation. Just as in other modes of chromatography, the efficiency of an SMB separation is negatively impacted by extracolumn volume, which in this case consists of the volume of the recycling pump, flow meter, pressure sensor, and valves. Antia (Chapter 5) describes a simple and effective approach to the design and control of SMB separations.

C. Practical Considerations for Modes of Interaction

As with modes of operation, there are many practical issues peculiar to each mode of interaction. The general features of each mode are discussed briefly in this section. Again, the relevance of these issues in the design of practical separations will be highlighted by the design choices made in the case studies described in the following chapters.

1. Ion-Exchange Chromatography

Separation in ion-exchange chromatography (IEC) takes place because of differential ionic interactions between the charged ligands on the stationary phase and the charged sample components in the feed in the presence of aqueous buffer solution. Elution is then performed with increasing salt concentration by altering the pH of the mobile phase, resulting in weakening of the ionic forces, with the components eluting in the order of increasing binding strength with the stationary phase.

Ion-exchange chromatography is the most widely used mode of chromatography for protein separation. This is due to the high dynamic capacities and low relative costs of the IEC resins, simple buffers used, high usable flow rates, and the robustness, scalability, and ease of operation of most IEC methods. Most IEC resins have large volumetric capacities, with operation limited only by the total concentration of the product and contaminants in the feed. Scale-up in IEC can be performed following the generic guidelines mentioned above, and it is the technique of choice as an early processing step due to the large qualities of material that have to be handled [50,51].

Watler et al. (Chap. 4) and Levison (Chap. 6) provide a clear and comprehensive discussion of the use of IEC in preparative separations.

2. Hydrophobic Interaction Chromatography

Separation in hydrophobic interaction chromatography (HIC) is based on differences in hydrophobicity of the different sample components. At high salt concentrations, the solubility of the hydrophobic product is reduced and the hydrophobic side chains of the sample component associate with the hydrophobic ligands on the stationary phase [52]. Elution can be performed by reducing the polarity of the mobile phase in a continuous or stepwise manner. Because of the differences in the mechanisms of separation of HIC and IEC, HIC is frequently used to process the eluant stream from an IEC column.

Scale-up in HIC can be performed by following the generic guidelines outlined above [53–55]. However, there are several considerations peculiar to HIC that must be kept in mind. Denaturation of the product molecule may occur at high salt concentrations during the separation process; the mobile phase may be too viscous and thus impair the accuracy and reproducibility of gradient formation, apart from limiting the usable bed height; and temperature variations may change column performance considerably.

3. Size Exclusion Chromatography

The separation in size exclusion chromatography (SEC) is due to differences in the sizes and shapes of the sample components as they are carried by the mobile phase through the three-dimensional porous structure of the stationary phase. In most cases the maximum bed height and diameter that can be used are limited by the physical stability of the gel. This is resolved by using a series of smaller columns instead of a single long column during scale-up. Silica and agarose matrices are most commonly used in SEC and often contain negatively charged moieties at the surface that may cause adsorption of positively charged proteins and the exclusion of negatively charged proteins at low ionic concentrations.

An Overview

In SEC, although the separation time increases linearly with column length as in other forms of chromatography, the resolution increases only as the square root of the length. Also, the resolution of the sample components is more sensitive to the width of the sample band than the flow velocity. One of the suggested strategies for SEC optimization and scale-up is to select the bed height and linear flow velocity to obtain the desired cycle time and satisfactory resolution [56,57]. Next, the width of the sample zone is varied to fine-tune resolution. Finally, the column diameter is changed to meet the productivity requirements for the column.

Feed volume in SEC normally varies between 1% and 5% of the total column volume to obtain satisfactory resolution for most large-scale separations and up to 30% for desalting. Resins used in SEC have low binding capacities and are often compressible, so the separations are generally performed at very low flow velocities, and all the issues mentioned earlier in Section IV.A.1 apply. Therefore, SEC is not considered by most an ideal technique for scale-up except for desalting or buffer exchange or as a final polishing step when the volumes and sample quantities are small enough (as in purification of human serum albumin).

4. Reversed-Phase Chromatography

Separation in reversed-phase chromatography (RPC) is based on the differences in the hydrophobicities of the different sample components. The hydrophobic groups on the surface of the sample molecules bind with the very hydrophobic stationary phase particles in the presence of an apolar mobile phase. Elution occurs when a mobile phase is used with increasing concentration of an organic phase such as acetonitrile, methanol, or isopropanol. The components elute in the order of their hydrophobicity, and a separation is achieved.

Reversed-phase chromatography is very popular in the separation of small molecules. Its limited use in protein separation, particularly at large scale, is due to the various issues that are associated with the use of the organic solvents in the purification process. These include protein denaturation and/or unfolding, waste handling, and the need for special explosion-proof handling due to solvent volatility. Scale-up in RPC follows the guidelines outlined above.

5. Affinity Chromatography

The separation of a target molecule from a mixture of species in affinity chromatography (AC) takes place by virtue of specific and reversible binding with a ligand that is immobilized on the matrix. This technique offers short process times and high specificity and resolution and is particularly useful when the target is present in very small quantities in the complex mixture. The primary issues in this technique are that the ligands may be expensive and unstable and may leach from the matrix and be present in the product. Fouling and regeneration of the ligand upon use may also offer challenges, so AC is generally used as the final polishing step in the purification scheme [58].

Scale-up in AC is carried out following the generic guidelines outlined in the previous section [59–62]. The bed height is kept constant, and column productivity is increased by increasing column diameter. However, resins used in AC are often compressible, so all the issues mentioned earlier in Section IV.A.1 apply and should be considered during scale-up.

Another approach to scaling up affinity separations is to increase the capacity of the adsorbent. This may be accomplished by increasing the concentration of the binding ligand that is coupled to the support particles. However, the binding capacity does not increase linearly with the ligand concentration and depends on the characteristics of the ligand as well as those of the binding product. This route for scale-up is therefore sensitive to the problems of leaching of ligands and fouling of resins. In addition, there is a partical limit on the maximum attainable ligand concentration on the resin surface. Thus, the final decision should depend on the comparison of economics and issues associated with the two approaches highlighted above.

6. Metal (Chelate) Chromatography

The separation and operation principles in metal (chelate) chromatography (MC) are very similar to those in affinity chromatography. Separation is based on interactions between the affinity tail on the sample component (such as dihistidyl tag) with the complexed heavy metals such as zinc and nickel on the stationary phase [9]. MC can be as efficient as affinity chromatography, particularly in the removal of endotoxins.

Scale-up issues for metal chromatography are very similar to those of affinity chromatography. The resins are generally custom-made and are relatively very expensive. The affinity tail often needs to be clipped before the end of the purification process. Care must be taken to avoid denaturation of protein feedstocks by the metals used in MC.

V. CONCLUSIONS

In this chapter we have provided an overview of the basic principles and practice of scale-up in the preparative chromatography. Modes of interaction and

An Overview

modes of operation were defined to clarify the options available upon scale-up. Because it is necessary to optimize a bench scale separation before attempting to scale it up, attention was first focused on method development at bench scale. Thermodynamic issues (resin screening) and kinetic issues (determination of operating conditions) were decoupled. A simple but generally applicable approach to resin screening was described. After a physically motivated introduction to scale-up, a straightforward calculation based on pore diffusion as the controlling kinetic contribution was presented and should suffice for many realistic separations. More detailed analyses of the fundamental chromatographic processes, modeling approaches to important modes of interaction and operation, and industrial case studies are given in the subsequent chapters.

REFERENCES

1. HK Rhee, R Aris, NR Amundson. First-Order Partial Differential Equations, Vol II: Theory and Application of Hyperbolic Systems of Quasilinear Equations. Englewood Cliffs, NJ: Prentice-Hall, 1989.
2. G Guiochon, S Golshan-Shirazi, AM Katti. Fundamentals of Preparative and Nonlinear Chromatography. Boston: Academic Press, 1994.
3. A Lee, A Velayudhan, Cs Horváth. Preparative HPLC. In: G Durand, M Bobichon, J Florent (eds.) Proceedings of the 8th International Biotechnology Symposium. Vol. 1. Paris: Société Française de Microbiologie, 1988, pp 593–610.
4. AS Rathore. Resin screening for optimization of chromatographic separations. LC-GC 19:616, 2001.
5. JC Janson, T Pettersson. Large-scale chromatography of proteins. In: G Ganetsos, PE Barker, eds. Preparative and Production Scale Chromatography. New York: Marcel Dekker, 1993, pp 559–590.
6. G Sofer, L Hagel. Handbook of Process Chromatography—A Guide to Optimization, Scale-up and Validation. New York: Academic Press, 1997, pp 27–113.
7. P Hedman, JC Janson, B Arve, JG Gustafsson. Large scale chromatography: optimization of preparative chromatographic separations with respect to column packing particle size, mass loading and flow rate. In: G Durand, L Bobichon, J Florent, eds. Proc 8th Int Biotechnol Symp, Vol 1. Paris: Soc Franc Microbiol, 1988, pp 623–643.
8. S Fulton, T Londo. Systematic development of chromatographic processes using perfusion chromatography technology. In: G Subramanian, ed. Bioseparation and Bioprocessing, Vol 1. New York: VCH, 1994, pp 41–64.
9. R Wisniewski, E Boschetti, A Jungbauer. Process design considerations for large-scale chromatography of biomolecules. In: KE Avis, VL Wu, eds. Biotechnology and Biopharmaceutical Manufacturing, Processing, and Preservation. Buffalo Grove, IL: Interpharm, 1996, pp 61–198.

10. MA Quarry, RL Grob, LR Snyder. Prediction of precise isocratic retention data from two or more gradient elution runs. Analysis of some associated errors. Anal Chem 58:901–917, 1986.
11. P Jandera, D Kromers, G Guiochon. Effects of the gradient profile on the production rate in reversed-phase gradient elution overloaded chromatography. J Chromatogr A 760:25–39, 1997.
12. P Jandera. Simultaneous optimization of gradient time, gradient shape, and initial composition of the mobile phase in the high-performance liquid chromatography of homologous and oligomeric series. J Chromatogr A 845:133–144, 1999.
13. S Yamamoto, E Miyagawa. Retention behavior of very large biomolecules in ion-exchange chromatography. J Chromatogr A 852:25–30, 1999.
14. S Yamamoto, E Miyagawa. Dynamic binding performance of large biomolecules such as γ-globulin, viruses, and virus-like particles on various chromatographic supports. Prog Biotechno 16:81–86, 2000.
15. E Groundwater. Guidelines for chromatography scale-up. Lab Pract 34:17–18, 1985.
16. EL Grushka, LR Snyder, JH Knox. Advances in band spreading theories. J Chromatogr Sci 13:25–33, 1975.
17. MR Ladisch, PC Wankat. Scale-up of bioseparations for microbial and biochemical technology. In: M Phillips, S Shoemaker, R Ottenbrite, R Middlekauf, eds. Impact of Chemistry on Biotechnology. ACS Symp Ser No. 362. Washington, DC: Am Chem Sco, 1988, pp 72–96.
18. DM Ruthven. Principles of Adsorption and Adsorption Processes. New York: Wiley, 1984.
19. S Yamamoto, M Nomura, Y Sano. Scaling up of medium-performance gel filtration chromatography of proteins. J Chem Eng Jpn 19:227–234, 1986.
20. S Yamamoto, M Nomura, Y Sano. Factors affecting the relationship between plate heights and the linear mobile phase velocity in gel filtration chromatography of proteins. J Chromatogr 394:363–371, 1987.
21. G Pieri, P Piccardi, G Muratori, L Caval. Scale-up for preparative chromatography of fine chemicals. Chim Ind 65:331–340, 1983.
22. MR Ladisch, A. Velayudhan. Scale-up techniques in bioseparation processes. In: RK Singh, SSH Rizvi, eds. Bioseparation Processes in Foods. New York: Marcel Dekker, 1995, pp 113–138.
23. CK Lee, Q Yu, SU Kim, NHL Wang. J Chromatogr 483:85, 1989.
24. JH Knox, HM Pyper. Framework for maximizing throughput in preparative liquid chromatography. J Chromatogr 363:1, 1986.
25. PC Wankat, YM Koo. Scaling rules for isocratic elution chromatography. AIChE J 34:1006–1012, 1988.
26. S Golshan-Shirazi, G Guiochon. Theory of optimization of the experimental conditions in preparative chromatography: optimization of the column efficiency. Anal Chem 61:1368, 1989.
27. S Golshan-Shirazi, G Guiochon. Optimization of the experimental conditions in preparative liquid chromatography with touching bands. J Chromatogr 517:229, 1990.

28. PC Wankat. Scaling rules and intensification of liquid chromatography: extension to gradient elution and displacement chromatography. Prep Chromatogr 1:303–322, 1992.
29. HP Jennissen. Evidence for negative cooperativity in the adsorption of phosphorylase b on hydrophobic agaroses. Biochemistry 15:5683, 1976.
30. A Velayudhan, Cs Horváth. Preparative chromatography of proteins: analysis of the multivalent ion-exchange formalism. J Chromatogr 443:13–29, 1988.
31. G Guiochon, AM Katti. Chromatographia 24:165, 1987.
32. C Heuer, P Hugo, G Mann, A Seidel-Morgenstern. Scale up in preparative chromatography. J Chromatogr A 752:19–29, 1996.
33. K Jones. Process scale high-performance liquid chromatography, Part I: An optimization procedure to maximise column efficiency. Chromatographia 25:437–446, 1998.
34. K Jones. A review of very large scale chromatography. Chromatographia 25:547–559, 1998.
35. H Colin. Large-scale high-performance preparative liquid chromatography. In: G Ganetsos, PE Barker, eds. Preparative and Production Scale Chromatography. New York: Marcel Dekker, 1993, pp 11–46.
36. E Bayer, E Banneister, U Tallarek, K Albert, G Guiochon. NMR imaging of the chromatographic process. J Chromatogr A 704:37, 1995.
37. DE Cherrak, G Guiochon. Phenomenological study of the bed-wall friction in axially compressed packed chromatography columns. J Chromatogr A 911:147, 2001.
38. S Katoh. Scaling-up affinity chromatography. Trends Biotechnol 5:328–331, 1987.
39. AW Mohammad, DG Stevenson, PC Wankat. Pressure drop correlations and scale-up of size exclusion chromatography with compressible packing. Ind Eng Chem Res 31:549–561, 1992.
40. JC Janson. Large-scale chromatography. GBF Monogr Ser 7:13–20, 1984.
41. JC Janson, P Dunhill. Factors affecting scale-up of chromatography. Fed Eur Biochem Soc Meet (Proc). In: B Spencer, ed. Industrial Aspects of Biochemistry, Part 1, Vol 30 1974, pp 81–105.
42. JC Janson, P Hedman. Large-scale chromatography of proteins. Adv Biochem Eng 25:43–99, 1982.
43. P Zeilon, L Stack, RE Majors. System design for process-scale chromatography. LC-GC 10:736–742, 1992.
44. PR Levison. Techniques in process-scale ion-exchange chromatography. In: G Ganetsos, PE Barker, eds. Preparative and Production Scale Chromatography. New York: Marcel Dekker, 1993, pp 617–626.
45. G Sofer. Ensuring the accuracy of scaled-down chromatography models. BioPharm 10:36–39, 1996.
46. R Hjorth, P Leijon, AB Frej, C Jägersten. Expanded bed adsorption chromatography. In: G Subramanian, ed. Bioseparation and Bioprocessing, Vol 1. New York: VCH, 1994, pp 199–228.

47. AB Frej. Expanded bed adsorption. In: MC Flickinger, SW Drew, eds. Encyclopedia of Bioprocess Technology: Fermentation, Biocatalysis, and Bioseparation, Vol 1. New York: Wiley-Interscience, 1999, pp 20–31.
48. J Strube, A Jupke, A Epping, H Schmidt-Traub, M Schultze, R Devant. Design, optimization, and operation of SMB chromatography in the production of enantiomerically pure pharmaceuticals. Chirality 11:440–450, 1999.
49. RM Nicoud, RE Majors. Simulating moving bed chromatography for preparative separations. LC-GC 18:680–687, 2000.
50. S Ostrove. Considerations for scaling up to process chromatography. LC-GC 7: 550–554, 1989.
51. S Cresswell. Scaling up in ion exchange chromatography. Biotech Forum Eur 9:446–452, 1992.
52. E Grund. Hydrophobic interaction chromatography of proteins. In: G Subramanian, ed. Bioseparation and Bioprocessing, Vol 1. New York: VCH, 1994, pp 65–88.
53. S Ishida, J Saeki, E Kumazawa, G Kawanishi, E Sada, S Katoh. Scale-up of hydrophobic interaction chromatography for purification of antitumor antibiotic SN-07. Bioprocess Eng 4:163–167, 1989.
54. P Kårsnäs. Hydrophobic interaction chromatography. In: MC Flickinger, SW Drew, eds. Encyclopedia of Bioprocess Technology: Fermentation, Biocatalysis, and Bioseparation, Vol 1. New York: Wiley-Interscience, 1999, pp 602–612.
55. K Vorauer, M Skias, A Trkola, P Schulz, A Jungbauer. Scale-up of recombinant protein purification of hydrophobic interaction chromatography. J Chromatogr 625:33–39, 1992.
56. JC Janson. Process scale size exclusion chromatography. In: G Subramanian, ed. Process Scale Liquid Chromatography. New York: VCH, 1995, pp 81–98.
57. CM Roth, ML Yarmush. Size exclusion chromatography. In: MC Flickinger, SW Drew, eds. Encyclopedia of Bioprocess Technology: Biocatalysis, and Bioseparation, Vol 1. New York: Wiley-Interscience, 1999, pp 639–650.
58. CR Goward. Affinity chromatography and its applications in large-scale separations. In: G Subramanian, ed. Process scale liquid chromatography. New York: VCH, 1995, pp 193–214.
59. HA Chase. Optimisation & scale-up of affinity chromatography. Makromol Chem, Macromol Symp 17:467–482, 1988.
60. J Pearson. Affinity chromatography. In: G Subramanian, ed. Bioseparation and Bioprocessing, Vol 1. New York: VCH, 1994, pp 113–124.
61. FH Arnold, JJ Chalmers, MS Saunders, MS Croughan, HW Blanch, CR Wilke. A rational approach to the scale-up of affinity chromatography. ACS Symp Ser, Purif Ferment Prod 271:113–122, 1985.
62. KA Kang, DDY Ryu. Studies on scale-up parameters of an immunoglobulin separation system using Protein A affinity chromatography. Biotechnol Prog 7:205–212, 1991.

2
Interaction of Mass Transfer and Fluid Mechanics

Edwin N. Lightfoot, John S. Moscariello, Mark A. Teeters, and Thatcher W. Root
University of Wisconsin–Madison, Madison, Wisconsin, U.S.A.

I. INTRODUCTION

For our purposes we define chromatography as a group of transient multicomponent separations resulting from differential rates of solute travel through a fixed sorbent bed via the flow of a percolating solvent. It is assumed here that the classification of chromatography is discussed elsewhere in this volume, and our emphasis is on the design and operation of chromatographic devices. The development of chromatography as a unified subject has been recently reviewed [1,2].

Although the success of chromatographic separations depends heavily upon finding a suitably selective adsorbent system, the cost and feasibility of operation tend to be dominated by interaction between fluid mechanics and mass transfer. In this respect chromatography is a rather typical mass transfer process, from overall systems aspects, through column and header design down to the detailed behavior at the level of individual adsorbent particles.

Organization of this chapter follows the top-down approach typical of design processes, and it begins with an overview of separation economics. Section II provides the framework for the more detailed discussions to follow and relates chromatography to other separations processes that can compete with or supplement it in a process system.

In Section III we provide discussions of mass transfer heuristics, with particular reference to chromatographic separations. Section IV shows that there are a limited number of column types in the chromatographic "family" and indicates the factors that can provide an economic selection among them for a particular application.

We then summarize the available knowledge of the behavior of an ideal differential chromatographic column with emphasis on the concept of ideal plates. Although strictly applicable only to differential chromatography, the plate concept provides useful insight into other operating modes as well.

We follow this discussion by describing some of the reasons columns fail to meet the predictions of idealized models, with emphasis on flow nonuniformities, both those introduced by packing irregularities and variations of solute residence time introduced by header design.

Our concluding section deals with diagnostics, how to determine whether or not a chromatographic device is meeting its potential, and, where possible, what can be done to improve performance.

II. OVERVIEW OF SEPARATION ECONOMICS

A. Generic Economic Characteristics

Whereas each separation problem has its own special characteristics, it is nonetheless helpful to take a general overview [3]. It has been shown [4] that for a wide range of biological products, ranging from commodity chemicals through antibiotics and related substances to injectable proteins, sales price is proportional to the mass of the feed stream to the separation chain needed to produce the product from a crude feed, e.g., a fermentation broth:

Cost per unit mass of product =
$$K \times \text{impurities per unit mass of product} \quad (1)$$

Here K is a species-independent constant, and it is particularly noteworthy that purity of the final product does not appear in this equation.

This simple observation has had a great impact on process design. In most cases of interest to us here, materials handling means fluid flow and requires a variety of considerations. Items to consider include column pressure drops and a variety of less dramatic factors. Solvent consumption, cost of pumps, lines and storage tanks, and even solvent disposal become major considerations in the plant, whereas in the laboratory the major cost is labor. Recognition of this cost structure is useful in both process and equipment

Mass Transfer and Fluid Mechanics

design, and we now briefly consider these two aspects of separation technology.

B. Process Design Strategy

Most separation processes involving chromatography are complex and do not lend themselves to true optimization. Rather, one must depend heavily upon heuristic considerations, and perhaps the simplest is to recognize that concentration and thermodynamic activity are independent of each other once a choice of solvents or adsorbents is permitted. Much can therefore be learned from plotting the process trajectory on an activity–concentration diagram such as the one pictured in Fig. 1. It is clear from the above discussion that the trajectory on such a diagram should be as concave as possible when viewed from above. Thus trajectory *FEP* is generally preferable to *FIP*, and here the difference between laboratory and process conditions becomes important.

In the laboratory it is common practice to concentrate a dilute protein feed by salting out, which first increases the activity for any given concentration and then results in a spontaneous concentration in the form of precipitation (trajectory *FIP* in Fig. 1). However, at the process scale the cost of adding large amounts of salt, typically ammonium sulfate, and the subsequent cost of disposal can be prohibitive. It is therefore preferable to begin with a spontaneous concentration process such as the adsorption of protein on an ion-exchange adsorbent. The spent process stream can then be disposed of at a reasonable cost, and the protein can be eluted with a modest volume of salt

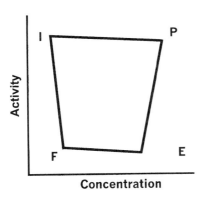

Figure 1. Representative separation trajectories.

solution, now at a concentration that competes with the protein for adsorption sites but does not precipitate it (trajectory *FEP*).

The recovery of high purity biologicals, the primary province of process scale chromatography, almost always requires a series of recovery operations, typically including membrane filtration and centrifugation in addition to one or more chromatographic steps. Proper sequencing is always among the first jobs facing the designer. In accord with the above-described importance of materials handling costs, it has been found useful to organize these recovery operations into three stages:

1. Capture or concentration
2. Fractionation
3. Polishing

Each has a different function and requires a different approach. We discuss these separately.

1. Capture or Concentration

The primary functions of the capture stage are to reduce the volume of the process stream and to remove substances that, though largely unrelated to the product of interest, can interfere with subsequent steps. It is here that solvent and capital costs are highest and the greatest amount of attention is needed to both equipment and process design improvements. Batch adsorption on a chromatographic column is well suited to such a concentration step. One can also concentrate a dilute feed using a membrane process such as nano- or ultrafiltration, but this inevitably concentrates many undesirable large impurities along with the desired product.

At this stage one seeks equipment and packings that permit high flow rates, preferably at modest pumping pressures, and provide a high volumetric capacity for the desired product. Packings must withstand the crushing pressure at the downstream end of the column and should also resist fouling and other forms of inactivation. In addition, packings should be easily regenerated, and they should be chosen to minimize pretreatment of the feed solution. The most common choices are an ion-exchange or hydrophobic interaction adsorbent.

Hydrogels, pioneered by Pharmacia, have the highest capacity of available adsorbents and have been widely used. They also exhibit high diffusion coefficients for adsorbed species. Their dynamic capacity is relatively insensitive to percolation velocity and usually increases with dilution of the feed stream. They do tend to be soft, and the addition of a rigid skeleton, as in the

Mass Transfer and Fluid Mechanics

HyperD adsorbents developed by BioSepra, can remedy this problem at a relatively modest loss of capacity. However, available skeletons tend to be damaged by repeated regeneration—there is no perfect adsorbent. We return to selection of adsorbents later, along with equipment morphology.

2. Fractionation

The primary function of the fractionation step is to remove species closely related to the desired product, and it is highly desirable to use a different mechanism from that employed during capture. Moreover, it is here that highly selective adsorbents, such as dye affinity, protein A, or even bioaffinity types, are most often used. In fact, these are sometimes used on the crude feedstream, thus combining capture and fractionation in one step. However, these adsorbents tend to be expensive, and nonselective adsorption is at its maximum in what can be very crude process streams. Such inadvertently adsorbed impurities not only reduce capacity but also can shorten adsorbent life and complicate regeneration.

Among the major engineering problems encountered here is that of obtaining high solute loading, for obvious economic reasons, and gradient elution [5,6] is commonly employed for this purpose. For polymers such as proteins a primary advantage of this technique is that it is possible to achieve selective elution [6], in addition to the more familiar selective migration, and this can lead in especially favorable situations to on–off operation for which one needs only a single stage, equivalent to a stirred tank of slurry. However, the downside of gradient elution is that columns can be too long—another aspect of the interaction of flow and mass transfer.

3. Polishing and Auxiliary Operations

The purpose of the final stage of purification is to remove traces of impurities and prepare the product stream for the drying and other steps needed for packaging. It may also be necessary to change the solvent between the above-discussed processing steps to maximize their performance, and this has historically been done by size exclusion chromatography. However, size exclusion chromatography tends to be expensive, because of both low volumetric capacity and a tendency to hydrodynamic instability at high solute loading. Hydrodynamic instability often takes the form of "fingering," as shown in Fig. 2 for native hemoglobin [7]. Here flow is from the top down, and it is the trailing end of the protein band that is eroded by the lower viscosity solvent. Protein concentration toward the base of the band is about 15 mg/mL. Size exclusion chromatography is being replaced more and more by membrane processes.

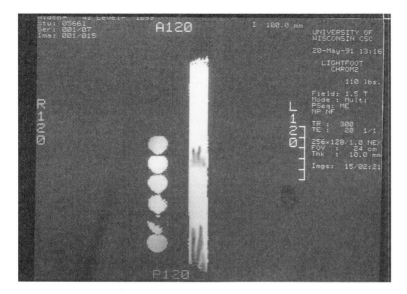

Figure 2. Fingering of native hemoglobin in a size exclusion column.

These latter are in fact becoming selective enough to encroach even on fractionation [8,9], and they owe their success to low materials handling costs.

Removal of pyrogens, nucleic acids, viruses, and microorganisms is also often provided by adsorption processes.

III. MASS TRANSFER HEURISTICS

At the start of any design process the critical task is to reduce the extent of parameter space that must be investigated to arrive at a satisfactory result, and this is inevitably a heuristic process. We are dealing here with the emergence characteristics [10] of transport phenomena and their complex supporting technology, and the number of possibilities is staggeringly large [11].

Moreover, it is seldom that one can find a single operational objective function to be optimized. Rather one must settle for an approximate objective, which normally involves a number of conflicting effects. Fortunately, mass transfer processes have been examined in considerable depth, and chromatographic processes are not very different from other and generally better understood mass transfer operations. In general the history of innovation [12] also suggests that it is highly desirable to limit initial consideration to two independent variables—our vision and the paper and computer monitors we use in our search are primarily two-dimensional. The multidimensional (and hypothetical) fitness landscapes of evolutionary theory [13] have so far not proven very useful.

We therefore begin our search for a design strategy with the guidelines of Table 1. The primary thermodynamic goal, in accord with Eq. (1), is to maximize solute capacity. However, here we are more interested in the transport behavior, and this means maximizing mass transport rates. This in turn normally means keeping diffusion paths short and providing a large mass transfer surface. However, experience teaches that one should also minimize momentum transport, which means minimizing pressure drop and in our case crushing pressure on the packing at the downstream end of the column. This goal is favored by long momentum transfer distances and small interfacial areas between phases. It is the interplay of these two conflicting requirements that provides the basic framework for the design of essentially all large-scale mass transfer devices. This means that we must find differences between mass and momentum transfer that we can exploit to achieve a satisfactory design. We shall find that there are only a few ways to do this and therefore that there are a limited number of basic morphologies for chromatographic devices.

However, chromatographic columns of interest here all depend upon essentially one-dimensional counterflow of two interacting phases, and there are therefore two additional, subsidiary but important, constraints that must be taken into account in the design process: dispersion of solute residence times across the direction of flow and nonuniformity of flow in this direction. These are strongly influenced by column morphology.

Finally, there are practical constraints imposed by the physical nature of the systems being studied and the existing state of technology. It is difficult

Table 1 Guidelines for Separator Design

Maximize solute capacity
Maximize lateral mass transport
   Small distances and large surface
Minimize pressure drop
   Large distances and small surface
Minimize axial dispersion
   Small diameters, especially tubes
Maintain flow uniformity

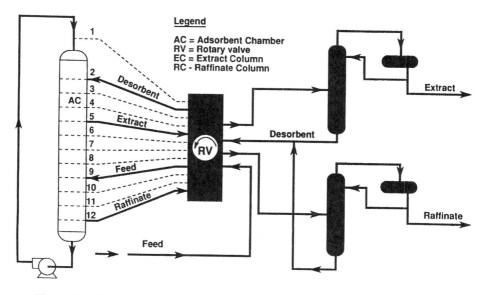

Figure 3. A simulated moving bed.

to systematize these, but the interplay of fluid mechanics and mass transfer plays a strong role here. Specifically it is only the difficulty of moving solid particulates relative to a fluid stream that makes the chromatographic operating mode attractive as opposed to the much more efficient [14] steady counterflow used in essentially all fluid–fluid contactors. It is for this reason that simulated moving bed systems, sketched in Fig. 3, are so cumbersome and expensive. These devices provide excellent separations, but they are so expensive that they are restricted to large-scale applications. In addition, their complexity is such that true feedback control is not feasible. As a result they have limited applicability for proteins, where fouling and other unpredictable variations in capacity and kinetics are common.

IV. ESTABLISHING THE CHROMATOGRAPHIC FAMILY

We now turn to the specific case of chromatographic design and begin by noting that we are dealing with a transient operation, typically concerned with unsteady diffusion in spheroidal packing particles and the accompanying momentum transfer in a granular bed. These two processes are compared in Table

Mass Transfer and Fluid Mechanics

Table 2 Mass vs. Momentum Transport

Mass transport to a sphere

c = mass of solute/volume

$$\partial c/\partial t \approx \frac{6\mathcal{D}_A}{R^2} \exp[-\pi^2 t \mathcal{D}_A / R^2]$$

Momentum transport to a (bed of) spheres

$$F = d(mv)/dt; \quad dp/dz = F/V$$

$$dp/dz = d(mv/V)dt$$

$$= \frac{150}{4} v_0 \frac{(1-\varepsilon)^2}{\varepsilon^3} \frac{\mu}{R^2}$$

Both mass and momentum transfer show the same dependence on particle size!

2, and for mass transfer we take advantage of the insensitivity of diffusional response times to boundary conditions. It is important to note that time t always occurs simultaneously with particle radius R and diffusivity $\mathcal{D}_A$ in the ratio $t\mathcal{D}_A/R^2$.

The expression for momentum transfer is just a commonly used form of the Blake–Kozeny equation for creeping flow through a bed of spheres [15]. We see that the pertinent transport property for momentum transfer, viscosity μ, appears only in the ratio μ/R^2 and therefore that the dependence of mass and that of momentum transfer on particle radius are identical: Changing radius has no effect on the ratio of mass to momentum transfer. This is one form of a dilemma facing all designers of mass transfer equipment.

However, we also see that viscosity is multiplied by three terms that have no counterpart in mass transfer:

The numerical coefficient, about 150/4 for spherical packings
The percolating fluid velocity v and the superficial velocity v_0
A term containing only the fluid void fraction of the column, ε

It is in these terms that we must seek exploitable differences between mass and momentum transport.

There appear to be five fundamental ways to improve mass transfer relative to momentum transfer:

1. Raise pressure drop.
2. Decrease momentum load.

3. Thicken momentum boundary layer.
4. Eliminate form drag.
5. Modify packing.

Conceptually the simplest is to raise the pressure drop as in the Prochrom dynamic axial compression columns, which are typically packed at 70 bar and maintained at 50 bar. This approach requires massive columns; strong, rigid packings; and high pressure pumps, but it is capable of producing very high resolution separations. In the differential chromatographic mode it can, for example, exhibit 40,000 or more plates.

Alternatively one can use a large ratio of column diameter to packed height, thus yielding a low percolation velocity for a given packed volume. This is a popular approach and permits the use of cheaper pumps and columns and softer packings; the latter can in turn be of higher volumetric capacity. Pressure drops can be further reduced simply by relaxing the requirement of high resolution, not actually needed in many applications. This is particularly true in the batch adsorption so widely used in the capture stage of the separation chain.

However, as discussed in connection with mass transfer modeling below, low velocities may result in unacceptably low resolution, and large diameter/length (D/L) ratios can cause problems in achieving acceptable flow distribution. A very promising special case is to use a packing of stacked adsorptive membrane sheets such as those produced by the Pall and Sartorius corporations. These also eliminate most lateral mass transfer resistance, and as a result resolution is very insensitive to percolation velocity. Early problems with volumetric solute capacity seem to have been solved, but some problems of flow distribution remain and fouling can be a problem.

Another approach is to eliminate form drag, for example by using bundles of hollow fibers coated on the inside with adsorptive material. Such devices can have much higher ratios of mass to momentum transfer and in particular offer both low pressure drops and resistance to fouling. They do, however, present two flow-related problems: those of obtaining low axial dispersion and good flow distribution. As will be seen in more detail below, axial dispersion is much more sensitive to characteristic packing diameters in cylindrical pores than in granular beds, and the practical result is that lumen diameters must be less than about 40 µm. In addition, lumen diameters must be highly uniform or the bulk of flow will be through large, diffusionally inefficient channels. However, as suggested in Fig. 4, these problems have been solved by the Alltech Corporation for gas chromatography, and there is no reason why their techniques cannot be extended to other applications.

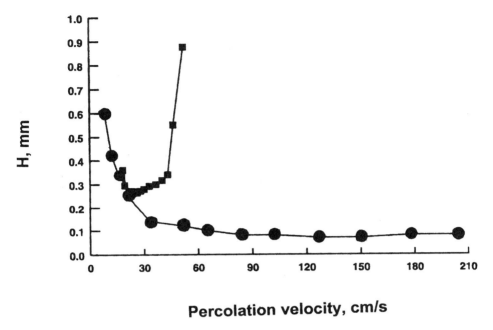

Figure 4. Van Deemter plot for the Alltech SE-54 multicapillary column (●) compared with a conventional granular packed bed apparatus (■). The capillary lumens are hexagonal with an effective diameter on the order of 40 μm.

A fourth approach is to thicken the momentum transfer boundary layers by increasing column void fraction. This approach depends on the fact that all momentum transfer takes place in the fluid phase but the high Schmidt numbers of all liquid-phase chromatography require that the great bulk of mass transfer resistance reside within the packing particles. This means operating in upflow so that the packing expands, and it results in both low pressure drop and the ability to handle particulate suspensions. Success is due to the work of Howard Chase [16] at the University of Cambridge, who showed that with dense particles of varying size and careful leveling of column supports one can avoid the very strong vertical particle mixing of typical fluidization operations. This approach has been implemented by Upfront Chromatography and on a larger scale by Pharmacia in their "streamline" system using their "big beads": large and dense packing particles. However, one cannot really approach the resolution of a well-packed fixed bed, and this approach is best suited to the capture stage, where processing cost is a major consideration.

Finally, much can be done by modification of the packings in granular columns. We have already talked about the high capacity and large diffusivity of hydrogels, and both appear to result from the distribution of fixed charges throughout the interior of the adsorbent (see Fig. 5a). This contrasts strongly with the composite structure of Figs. 5b and 5c, where the adsorbent sites are

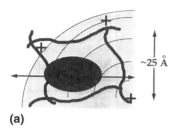
(a)

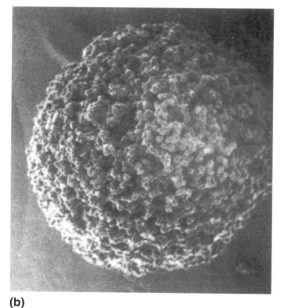
(b)

Figure 5. Representative adsorbent structures. (a) Schematic view of a hydrogel. (b) Scanning electron micrograph of Amberlite IRA-938 at 100× magnification. (c) Scanning electron micrograph of Amberlite IRA-938 at 300× magnification. [(b) and (c) courtesy of Robert L. Albright.]

Mass Transfer and Fluid Mechanics

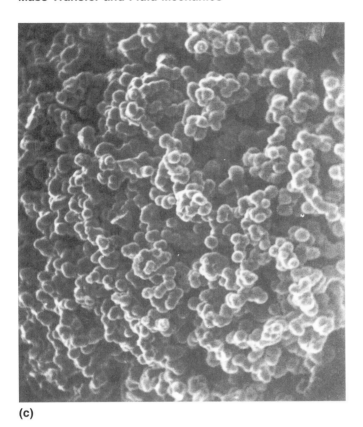

(c)

on the surfaces of very small spheroids that form the basic structure of the adsorbent. These latter structures are, however, quite strong and chemically stable. Perseptive Biosystems claims that their Poros adsorbents permit internal convection that speeds equilibrium between the sorbent and percolating fluid. Another promising approach is that of homogeneous beds pioneered independently by Frantisec Svec and his coworkers [17] and by Stellen Hjerten et al. [18]. These are preformed cylindrical structures with internal pores providing internal flow and good access to adsorbent sites. They are marketed by Bio-Rad Laboratories under the trade name UNO beds. There is little doubt that further innovations will be made in packing design.

We shall have more to say about equipment performance later, but first we need to introduce quantitative measures of effectiveness.

V. DIFFERENTIAL CHROMATOGRAPHY AND HEIGHTS OF A THEORETICAL PLATE

Differential chromatography is widely used at an analytical level and for determining the efficiency of process scale columns. In addition, much of the nomenclature of chromatography was developed for this simple process, and we will need to use some of the most widely used terms in our subsequent discussions. It is commonly assumed that for this situation the effluent concentration of a given solute i is given by a Gaussian distribution

$$c_i(L, t) = \frac{m_0}{A\varepsilon v}\left(\frac{1}{\sqrt{2\pi\sigma^2}}\right) \exp\left[\frac{-(t - T_i)^2}{2\sigma^2}\right] \qquad (2)$$

or, in scaled terms,

$$C_i(L, t) = \frac{1}{\sqrt{2\pi s^2}} \exp\left[\frac{-(t/T_i - 1)^2}{2s^2}\right] \qquad (3)$$

Here m is the mass of solute fed to a column length, v is interstitial solvent velocity, ε is column void fraction, t is time, T_i is the mean residence time of solute i, and σ is the standard deviation of the distribution. The scaled solute concentration is

$$C_i = c_i \frac{A\varepsilon v T_i}{m_0} \qquad (4)$$

and

$$s^2 \equiv (\sigma/T_i)^2 \qquad (5)$$

is the *variance* of the distribution scaled with respect to mean solute residence time. The variance, in turn, is the scaled second moment of the solute concentration with respect to time,

$$s^2 = \int_{-\infty}^{\infty} (\tau - 1)^2 \, C_i \, d\tau \qquad (6)$$

We use all of these means for expressing solute distribution in the discussions below.

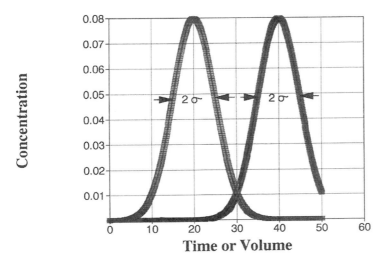

Figure 6. Scaled Gaussian effluent curves.

We begin with Fig. 6, which shows a two-solute effluent curve with normalized Gaussian distributions of solute concentration about the peaks at mean residence times T_1 (left) and T_2 (right). The respective standard deviations about these mean times—σ_1 and σ_2, respectively—are the half-widths at the inflection points of the distributions. The degree of separation of the two solutes is usually expressed in terms of resolution defined as

$$R_{12} \equiv \frac{1}{2}\left(\frac{|T_2 - T_1|}{\sigma_1 + \sigma_2}\right) \tag{7}$$

The degree of separation of such normalized peaks can then be determined from the shape of a Gaussian distribution. For a resolution of unity, each fraction is 97.73% pure if the cut is made at the point of overlap in these normalized curves.

We next look a little more closely at the Gaussian distribution in Fig. 7, with the time scale now normalized with respect to mean residence time. It may be seen that this distribution is determined by a single parameter s^2, the normalized variance, and it defines the number of theoretical plates, N,

$$N_i \equiv 1/s_i^2 \equiv L/H_i \tag{8}$$

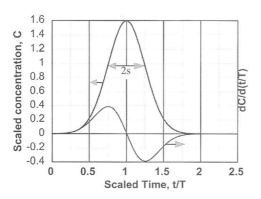

Figure 7. A Gaussian distribution with a plate number of 16.

or

$$N_i \equiv L/H_i \qquad (9)$$

as well as the height H_i of a theoretical plate. Here L is the packed length of the column. The subscript i is a reminder that each species will in general exhibit a different value for plate height. Henceforth this subscript is omitted for convenience. It may be seen on careful examination of this figure that s, the standard deviation, is the half-width at the inflection point of the distribution and that its magnitude is $1/4$. It follows that the number of plates represented here is 4^2, or 16.

It is very important to note that H is defined only for a Gaussian solute distribution, and it is useful to recall that Gaussian distributions are typically the result of adding numerous small random deviations about a mean value [19]. For a granular bed these deviations result from the successive interaction of the percolating solvent with a very large number of adsorbent particles.

For differential chromatography H and T_i are the only parameters needed to characterize column effectiveness, and T_i depends only upon thermodynamic parameters. We therefore concentrate our attention here on estimation of H. This problem is discussed at length in many sources, for example, Athalye et al. [20], which we use as a basic reference. The reader is referred to Athalye et al. for details, but this and essentially all other accessible references are based upon two assumptions that we question later:

1. Flow is rectilinear on size scales large compared to the diameters of packing particles.

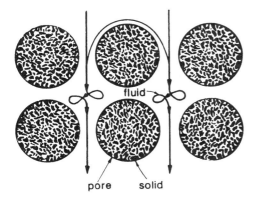

Figure 8. Schematic view of the convective mixing of the percolating fluid in the particle interstices and the highly nonuniform concentration boundary layer around each bead.

2. Changes in local concentration are slow enough that lumped parameter approximations are valid for calculating contribution to plate height.

Referring to Fig. 8, these contributing factors are finite kinetics of the adsorption/desorption process on the adsorption sites within the granule, diffusion to and from these sites through the bulk of the granule, localized mass transfer resistance in the diffusional boundary layer about the particle, and convective dispersion resulting from local flow disturbances between particles.

For the conditions leading to Gaussian concentration distributions these contributions may be considered additive so that

$$H = H_{rxn} + H_{diff} + H_{bl} + H_{disp} \tag{10}$$

Here the subscripts rxn, diff, bl, and disp refer, respectively, to reaction kinetics, internal diffusion, boundary layer resistance, and convective dispersion contributions to band spreading. Equation (10) and the expressions used to calculate H have been tested by Athalye et al. using commercial size exclusion columns, and representative results are shown in Fig. 9. Here the reduced plate height h is the ratio of plate height to packing diameter d_P:

$$h \equiv H/d_P \tag{11}$$

The abscissa is a dimensionless percolation velocity, defined by

$$\text{Re Sc} \equiv d_P v / Đ_F \tag{12}$$

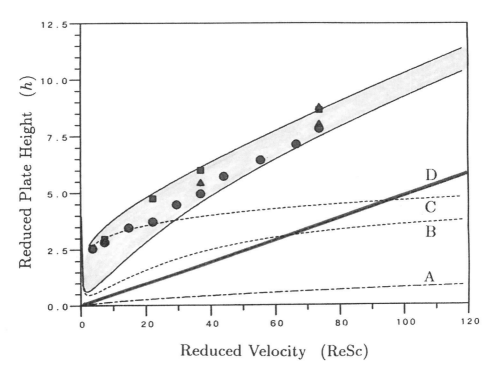

Figure 9. Comparison of actual behavior of a size exclusion column with predictions of a one-dimensional lumped parameter model. Bovine hemoglobin in a 1 in. Toyopearl HW65C column. See text discussion.

Here Re denotes the Reynolds number, Sc is the Schmidt number, v is interstitial fluid velocity, and $Ð_F$ is the effective solute diffusivity in the fluid phase. For size exclusion there is no adsorption and therefore no contribution from adsorption kinetics.

In this figure the labeled lines show estimates of the indicated contributions to h determined independently from the chromatographic process, and it should be noted that

1. The boundary layer contribution, line A, is obtained from a quite reliable correlation, but it is of only minor importance.
2. There is enough uncertainty in the dispersion contribution that a range, between lines B and C, is given rather than a single correlation. This range was arrived at from an examination of often con-

Mass Transfer and Fluid Mechanics

flicting published literature. In practice it is often assumed that the dispersion contribution is a velocity-independent constant. One can find values from about 2 to 5 or even higher depending on how well the bed packs. HyperD granules, for example, act as if they were sticky and can often show values on the order of 5.

3. The contribution of internal diffusion, line D, must be measured for each adsorbent–solute system. One can sometimes predict these diffusivities [21], but such circumstances are still exceptional. This contribution tends to be the largest of the group, especially for the high percolation velocities of modern practice.

The upper band represents the computed plate height as the sum of estimates for the three pertinent contributions. It may be seen to include all of the experimental data, and this finding strongly suggests that the mechanism-based correlations of Athalye et al. do have real predictive power. However, some caveats are in order.

First there is the uncertainty of the convective dispersion contribution, which can be important at the very low percolation velocities needed to minimize plate height. More important is the fact that a lumped-parameter approximation,

$$h_{\text{diff}} \approx \frac{(1-u)^2}{3(1-\varepsilon_b)} \left(\frac{d_p v}{\mathcal{D}_F}\right)\left(\frac{m}{10}\right) \tag{13}$$

is used for the diffusive contribution:

$$u \equiv \frac{1}{1 + K[(1-\varepsilon_b)/\varepsilon_b]} \tag{14}$$

$$m \equiv \frac{\mathcal{D}_F}{\mathcal{D}_A}\left(\frac{1}{K}\right) \tag{15}$$

Here ε_b is the volume fraction of interstitial fluid in the bed and K is the equilibrium ratio of observable (volume-average) solute concentration in the adsorbent to that in the external fluid. The subscripts A and F represent the adsorbent and percolating fluid, respectively.

From a physical standpoint, u is the equilibrium fraction of solute in the percolating fluid in the fluid–adsorbent mixture or the ratio of average solute migration velocity to interstitial fluid velocity. Then the mean distance traveled by the solute in time t is

$$\bar{z} = uvt \tag{16}$$

The factor 10 in Eq. (13) is the Sherwood number for mass transport in the adsorbent granule defined by using the fluid-phase diffusivity:

$$\text{Sh} \equiv d_p k_c / Ð_F = 10 \tag{17}$$

whereas the mass transfer coefficient k_c is defined by

$$V \frac{dq_b}{dt} = k_c S(q_s - q_b) \tag{18}$$

Here V is the volume of a packing particle, S is its surface area, q_b is the average solute concentration within the particle, and q_s is the solute concentration just inside the particle surface.

Now, the Sherwood number of 10 can be obtained as the long-time asymptote for diffusion into a sphere with a linearly rising surface concentration,

$$q_s = gt \tag{19}$$

where g is an arbitrary constant [22]. It has also been shown that this is a correct first approximation for differential chromatography [23]. However, it is a good approximation only when the time scale over which g changes is long with respect to the diffusional response time of the particle,

$$T_{\text{diff}} \approx d_p^2 / 24 Ð_A \tag{20}$$

The time scale over which g changes is of the order of the standard deviation of the effluent curve, $s_i T_i$, and as a pratical matter Eq. (13) should not be used unless

$$s_i T_i \gg d_p^2 / 24 Ð_A \tag{21}$$

This effect has, however, not been systematically investigated, and it is not yet possible to quantify this requirement.

If changes in g take place too fast, effluent curves become sharply skewed, the peak appears significantly earlier than the true mean solute residence time, and the numerous short-cut methods [24] for estimating plate height give values that are too low (see, e.g., Ref. 25).

Another deviation from Gaussian behavior results from the fact that time, and hence degree of solute dispersion, changes as the solute peak passes the end of the column. This also causes the effluent distribution to be "left skewed," i.e., sharper as the effluent concentration rises and more gradual as it falls. This will be the case unless

$$s_i \ll 1 \tag{22}$$

and where this inequality is not satisfied one should replace s_i^2 by $s_i^2(t/T_i)$. In practice the observed effluent curve will be further distorted by mixing in the auxiliary tubing and in the flow cell used to measure its concentration. We return to this point later.

It is therefore desirable to have some means of dealing with non-Gaussian effluent curves. The most rigorous way to do this is to solve the defining equations completely, including the complete statement of Fick's second law for the adsorbent particles. If they may be considered homogeneous this takes the form

$$\frac{\partial q}{\partial t} = \mathcal{D}_A \frac{1}{r^2} \frac{\partial}{\partial r} r^2 \frac{\partial q}{\partial r} \tag{23}$$

Even here one is neglecting the effect of particle interactions on the symmetry of the solute distribution, but it can be shown that this effect is not very important [26].

It has been shown that one can obtain directly a closed-form solution for s^2 and hence N from Eq. (9) from this more complete description [27]. The result of such a calculation is shown in Fig. 10, for actual data, compared

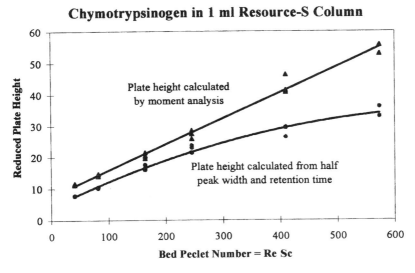

Figure 10. Comparison of plate heights calculated with Eqs. (24) and (26).

with a widely used approximation based on the assumption of a Gaussian distribution:

$$\frac{1}{N} = \frac{H}{L} = \frac{(W_{0.5}/T_{max})^2}{5.54} \tag{24}$$

Here $W_{0.5}$ is the effluent peak width at half the peak concentration and T_{max} is the time of the peak appearance.

It may be seen that use of the variance s^2 yields a correlation that is almost a straight line, with an intercept corresponding to a first approximation to the convective dispersion contribution and a slope corresponding to $dh_{diff}/d(\text{Re Sc})$. Calculations based on Eq. (24), on the other hand, show a steadily decreasing slope that seems to be approaching an asymptotic value. This curvature suggests to the unwary that there is an advantage to using very high percolation velocities, but examination of actual effluent curves shows increasingly serious tailing and progressively poorer separability as well as more dilute products.

Closed-form solutions incorporating the complete solution of Fick's second law are also available for the third central moment of concentration, and these can be used along with available alternatives [28] to the Gaussian distribution for correlating experimental data.

Recognizing the fact that Eq. (24) introduces what can be serious errors in interpretation is important, because if the h vs. Re Sc curve in Fig. 9 (a *van Deemter plot*) is convex one can improve separation by increasing percolation velocity. This is often desirable on scale-up because laboratory columns are normally quite short. To show this, note that

$$v_0 = Q/S \tag{25}$$

and

$$H = L/N = Ls^2 \tag{26}$$

It follows that the number of plates is given by

$$N = \frac{Q}{V_{col}}\left(\frac{H}{v_0}\right) \tag{27}$$

where Q is the volumetric flow rate of the percolating solvent, V_{col} is the packed column volume, and H/v_0 is the slope of a line from the origin to any point on a plot of H vs. v_0.

Because even a properly calculated van Deemter plot is always convex, it follows that *for a given ratio of packed volume to solvent flow rate* it is

Mass Transfer and Fluid Mechanics

always safe to increase percolation velocity. This is important for scale-up purposes because the laboratory columns used to obtained preliminary design data tend to be quite short. It also follows from this simple analysis that the common tendency to scale up at constant packed height is not conservative. In fact, this can lead to very high diameter-to-length ratios, which make it difficult to obtain uniform flow distribution.

It may also be seen that where H is properly calculated as Ls^2 this ratio of H to v_0 approaches an asymptotic value. Where this is the case, boundary layer resistance and convective dispersion are completely dominated by internal diffusion resistance of the packing particles. Because the behavior of individual particles depends only upon the time course of fluid concentration at their surface, the only external factor influencing system behavior is solvent holdup time in the bed. In modern industrial practice this high-velocity limit tends to be closely approached.

We find, then, that as long as we are operating close to this asymptotic limit system behavior depends upon only two parameters: the diffusional response time T_{diff} of the packing particles and the solvent holdup time L/v. Note, however, that all discussion so far has been based on the implicit assumption of very fast adsorption kinetics. This assumption is generally acceptable for ion exchange and hydrophobic interaction but not for the various types of affinity columns.

VI. TIME CONSTANTS AND PROCESS OPERATIONS

Most process chromatography is based upon either batch adsorption or gradient elution, and the concept of plates has limited relevance. However, the concepts of time constants and high-flow limits are highly useful, and we now briefly discuss their application. We begin by defining four dynamic time constants that fix the behavior of a chromatographic system: solvent holdup time T_{vel}, reaction time T_{rxn}, the dynamic response time of the packing T_{diff} to changes in local environment, and finally T_{prc}, the time required to complete the desired process. These are factors under the control of the designer once the chemical nature of the system has been fixed.

The first is easily and unambiguously defined:

$$T_{vel} \equiv V_0/Q \tag{28}$$

Here V_0 is the volume of the packed column accessible to the solvent, and Q is the volumetric solvent flow rate. Thus T_{vel} is the mean solvent holdup time.

The second is a bit harder to pin down, but we suggest here the definition

$$T_{rxn} \equiv q_0/\langle R_0 \rangle \tag{29}$$

Here q_0 is the average solute concentration with the packing particle in some arbitrarily chosen standard state, for example in equilibrium with feed for batch adsorption. The specific effect of T_{rxn} will depend upon the specific kinetic behavior of the system under consideration, but experience with heterogeneous catalysis [29] suggests that this is not a big problem.

We shall define the third time constant as the diffusional response time of the packing,

$$T_{diff} \equiv d_p^2/24 \mathcal{D}_A \tag{30}$$

making the implicit assumption that local dynamic response is dominated by internal diffusional resistance. The coefficient 1/24 is arbitrary, and others are commonly used. We are concerned now only with the effect of changing conditions.

Equation (30) is satisfactory under most present-day conditions in process chromatography, but one can modify this expression to account for boundary layer resistance and convective dispersion through use of an alternative time constant,

$$T_{diff}^{tot} \equiv T_{diff} \frac{s_{tot}}{s_{diff}} \tag{31}$$

if sufficient information is available. Here s_{tot} is the square root of the variance of the effluent curve under conditions of differential chromatography, and s_{diff} is the contribution of internal diffusion. The use of time constants to characterize a chromatographic system prove beneficial in many ways, but we shall consider only two here by way of example:

1. Ensuring that scale-up of a laboratory process will maintain product yield and quality
2. Explaining the performance of a system so as to gain insight into possible further improvement

In the first of these examples we attempt to show that maintaining time constant ratios on scale-up will accomplish the stated objective.

For both of these examples we will use a commercial adsorbent, S HyperD, developed by the BioSepra corporation, and examine its performance

Mass Transfer and Fluid Mechanics

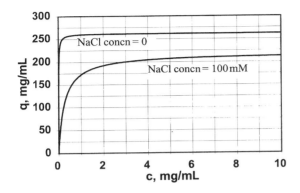

Figure 11. Nonlinear adsorption equilibria: lysozyme on S HyperD M from a 10 mM phosphate buffer at pH 6.5 with indicated NaCl concentrations.

in batch adsorption. It may be seen from Fig. 11 that the equilibrium distribution of solute is highly nonlinear, and we emphasize that the use of time constants is especially appropriate for such systems.

We begin in Tables 3 and 4 by testing our scale-up criteria through use of a simulator, the BioSepra Calculator, distributed by the company for predicting the performance of their HyperD family of adsorbents as well as those of selected competitors. In Table 3 we probe the effect of particle diameter and find very good agreement with the stated criterion for bed capacity at values of vd_p^2 less than 5 cm^3/h for both the 50 μm and 100 μm particles. In Table 4 we find excellent agreement with the effects of bed length and velocity over the entire range tested.

Table 3 Dimensional Analysis of Batch Adsorption: Effect of Particle Diameter[a]

	$d_P = 50$ μm			$d_P = 100$ μm		
$vd_p^2 \times 10^{-6}$	v cm/h	% Capacity	Pressure drop (bar)	v cm/h	% Capacity	Pressure drop (bar)
40				4000	64	136
10	4000	91	544	1000	91	34
5	2000	95.5	272	500	95.5	17
2.5	1000	97.8	136	250	97.8	8
1.25	500	98.9	68	125		
0.625	250	99.4	34			

[a] As determined by the BioSepra Calculator for HyperD adsorbents.

Table 4 Dimensional Analysis of Batch Adsorption: Effect of Column Length[a]

Length (cm)	Diameter (cm)	Velocity (cm/h)	% Capacity	Pressure drop (bar)
25	1	250	95.9	1
100	0.5	1000	96.3	8
400	0.25	4000	96.4	136
1600	0.125	16,000	96.4	2176

[a] As determined by the BioSepra Calculator for HyperD adsorbents.

The remaining time constant, T_{prc}, is tied to T_{vel} because we are dealing with the same chemistry and loading in all cases. Moreover, S HyperD is a cation exchanger where reaction kinetics are very fast so we do not need to consider T_{rxn} [30]. Where adsorption is very slow, as is often the case in affinity chromatography, it is customary just to keep T_{vel} constant on scale-up.

We now turn our attention to Fig. 12, where we see that dynamic capacity in batch adsorption is higher with a more dilute feed. This is highly desirable and also unusual. The reasons for this are believed to be as follows:

1. Because the adsorbent solute concentration q is nearly independent of solution concentration c, the internal diffusional behavior is all but independent of external concentration. Moreover, because all adsorbed species can easily move from one charge site to the next, because the latter are so close, internal diffusivity, hence T_{diff}, is insensitive to external concentration.

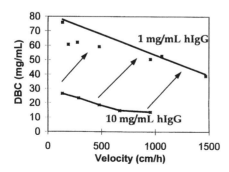

Figure 12. Effect of feed solute concentration on dynamic bed capacity (DBL) for hydrogel adsorbents.

Mass Transfer and Fluid Mechanics

2. Because process time, here time to saturate the column, increases in roughly inverse proportion to feed concentration, T_{prc} increases rapidly with dilution.
3. Finally, T_{vel} is identical for the two at any flow rate, and T_{rxn} is unimportant.

Thus the ratio T_{prc}/T_{diff} becomes larger with dilution, and this is favorable: More time is available for diffusion, and the effectiveness of diffusional processes is unchanged.

Time constants and their extension to characteristic lengths and other scale factors are useful for many other purposes, for example, in designing gradient elution systems [6]. We use them in the next section to examine some important flow-related problems.

VII. DEVIATIONS FROM SIMPLE MODELS

We now turn our attention to deviations from the predictions of simple one-dimensional chromatographic theories. Among the simplest of these are the departures of differential chromatographic effluent curves from Gaussian distributions, resulting either from the finite time required for solute bands to leave a column or from the use of lumped parameter diffusional resistance, both already touched on above. However, the most serious deviations are usually those resulting from maldistribution of flow. These latter can be caused either by inadequate flow distribution in column headers or by nonuniform packing, and they can have quite serious effects. Even very small flow perturbations, resulting, for example, from the existence of a bit of extra space between a set of adjacent packing particles, can suffice to "trigger" the kind of fingering shown in Fig. 2. Larger scale irregularities can increase dispersion, increasing plate height in differential chromatography or decreasing dynamic capacity in batch adsorption.

However, they can also have unexpected qualitative effects that can invalidate scale-up rather seriously. For example, sufficiently large scale variations in percolation velocity will cause bandwidth or breakthrough curves to broaden with the first power of effluent volume. This is clearly at odds with the one-half power typical of differential chromatography or the zero power ("constant-form fronts") of batch adsorption. It is thus important to determine when such variations occur and what to do about them if they are detected.

Many methods have been developed for detecting nonuniform packing and flow [31–39]. However, most of these are limited to laboratory scale columns, and we are concerned here only with process scale operations. A technique known as reverse-flow examination [40] has proven uniquely adapted to large columns, and we shall limit ourselves to it here. We begin by reviewing the effect of flow direction on pertinent chromatographic parameters.

A. Reversibility vs. Irreversibility and Detection of Flow Maldistribution

Here we examine the flow and mass transfer phenomena introduced above with an eye to their characteristic magnitudes and their symmetry with respect to time. We shall find in the next subsection that both characteristics are useful both for understanding these phenomena and for diagnostic purposes.

We first look at fluid flow through chromatographic devices and begin by noting that under normal process conditions it is steady, incompressible, and characterized by very low Reynolds numbers,

$$\text{Re} \equiv \frac{d_P v \rho}{\mu} \ll 1 \tag{32}$$

where ρ is fluid density. Such flows are described by the *creeping flow* equations of continuity and motion [15]:

$$(\nabla \cdot v) = 0 \tag{33}$$
$$(\nabla^2 [\nabla \times v]) = 0 \tag{34}$$

These equations are both linear in velocity, and it follows that these flow are *reversible*, i.e., if one changes the direction of flow, for example, by reversing the driving pressure, each fluid element will retrace the path originally made in the opposite direction. Moreover, changing the magnitude of velocity will change only the speed at which this retracing takes place. This expectation is borne out by the Blake–Kozeny equation, introduced to describe momentum transport in Table 2.

$$\frac{dp}{dz} = \frac{150}{4} v_0 \frac{(1-\varepsilon)^2}{\varepsilon^3} \left(\frac{\mu}{R^2}\right) \tag{35}$$

However, in the interstices between particles there are small eddies where inertial effects are appreciable, and these small-scale flows are irreversible: They depend in part on inertial forces that are quadratic in velocity. It is these small flow nonuniformities that give rise to the convective dispersion contribution to plate height.

Mass Transfer and Fluid Mechanics

Convective dispersion and all of the other above-defined contributions to plate height are also irreversible, so we arrive at a conclusion of considerable utility: *All contributions to plate height in one-dimensional models, all localized to the size scale of the packing particle diameter, are irreversible, whereas fluid motion on a size scale large relative to particle diameter is reversible.* It is also important to note that under the conditions of differential chromatography, mass transfer is governed by equations that are linear with respect to solute concentration; the trajectory of any particle is independent of its neighbors.

Now consider an experiment carried out in the apparatus shown in Fig. 13 that permits reversal of flow. After injection of a small pulse of tracer, the column is operated normally until the center of the tracer mass is exactly at the center of the packing. The flow is then reversed, and the solute concentration is measured at the initial inlet, which is now the outlet. If the measured solute distribution is compared with that for normal operation, it will have approximately the same peak appearance time, but it will be sharper and will more closely resemble the ideal Gaussian response than that for normal operation. Such a comparison is made in Fig. 14.

Any broadening caused by macroscopic flow maldistribution will now have disappeared, and the plate height will be that of a column with a macroscopically flat velocity profile but otherwise identical to the one under observation. Moreover,

$$s^2_{\text{normal}} - s^2_{\text{reversed}} = s^2_{\text{flow}} \qquad (36)$$

where the term on the right-hand side represents the dispersion caused by flow maldistribution. This conclusion follows from the additivity of contributions to the variance already discussed.

However, the validity of Eq. (36) depends upon the assumption that diffusion or dispersion of solute across fluid streamlines is negligible. The validity of this assumption can be tested at an order-of-magnitude level by noting that the distance δ_i traveled by a solute molecule in time t is

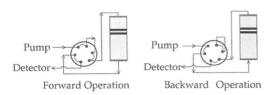

Figure 13. Schematic view of a reverse-flow setup.

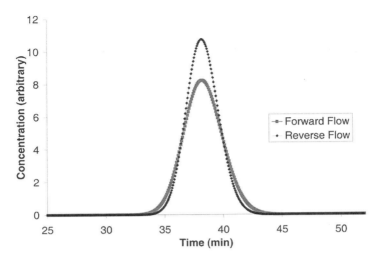

Figure 14. Comparison of a conventional run under normal operation and a reverse-flow run, where the direction of the solvent flow was reversed halfway through the column. Acetone pulse in a Millipore IsoPak column.

$$\delta_i \sim 2\sqrt{\mathcal{D}_{\text{eff}} T_i} \tag{37}$$

where $\mathcal{D}_{\text{eff}}$ is the effective diffusivity, which for process columns will be primarily a lateral dispersion coefficient, not significantly influenced by true or molecular diffusivity. It may be estimated from the correlation [41]

$$\frac{d_p v}{\mathcal{D}_{\text{eff}}} \approx \frac{17.5}{\text{Re}^{0.75}} + 11.4 \tag{38}$$

Because these particle-based Reynolds numbers tend to be very small, the lateral effective diffusivities, or dispersion coefficients, are far smaller than their axial counterparts.

B. Modeling of Actual Systems

We now apply the above discussion to the examination of commercial columns by way of example. We begin with a large packed column that is already in widespread industrial use and then go on to a promising new design just entering commercial operation at the time of writing.

Mass Transfer and Fluid Mechanics

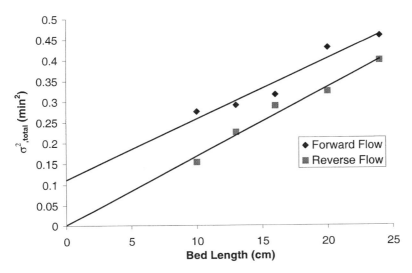

Figure 15. Examination of the effects of flow nonuniformity for the Millipore Iso-Pak chromatographic system.

1. Behavior of a Representative Packeded Column

Application of this technique is illustrated in Fig. 15 and Table 5 for the Millipore IsoPak system [42]. This is an industrial-scale chromatographic column, the unique loading apparatus of which permits control of packed depth. This characteristic permits measuring the degree of maldistribution caused by the header from that resulting from nonuniform packing. The effect of the packing is determined by the slope of the curves in Fig. 15, and the effect of the header is determined by the intercept. That is, it has been assumed that

$$\sigma^2_{\text{total}} = \sigma^2_{\text{header}} + \frac{H_{\text{bed}}}{v_0^2} L \tag{39}$$

Table 5 Comparison of Forward Flow and Reverse-Flow Parameters Determined by Eq. (39)

Flow direction	$\sigma^2_{\text{header}}(\text{min}^2)$	$H_{\text{bed}}(\text{cm})$
Forward	0.11 ± 0.04	0.010 ± 0.002
Reverse	0.00 ± 0.03	0.011 ± 0.001

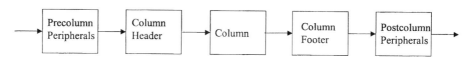

Figure 16. Postulated model for the additivity of the individual contributions of each subsystem in an overall chromatographic system.

This equation fits the data for both normal and reversed flow quite well and suggests that the packing technique is quite reproducible and insensitive to final packing depth.

It may be seen that the packing appears to be quite uniform: The slopes of the curves for normal and reversed flow are statistically equivalent. All flow nonuniformity is therefore contributed by the header, and it may be seen to have only a modest effect.

Making this distinction between the two causes of dispersion requires assuming additivity of the variances contributed by the header and packing (see Appendix A). This in turn requires that there be no diffusional transport or lateral variation in flow across their joint interface as suggested in Fig. 16. It is common practice to make this assumption, but critical examination is in order.

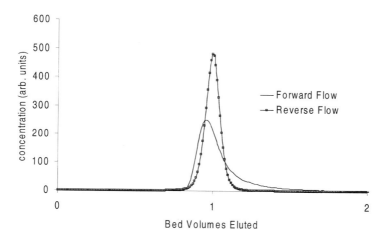

Figure 17. Comparison of conventional and reverse-flow elution profiles on a 10 mL Mustang using a tryptophan (nonadsorbing) tracer.

2. An Adsorptive Membrane Column

Header maldistribution is particularly serious for stacked membrane columns, both because they typically have very large diameters relative to packed height and because the intrinsic resolution of the packing tends to be very high.

Representative behavior for a column of this type is shown in Fig. 17. Here normal- and reverse-flow behavior is contrasted for the Pall Corporation 10 mL Mustang stacked membrane chromatograph. These results were obtained with a nonadsorbing ($k' = 0$) tracer, tryptophan, where

$$k' \equiv \frac{T_t - T_0}{T_0} \tag{40}$$

where k' is the retention factor, T_t is the mean residence of the tracer, here tryptophan, and T_0 is typically that of the solvent, a small nonadsorbing reference species. It may be seen that the reverse-flow data show a very high plate

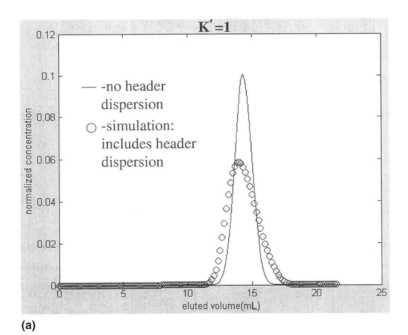

(a)

Figure 18. Simulated elution curves for adsorbing tracers on a 10 mL Mustang unit (circles). (a) $k' = 1$; (b) $k' = 3$; (c) $k' = 5$. Solid lines represent the elution curves in the absence of any header dispersion determined by reverse-flow experiment.

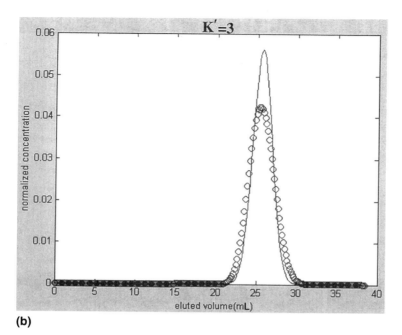

(b)

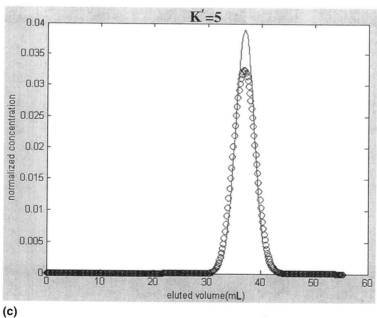

(c)

Figure 18. Continued

count, 480, which is quite impressive when one realizes that the stack is only 0.97 cm thick. However, maldistribution in the header causes a significant spreading and skewing of the tracer effluent curve.

Use of a nonadsorbing tracer is misleading, however, in that significantly better behavior is observed with solutes of separative significance, because these by definition do adsorb and thus exhibit larger mean residence times. The reason for this lies in the additivity of both residence times and variance, according to the expression

$$N_{observed} = \frac{(T_{bed} + T_{header})^2}{\sigma_{bed}^2 + \sigma_{header}^2} \qquad (41)$$

Simulations based on the reverse-flow data represented in Fig. 17 and dimensions of the Mustang unit are shown in Fig. 18 for k' values of 1, 3, and 5. It may be seen here that these predictions are much more favorable and that

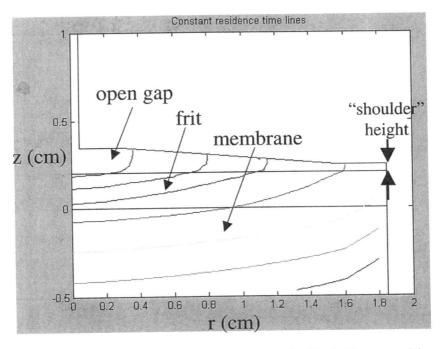

Figure 19. Simulation of macroscopic flow through the 10 mL Mustang module. Each line represents the position of the solute band at a constant residence time.

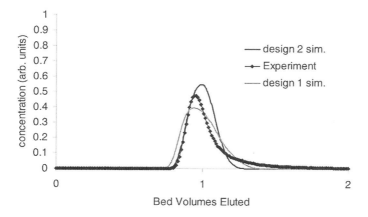

Figure 20. Comparison of experimentally determined elution curve for tryptophan with simulated curves that use a range of expected header configurations. The experimental and simulated profiles are for the 10 mL Mustang using a nonadsorbing tracer.

separability is already only modestly affected by header distribution for $k' = 3$, which is representative of practice.

Moreover, header design is amenable to even rather simple modeling procedures, and in Fig. 19 the results of a MATLAB simulation for the above Mustang unit are shown. By way of example, the calculated effect of shoulder height is shown in Fig. 20. Here predictions for heights of 0.44 mm (design 1 in Fig. 20) and 0.22 mm (design 2 in Fig. 20) are contrasted with actual operating data. This figure is useful from at least two standpoints:

1. It shows that the relatively simple simulation procedure used, though not quantitatively precise, is capable of capturing the essential features of the system.
2. It demonstrates the importance of controlling shoulder height, a relatively difficult variable to control in the manufacturing process.

Simulation procedures can be improved by using available programs if the engineering cost is warranted.

VIII. CONCLUDING REMARKS

Chromatographic equipment and process design have improved greatly over recent years [1], thanks in large part to their success in biotechnology. The

Mass Transfer and Fluid Mechanics

interplay of fluid mechanics and mass transfer has been a key factor, and organization of the "family" of available equipment can be clarified by thinking in these terms. This interplay will continue to be important in continuing improvement of these devices, and success in such an effort is particularly important in view of the increasing effectiveness and selectivity of membrane processes, which have an intrinsic cost advantage.

Increasingly, improvements will be concentrated on the departures of available equipment from the one-dimensional lumped parameter models still in widespread use. These improvements in packing uniformity and header design are basically fluid mechanical in nature. Packing presents the greater challenge, but this is a complicated issue just beginning to be understood. Header design is amenable to detailed modeling, and refinements can be expected in the near future.

NOMENCLATURE

A	column cross-sectional area, Eq. (2)
c_i	solute concentration in the moving fluid phase, Eq. (2)
C_i	scaled solute concentration, defined in Eq. (4)
d_P	diameter of a stationary-phase particle, Eq. (11)
D	column diameter
D_A	effective diffusivity of the solute of interest within the adsorbent particle, Eq. (15)
D_{eff}	effective solute diffusivity, Eq. (37)
D_F	effective solute diffusivity in unbounded solution, Eq. (12)
g	arbitrary constant, Eq. (19)
h	reduced plate height, defined in Eq. (11)
H	height equivalent to a theoretical plate (HETP) or plate height, defined in Eq. (9)
k'	capacity factor, defined in Eq. (40)
k_c	concentration-based fluid-phase mass transfer coefficient, Eq. (17)
K	ratio of masses of adsorbed and pore-liquid solutes at equilibrium, Eq. (14)
L	column length, Eq. (9)
m	parameter in the reduced plate height expression, defined in Eq. (15)
m_0	mass of solute fed to the column in a pulse, Eq. (3)

N	number of theoretical plates of a packed bed, defined in Eq. (8)
p	pressure, Eq. (35)
q_b	average solute concentration within a stationary-phase particle, Eq. (18)
q_s	solute concentration just inside a stationary-phase particle, Eq. (18)
q_0	average solute concentration in the packing particle in some arbitrarily chosen standard state, Eq. (29)
Q	volumetric flow rate of the percolating solvent, Eq. (25)
R	radius of a stationary-phase particle, Eq. (35)
R_{xy}	degree of separation of peaks x and y, Eq. (7)
R_0	average volumetric rate of reaction, Eq. (29)
Re	Reynolds number, defined in Eq. (32)
Re Sc	reduced velocity or diffusion Péclet number, defined in Eq. (12)
s	scaled standard deviation of the solute distribution with respect to the mean solute residence time, defined in Eq. (5)
S	surface area of a stationary-phase particle, Eq. (18)
Sh	Sherwood number for mass transport in the adsorbent granule, defined in Eq. (17)
t	time, Eq. (2)
T_{bed}	mean residence time of a solute in the packed bed alone, Eq. (41)
T_{header}	mean residence time of a solute in the column header alone, Eq. (41)
T_{diff}	diffusional response time of a stationary-phase particle, Eq. (20)
$T_{\text{diff}}^{\text{tot}}$	total diffusional response time of a stationary-phase particle, defined in Eq. (31)
T_i	mean residence time of a solute, Eq. (2)
$T_{\max}$	time of the appearance of the maximum peak concentration, Eq. (24)
T_{prc}	process completion time
T_{rxn}	reaction time, Eq. (29)
T_t	mean residence time of a solute, Eq. (40)
T_{vel}	mean solvent holdup time, Eq. (28)
T_0	mean residence time of a nonadsorbing solute, Eq. (40)
u	fraction of solute in the moving phase at long times, defined in Eq. (14)

Mass Transfer and Fluid Mechanics

v	interstitial solute velocity, Eq. (2)
v_0	superficial fluid velocity, Eq. (27)
V	volume of a stationary-phase particle, Eq. (18)
V_{col}	volume of the packed bed, Eq. (27)
V_0	volume of the packed bed accessible to the solvent, Eq. (28)
$W_{0.5}$	effluent peak width measured at half the maximum peak concentration, Eq. (24)
$\bar{z}$	mean position of a solute at a given time, defined in Eq. (16)

Greek Symbols

δ_i	distance traveled by a solute molecule, defined in Eq. (37)
ε	column void fraction, Eq. (2)
ε_b	volume fraction of interstitial fluid, Eq. (13)
μ	fluid viscosity, Eq. (32)
ρ	fluid density, Eq. (32)
σ	standard deviation of the solute distribution, Eq. (2)
σ_{bed}	standard deviation of the solute distribution in the packed bed alone, Eq. (39)
σ_{header}	standard deviation of the solute distribution in the column header alone, Eq. (39)

APPENDIX A ADDITIVITY OF MOMENTS

Consider two systems in series with a sharp tracer pulse introduced to the first at time zero. Define a response function for each of the two systems to such a pulse input as $h_i(t)$, with $i = 1$ or 2. Now consider introducing a unit pulse to systems 1 and 2 in series, so that the input to system 2 is the output from system 1.

Because the behavior of both systems is linear in concentration (each described by the diffusion equation with concentration-independent diffusivity and velocity) and the flow rate through the two systems is constant at Q, one may write the outlet concentration from the second system as

$$c(t) = \int_0^t h_1(t - \tau) h_2(\tau)\, d\tau \equiv h_1 * h_2 \qquad (A1)$$

This convolution integral is just a mathematical way of saying that the exit concentration from system 2 is just the sum of small differential inputs from system 1 and that each small input pulse is broadened to the extent described by $h_1(t)$. Moreover, the distribution of such pulses in time is given by $h_1(t)$.

For an outside observer, the various pulses enter system 2 at time $(t - \tau)$ and remain in system 2 for a time τ.

It is now convenient to note that if one takes the Fourier transform of the convolution integral and its moments one finds

$$\frac{M_1}{M'_2} = \langle t^2 \rangle \equiv \frac{\int_{-\infty}^{\infty} t^2 f(t)}{\int_{-\infty}^{\infty} f(t)} = \frac{F''(0)}{4\pi^2 F(0)} \tag{A2}$$

where $F(s)$ is the Fourier transform of $f(t)$ and the double prime denotes second-order derivatives with respect to s. Then, for our system,

$$M_2 = \int_{-\infty}^{\infty} t^2 c(t)\, dt = \int_{-\infty}^{\infty} t^2 (h_1 h_2)\, dt = \frac{(H_1 H_2)''|_{s=0}}{4\pi^2} \tag{A3}$$

where $H_i(s)$ is the Fourier transform of $h_i(t)$ and

$$(H_1 H_2)'' = H''_1 H_2 + 2H'_1 H'_2 + H_1 H''_2 \tag{A4}$$

It follows that

$$M_2(h_1 h_2) = M_2(h_1) + M_2(h_2) - \frac{2H'_1(0) H'_2(0)}{4\pi^2 H_1(0) H_2(0)} M_1(h_1 h_2) \tag{A5}$$

Now the variance is just the second moment taken about the mean residence time. For this special case, the last term of the above equation is zero, and we obtain

$$\sigma_{\text{tot}}^2 = \sigma_1^2 + \sigma_2^2 \tag{A6}$$

That is, the normalized variances are additive. This line of argument can obviously be extended to any number of subsystems in series, and the order in which they are connected has no effect on the overall variance. Note also that the range of the above integrals is not a problem, because for the system being considered there is no signal before time zero.

The third normalized central moment can be shown to be additive in the same way. Here,

$$M_3 = \frac{(H_1 H_2)'''|_{s=0}}{8\pi^3 i} \tag{A7}$$

$$= \frac{(H'''_1 H_2 + 2H''_1 H'_2 + 2H'_1 H''_2 + H_1 H'''_2)}{8\pi^3 i}$$

Mass Transfer and Fluid Mechanics

Again, the first derivative terms are zero, and additivity is proven. Note that this additivity does not hold for fourth moments and above.

REFERENCES

1. EN Lightfoot. The invention and development of process chromatography: Interaction of mass transfer and fluid mechanics. Am Lab 31:13–23, 1999.
2. EN Lightfoot. Speeding the design of bioseparations: a heuristic approach to engineering design. Ind Eng Chem Res 38:3628–3634, 1999.
3. EN Lightfoot, MCM Cockrem. What are dilute solutions? Sep Sci Technol 22: 165–189, 1987.
4. B Atkinson, F Mavituna. Biochemical Engineering and Biotechnology Handbook. 2nd ed. New York: Stockton Press, 1991, pp 96–99, 873–904.
5. S Yamamoto. Plate height determination for gradient elution of proteins. Biotechnol Bioeng 48:444–451, 1995.
6. JL Coffman, DK Roper, EN Lightfoot. High-resolution chromatography of proteins in short columns and adsorptive membranes. Bioseparation 4:183–200, 1994.
7. AM Athalye. Effects of high sample loading in nonadsorptive protein chromatography. PhD dissertation, University of Wisconsin-Madison, Madison, WI, 1993.
8. R Van Reis, AL Zydney. Protein ultrafiltration. In: MC Flickinger, SW Drew, eds. Encyclopedia of Bioprocess Technology: Fermentation, Biocatalysis, and Bioseparation. New York: Wiley, 1999, pp 2197–2214.
9. LJ Zeman, AL Zydney. Microfiltration and Ultrafiltration: Principles and Applications. New York: Marcel Dekker, 1996.
10. JH Holland. Hidden Order: How Adaption Builds Complexity. Reading, MA: Addison-Wesley, 1995, p 185.
11. EN Lightfoot. What chemical engineers can learn from Mother Nature. Chem Eng Prog 94:67–74, 1998.
12. AI Miller. Insights of Genius. New York: Springer-Verlag, 1996.
13. S Kauffman. At Home in the Universe. London: Oxford Univ Press, 1995.
14. DK Roper, EN Lightfoot. Comparing steady counter-flow and differential chromatography. J Chromatogr A 654:1–16, 1993.
15. RB Bird, WE Stewart, EN Lightfoot. Transport Phenomena. New York: Wiley, 2nd edition, 2002.
16. HA Chase. Development and operating conditions for protein purification using expanded bed techniques: the effect of the degree of bed expansion on adsorption performance. Biotechnol Bioeng 49:512–516, 1995.
17. M Petro, F Svec, JMJ Frechet. Molded continuous poly(styrene-co-divinylbenzene) rod as a separation medium for the very fast separation of polymers. Comparison of the chromatographic properties of the monolithic rod with columns

packed with porous and non-porous beads in high-performance liquid chromatography of polystyrenes. J Chromatogr A 752:59–66, 1996.
18. JL Liao, R Zhang, S Hjerten. Continuous beds for standard and micro high-performance liquid chromatography. J Chromatogr A 586:21–26, 1991.
19. A Klinkenberg, F Sjenitzer. Holding time distributions of the Gaussian type. Chem Eng Sci 5:258–270, 1956.
20. AM Athalye, SJ Gibbs, EN Lightfoot. Predictability of chromatographic protein separations: study of size-exclusion media with narrow particle-size distributions. J Chromatogr 589:71–85, 1992.
21. JL Coffman, EN Lightfoot, TW Root. Protein diffusion in porous chromatographic media studied by proton and fluorine PFG-NMR. J Phys Chem B 101:2218–2223, 1997.
22. HS Carslaw, JC Jaeger. Heat Conduction in Solids. 2nd ed. London: Oxford Univ Press, 1959, p 235.
23. JFG Reis, PT Noble, AS Chiang, EN Lightfoot. Chromatography in a bed of spheres. Sep Sci Technol 14:367–394, 1979.
24. Standard Practice for Liquid Chromatography Terms and Relationships. ASTM Publ E, 2000, pp 682–692.
25. MS Jeansonne, JP Foley. Improved equations for the calculation of chromatographic figures of merit for ideal and skewed chromatographic peaks. J Chromatogr 594:1–8, 1992.
26. SJ Gibbs, SJ Karrila, EN Lightfoot. Mass-transfer to a sphere with a nonuniform boundary-layer: implications in chromatographic analysis. Chem Eng J Biochem Eng J 36:B29–B37, 1987.
27. P Schneider, JM Smith. Adsorption rate constants from chromatography. AIChE J 14:762–771, 1982.
28. M Abramowitz, I Stegun. Section 26: Probability Functions. Handbook of Mathematical Functions. NBS Appl Math Ser Natl Bur Stand, New York, 1974.
29. PB Weisz. Diffusion and chemical reaction: an interdisciplinary excursion. Science 179:433–440, 1973.
30. MA Fernandez, G Carta. Characterization of protein adsorption by composite silica-polyacrylamide gel anion exchangers. I. Equilibrium and mass transfer in agitated contactors. J Chromatogr A 746:169–183, 1996.
31. FG Lode, A Rosenfeld, QS Yuan, TW Root, EN Lightfoot. Refining the scale-up of chromatographic separations. J Chromatogr A 796:3–14, 1998.
32. EN Lightfoot, JL Coffman, F Lode, QS Yuan, TW Perkins, TW Root. Refining the description of protein chromatography. J Chromatogr A 760:139–149, 1997.
33. DK Roper, EN Lightfoot. Separation of biomolecules using adsorptive membranes. J Chromatogr A 702:3–26, 1995.
34. RA Shalliker, BS Broyles, G Guiochon. On-column visualization of sample migration in liquid chromatography. Anal Chem 72:323–332, 2000.
35. BS Broyles, RA Shalliker, G Guiochon. Visualization of solute migration in chromatographic columns: quantitation of the concentration in a migrating zone. J Chromatogr A 867:71–92, 2000.

36. E Bayer, E Baumeister, U Tallarek, K Albert, G Guiochon. NMR imaging of the chromatographic process: deposition and removal of gadolinium ions on a reversed-phase liquid chromatographic column. J Chromatogr A 704:37–44, 1995.
37. U Tallarek, E Baumeister, K Albert, E Bayer, G Guiochon. NMR imaging of the chromatographic process: migration and separation of bands of gadolinium chelates. J Chromatogr A 696:1–18, 1995.
38. EJ Fernandez, CA Grotegut, GW Braun, KJ Kirschner, JR Staudaher, ML Dickson, VL Fernandez. The effects of permeability heterogeneity on miscible viscous fingering: a 3-dimensional magnetic-resonance-imaging analysis. Phys Fluids 7:468–477, 1995.
39. E Bayer, W Muller, M Ilg, K Albert. Visualization of chromatographic separations by NMR imaging. Angew Chem Int Ed Engl 28:1029–1032, 1989.
40. DK Roper, EN Lightfoot. Estimating plate heights in stacked-membrane chromatography by flow reversal. J Chromatogr A 702:69–80, 1995.
41. JB Butt. Reaction Kinetics and Reactor Design. 2nd ed. New York: Marcel Dekker, 1999, p 351.
42. JS Moscariello, G Purdom, JL Coffman, TW Root, EN Lightfoot. Characterizing the performance of industrial-scale columns. J Chromatogr A 908:131–141, 2001.

3
Optimization of Preparative Separations

Attila Felinger
University of Veszprém, Veszprém, Hungary

I. INTRODUCTION

Several studies have focused recently on the determination of the optimum experimental conditions and column design parameters of separations in preparative liquid chromatography. The nonlinear nature of preparative chromatography complicates the separation process so much that the derivation of general conclusions regarding the optimum conditions is a rather difficult, if not impossible, task. The optimization of preparative chromatography is further complicated by the fact that the choice of the objective function is not simple. In industrial applications, the production cost would be the major factor to consider. However, many components of the production cost are beyond the scope of the separation process itself. Accordingly, a more straightforward approach is chosen, and usually simply the production rate is maximized [1,2].

Some economic consequences of the conditions under which a separation is carried out have also been discussed. Optimum experimental conditions were determined in situations in which the cost of the solvent—a major cost factor in certain applications of preparative liquid chromatography—was also taken into account [3–6]. A hybrid objective function was introduced in order to weigh the importance of both the production rate (which should be as high as possible) and the solvent consumption (which should be as low as possible) [3,4]. Because all the modes of operation considered are usually applied as batch processes, the recovery yield during each run is lower than unity. Some optimization for maximum production rate was undertaken with a minimum yield constraint [1,2,7]. A very beneficial objective function was introduced: the product of the production rate and the recovery yield [8,9]. It was shown that, by means of that objective function, experimental conditions can be found such that the production rate is only slightly lower than when the production rate is the objective function whereas the recovery yield is significantly improved. This trade-off of a slight decrease in the production rate for a considerable yield improvement would be most economical.

The optimization of the different modes of preparative chromatography allowed the comparison of elution and displacement chromatography [7,10,11], revealing the relative advantages of either mode of separation. These studies suggested that elution can offer a higher production rate than displacement chromatography but delivers less concentrated fractions, which may significantly increase the cost of downstream processing.

The practical interest of these investigations of the optimization of the experimental conditions arises from the current availability of powerful personal computers. These machines allow the convenient calculation of the optimum conditions by the use of accurate models of the nonlinear separation process. The a priori knowledge of the competitive isotherms for the mixture of interest and of the parameters of the Knox plate height equation allows rapid calculation of the band profiles of each component, their integration, the determination of the positions of the cut points, the production rate, and the recovery yield. Combining the algorithms that perform these tasks with an optimization algorithm permits the determination of the experimental conditions that maximize the relevant objective function. Then, the experimental conditions obtained by computation can be fine-tuned experimentally. The introduction of the modeling step just described into the optimization process highly reduces its time and cost demands [12].

Although in several cases multicomponent mixtures are purified, the optimization problem can usually be reduced to the investigation of a binary mixture because there always exists a *limiting impurity* that elutes closest to the major component. The limiting impurity sets the constraint for the throughput, production rate, recovery yield, and other parameters [13].

II. DEFINITIONS

To facilitate a discussion of the optimization of the experimental and column design parameters, we need to introduce the following definitions of the parameters used in this study.

Optimization of Preparative Separations

The *loading factor* is the ratio of the total amount of the components in the sample to the column saturation capacity. For one component of the mixture the loading factor is

$$L_{f,i} = \frac{V_s C_i^0}{(1-\varepsilon) S_A L q_{s,i}} \tag{1}$$

where V_s is the sample volume, S_A is the column cross-sectional area, ε is the total porosity of the column, C_i^0 is the injected concentration of component i, L is the column length, and $q_{s,i}$ is the saturation capacity of the isotherm.

The *cycle time* is the time between two successive injections. To achieve uniform handling of the different modes of chromatographic separation, the cycle time should be introduced in such a way that the definition is applicable to all modes. The column must be regenerated after each run in both displacement and gradient elution. The same operation must also be performed occasionally in isocratic elution. The cycle time should account for the time required by this column regeneration step. Unless we introduce a new parameter, which would markedly complicate the issue, an arbitrary decision must be made. Therefore, the cycle time can be defined as the sum of the retention time of the more retained component under analytical (i.e., linear) conditions and the time needed for the regeneration of the column with j column volumes of mobile phase [11,14,15]:

$$\Delta t_c = t_{R,2} + j t_0 = t_0 (j + 1 + k_2') \tag{2}$$

Calculations indicate that under optimum conditions the column must be operated under the maximum pressure allowed. Thus, the regeneration time cannot be reduced by pumping the solvent faster.

Alternative definitions for the cycle time are also possible. Another option for defining it is the time between the emergence of the less retained component and the disappearance of the second one [1,2,4,7]. This latter definition is applicable only if there is no need to regenerate the column after each run.

The *recovery yield* of component i, Y_i, is defined as the ratio of the amount of component i recovered in the collected fraction to the amount of the same component in the sample.

The *purity* of a component is its concentration in the collected fraction. A typical value usually considered is 99%.

The *production rate* is the amount of a purified component produced per unit column cross-sectional area per unit time.

$$\mathrm{Pr}_i = \frac{V_s C_i^0 Y_i}{\varepsilon S_A \Delta t_c} \tag{3}$$

The *specific production* (SP) is defined as the amount of purified product obtained by using a unit amount of solvent.

$$SP = \frac{\text{amount of pure product}}{\text{amount of solvent}} \quad (4)$$

The above-defined specific production is the reciprocal of the solvent consumption.

In gradient elution chromatography, the rate of change of the mobile phase composition during gradient elution can be described by the *gradient slope*, $\Delta\varphi/t_G$, i.e., the ratio of the change in the modifier volume fraction to the gradient time [16].

The *gradient steepness* G is proportional to the gradient slope:

$$G = \frac{\Delta\varphi}{t_G} t_0 S \quad (5)$$

where φ is the volume fraction of the modifier, t_G is the gradient time, and S is the solvent strength parameter (the slope of the ln k' vs. φ plot).

III. COLUMN CHARACTERISTICS

In order to model—under linear conditions—the effect of mobile phase velocity on band broadening, a proper model should be employed. For this purpose, the Knox plate height equation is the most straightforward choice [17]:

$$h = \frac{A}{\nu} + B\nu^{1/3} + C\nu \quad (6)$$

where A, B, and C are numerical coefficients, $h = H/d_p$ is the reduced plate height, H is the actual height equal to a theoretical plate, d_p is the average particle size, and $\nu = ud_p/D_m$ is the reduced mobile phase velocity or particle Peclet number.

The pressure drop between the column inlet and outlet can be calculated using the equation

$$\Delta P = \frac{u\eta L}{k_0 d_p^2} \quad (7)$$

where ΔP is the pressure drop, k_0 the specific column permeability (approximately 1×10^{-3}), and η the mobile phase viscosity.

A. Adsorption Isotherms

The isotherms of the sample components should be known prior to the optimization. It is often reasonable to assume that the components of the sample behave according to the competitive Langmuir isotherm model. For a binary mixture the competitive Langmuir isotherm is written as

$$q_i = \frac{a_i C_i}{1 + b_1 C_1 + b_2 C_2}, \qquad i = 1,2 \tag{8}$$

where a_i and b_i are numerical coefficients. For thermodynamically consistent calculations, the isotherm parameters should be chosen such that the column saturation capacity, $q_s = a_i/b_i$, is the same for all components [18].

The use of the competitive Langmuir model is a simplification; in practice most competitive isotherms do not follow it exactly [19]. This model remains a good choice for theoretical studies, however, because it is simple and deviations from competitive Langmuir isotherm behavior are often small and are properly accounted for by the consequences of the difference in the column saturation capacities for the components studied [20].

In reversed-phase gradient elution chromatography, the linear solvent strength model connects the mobile phase composition with the retention factor through

$$\ln k'_i = \ln k'_{0,i} - S_i \varphi \tag{9}$$

where φ is the volume fraction of the modifier in the mobile phase. In this case, the strong solvent is assumed to be nonadsorbed on the stationary phase, which seems to be a reasonable assumption [21].

Because the retention factor at infinite dilution and the a parameter of the Langmuir isotherm are related through $k' = aF$, where F is the phase ratio, $F = (1 - \varepsilon)/\varepsilon$, the isotherm parameters can be determined as a function of the mobile phase composition, utilizing the equation above rewritten as

$$a_i(\varphi) = a_{0,i} \exp(-S_i \varphi) \tag{10}$$

It was shown by El Fallah and Guiochon [21] that in many cases, especially during the separation of small molecules, the saturation capacity, $q_{s,i} = a_i/b_i$, of the stationary phase remains practically unchanged when the composition of the mobile phase varies. If this statement holds true, the dependencies of the coefficients a_i and b_i on the mobile phase composition are the same or very similar.

IV. OBJECTIVE FUNCTIONS

As discussed above, the choice of the objective function is somewhat arbitrary during the optimization of preparative liquid chromatography. The objective functions discussed here focus merely on the separation process regardless of the economics of the industrial environment.

It is simplest to maximize the productivity or the *production rate* itself. This, however, often results in unacceptable recovery yields. A low recovery yield requires recycling of the mixed fractions. The recovery yield at the maximum production rate strongly depends on the separation factor. In the cases of difficult separations, when the separation factor under linear conditions is around or lower than $\alpha = 1.1$, the recovery yield is not higher than 40–60%. Even in the case of $\alpha = 1.8$, the recovery yield at the maximum production rate is only about 70–80% [1]. The situation is still less favorable in displacement chromatography, particularly if the component to be purified is more retained than the limiting impurity. In this case, from one side the impurity contaminates the product, whereas from the other side the displacer contaminates it.

When the *production rate* and the *recovery yield* are simultaneously maximized via their product Pr Y, the optimum conditions usually result in a configuration in which the recovery yield is much higher with a small sacrifice in production rate.

The characterization of the amount of solvent needed for a particular separation is conveniently described when using the term *specific production*. The minimum solvent consumption and the maximum specific production demand identical experimental conditions. Furthermore, the definition of specific production allows consistency in the use of hybrid objective functions consisting of both the production rate and the reciprocal of the solvent consumption.

The amount of solvent pumped through the column during one cycle is proportional to the mobile phase flow rate and to the cycle time. The amount of purified product made in one cycle is the product of the amount injected and the recovery yield. Therefore the specific production can be written as

$$SP_i = \frac{V_s C_i^0 Y_i}{\Delta t_c F_v} = \frac{V_s C_i^0 Y_i}{\Delta t_c \varepsilon S u} \tag{11}$$

Combining Eqs. (3) and (11) gives

$$SP_i = \frac{\Pr_i}{u} \tag{12}$$

Optimization of Preparative Separations

This simple relationship between production rate and specific production shows that in order to operate chromatographic columns at low levels of solvent consumption we need to achieve a high production rate at a low mobile phase linear velocity.

Because the feed volume injected is the product of the injection time and the flow rate, Eq. (12) can be rewritten as

$$SP_i = \frac{t_s}{\Delta t_c} C_i^0 Y_i \tag{13}$$

V. THEORETICAL CONSIDERATIONS

Several conclusions regarding the determination of the optimum experimental conditions can be simply derived by means of either the ideal or the equilibrium-dispersive model. Although there are more detailed models of nonlinear chromatography, the equilibrium-dispersive model accounts for the majority of the band profiles observed in preparative chromatography [12].

A. The Ideal Model of Chromatography

The ideal model has the advantage of supplying the thermodynamical limit, an estimate of the highest production rate and recovery yield allowed. These values could be achieved only with a column of infinite efficiency. In the case of a binary separation, the ideal model allows the calculation in algebraic form of all the attributes of the band profiles, except for the retention time of the shock of the less retained component of the pair. The optimum experimental conditions were reported previously for both components [22]. The following mass balance equation is written for each component:

$$\frac{\partial C_i}{\partial t} + F \frac{\partial q_i}{\partial t} + u \frac{\partial C_i}{\partial z} = 0 \tag{14}$$

where C_i and q_i are the concentrations of component i in the mobile and stationary phases, respectively; z is the length, t the time.

The recovery yield reported by Golshan-Shirazi and Guiochon [22] for the more retained component is

$$Y_2 = \begin{cases} 1 & \text{if } L_{f,2} \leq \dfrac{1}{1+r_1/\alpha}\left(\dfrac{(\alpha-1)/\alpha}{1-x}\right)^2 \\ \dfrac{1}{L_{f,2}(1+r_1/\alpha)}\left(\dfrac{(\alpha-1)/\alpha}{1-x}\right)^2, & \text{otherwise} \end{cases} \tag{15}$$

where r_1 is the positive root of the equation

$$b_2 C_2^0 r^2 - [\alpha - 1 + b_2(C_1^0 - C_2^0)]r - b_2 C_1^0 = 0 \tag{16}$$

x is calculated from the required degree of purity of the more retained component, Pu_2,

$$x = \left(\frac{1 - Pu_2}{\alpha r_1 Pu_2}\right)^{1/2} \tag{17}$$

For the definition of the cycle time, in the following we take into account the time required for regeneration and re-equilibration of the column after each run. For this reason, we must estimate the amount of washing required. Assuming that j column volumes of solvent is needed to regenerate the column, the cycle time will be defined as the analytical retention time of the more retained component plus j times the void time. With this definition, the maximum production rate can be expressed as

$$Pr_2 = \frac{uk_2'}{(1 + r_1/\alpha)b_2(k_2' + j + 1)} \left(\frac{(\alpha - 1)/\alpha}{1 - x}\right)^2 \tag{18}$$

This maximum production rate is reached when the two bands just touch each other. It remains constant when the loading factor is increased further. The general expression for production rate at any load is thus

$$Pr_2 = \begin{cases} \dfrac{uk_2' L_{f,2}}{b_2(k_2' + j + 1)} & \text{if } L_{f,2} \leq \dfrac{1}{1 + r_1/\alpha}\left(\dfrac{(\alpha - 1)/\alpha}{1 - x}\right)^2 \\ \dfrac{uk_2'}{(1 + r_1/\alpha)b_2(k_2' + j + 1)}\left(\dfrac{(\alpha - 1)/\alpha}{1 - x}\right)^2 & \text{otherwise} \end{cases} \tag{19}$$

Similar calculations can be carried out for the optimization of the production rate of the less retained component. For the sake of simplicity, we assume that 100% purity of the collected fraction is required (which is possible with the ideal model but is not with other models or in actual practice). In this case, the recovery yield of the less retained component is [22]

$$Y_1 = \begin{cases} 1 & \text{if } L_{f,2} \leq \dfrac{[(\alpha - 1)/\alpha]^2}{1 + r_1/\alpha} \\ 1 - \dfrac{\alpha r_1}{L_{f,1}(1 + r_1/\alpha)}\left[\dfrac{\alpha - 1}{\alpha} - \left(1 + \dfrac{r_1}{\alpha}\right)^{1/2}\sqrt{L_{f,2}}\right]^2 & \text{otherwise} \end{cases} \tag{20}$$

Optimization of Preparative Separations

With the definition of the cycle time given above, the production rate of this component is

$$\text{Pr}_1 = \frac{uL_{f,1}Y_1 k_2'}{b_2(k_2' + j + 1)} \tag{21}$$

The optimum conditions for the purification of the less retained and more retained components are entirely different. Because of the displacement effect, the intensity of which increases with increasing loading factor for the more retained component, the maximum production rate of the less retained component cannot be observed except by using values of the loading factor that are so high that the recovery yield becomes unacceptably poor. The production rate of the more retained component reaches a plateau when bands start to overlap.

Because of this phenomenon, the numerical method of optimization based on the calculation of solutions of either the ideal or the equilibrium-dispersive model is cumbersome in many instances. The surface of the objective function is rather flat. Some improvement could be observed by defining a cycle time that depends on the loading factor [1,2,4,7].

In Fig. 1, we compare the production rates that can be achieved for the two components. However, except at very high separation factors, high recovery yield can be reached only by constraining the optimization by introducing a minimum yield, usually set at 90%. Instead of maximizing the production rate with or without a yield constraint, the simultaneous maximizations of both the production rate and the recovery yield appears to be advantageous. A convenient way to achieve this goal is to choose an objective function that is the product of the production rate and the recovery yield. This objective function gives the following result when the ideal model is assumed.

$$\text{Pr}_2 Y_2 = \begin{cases} \dfrac{uk_2' L_{f,2}}{b_2(k_2' + j + 1)} & \text{if } L_{f,2} \leq \dfrac{1}{1 + r_1/\alpha}\left[\dfrac{(\alpha - 1)/\alpha}{1 - x}\right]^2 \\ \dfrac{uk_2'}{(1 + r_1/\alpha)^2 b_2(k_2' + j + 1)L_{f,2}}\left[\dfrac{(\alpha - 1)/\alpha}{1 - x}\right]^4 & \text{otherwise} \end{cases} \tag{22}$$

for the more retained component, and

$$\text{Pr}_1 Y_1 = \begin{cases} \dfrac{uL_{f,1} k_2'}{b_2(k_2' + j + 1)} & \text{if } L_{f,2} \leq \dfrac{[(\alpha - 1)/\alpha]^2}{1 + r_1/\alpha} \\ \dfrac{uL_{f,1} k_2'}{b_2(k_2' + j + 1)}\left\{1 - \dfrac{\alpha r_1}{L_{f,1}(1 + r_1/\alpha)}\left[\dfrac{\alpha - 1}{\alpha} - \left(1 + \dfrac{r_1}{\alpha}\right)^{1/2}\dfrac{1/2}{L_{f,2}}\right]^2\right\}^2 & \text{otherwise} \end{cases} \tag{23}$$

for the less retained component.

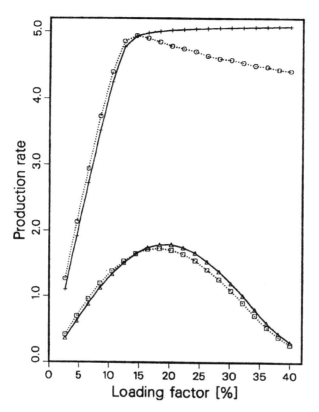

Figure 1 Effect of the definition of cycle time on the production rate of the two components of a binary mixture. Solid lines: Production rates of the first (△) and second (+) components if cycle time is $\Delta t_c = t_{R,2} - t_0$. Dotted line: Production rates of the first (□) and second (○) components if cycle time is the length of time between the emergence of the less retained component and the vanishing of the more retained component. (Reprinted with permission from Ref. 1.)

Equations (22) and (23) are the solutions of the ideal model. If we use these two objective functions for the optimization of the experimental conditions in preparative chromatography, we will find optimum conditions under which the production rate is somewhat lower than it would be if the production rate itself were the objective function, but the recovery yield will be much higher.

Optimization of Preparative Separations

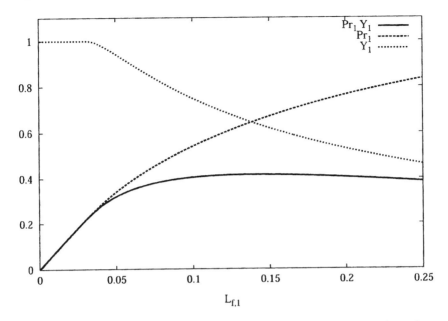

Figure 2 Plot of the calculated production rate, recovery yield, and their product for the less retained component against the loading factor, based on the ideal model. $\alpha = 1.2$; $k'_2 = 2$, $C_1^0 = 100$ mg/mL, $C_2^0 = 300$ mg/mL, $q_s = 260$ mg/mL. (Reprinted with permission from Ref. 8.)

Figures 2 and 3 show the results calculated with the ideal model. When the purification of the more retained component is optimized, the recovery yield is 100% until the bands of the two components touch each other. The production rate increases linearly with increasing value of the loading factor. When the two bands begin to overlap, the recovery yield begins to decrease with increasing value of $L_{f,2}$ whereas the production rate remains constant. Choosing the product of these two quantities as a Pr Y objective function, we observe a sharp maximum at the loading factor where overlap begins. The yield achieved is 100% and the production rate is maximum, an ideal situation.

On the other hand, when the separation of the less retained component is optimized, the ideal model fails to identify an optimum value of the loading factor for the maximum production rate. The production rate increases monotonously with increasing loading factor whereas the recovery yield decreases with increasing value of $L_{f,1}$ beyond the value at which a touching band is

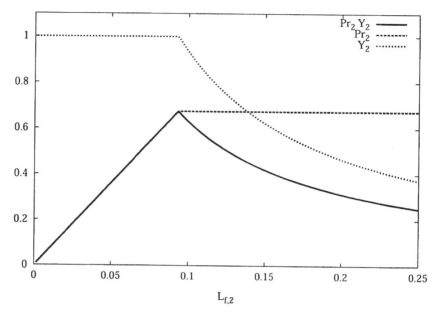

Figure 3 Same as Fig. 2, except that calculations were made for the more retained component. (Reprinted with permission from Ref. 8.)

reached. With the Pr Y objective function we are able to find an optimum value of the loading factor, although the maximum is very flat in most instances.

Figure 4 shows plots of Pr Y for the less retained component versus its loading factor at $\alpha = 1.5$ and at various values of the retention factor between 1 and 6. When the separation factor is not small, the ideal model predicts the existence of a relatively sharp maximum of the Pr Y objective function for the less retained component at tolerable experimental conditions. The recovery yield at the optimum is smaller than it is for the more retained component, but it is still approximately 75% in the entire range of retention factor studied. Note that it was shown in a previous publication that the optimum conditions for maximum production rate were at low values of the retention factor [1,2]. Figure 4 shows that this is also true for the Pr Y objective function.

B. Equilibrium-Dispersive Model

In the equilibrium-dispersive model we assume instantaneous equilibrium between the stationary and mobile phases and use an apparent dispersion term

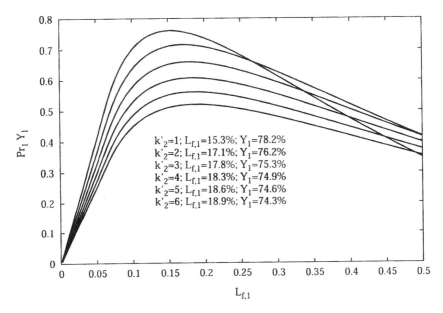

Figure 4 Plot of the Pr Y objective function against the loading factor. Calculations for the less retained component on the basis of the ideal model. $\alpha = 1.5$; $C_1^0 = 100$ mg/mL, $C_2^0 = 100$ mg/mL, $q_s = 260$ mg/mL. (Reprinted with permission from Ref. 8.)

to account for the smoothing effects of the axial dispersion and the finite rate of mass transfers [12]. A mass balance equation is written for each component of the sample:

$$\frac{\partial C_i}{\partial t} + F \frac{\partial q_i}{\partial t} + u \frac{\partial C_i}{\partial z} = D_a \frac{\partial^2 C_i}{\partial z^2} \tag{24}$$

where D_a is the apparent dispersion coefficient.

We can rewrite Eq. (24), using reduced time and length variables ($\tau = t/t_0$; $x = z/L$). Using these variables and substituting the apparent dispersion coefficient by

$$D_a = \frac{Hu}{2} = \frac{L^2}{2Nt_0} \tag{25}$$

the above-written mass balance equation can be transformed into

$$\frac{\partial C_i}{\partial \tau} + F\frac{\partial q_i}{\partial \tau} + \frac{\partial C_i}{\partial x} = \frac{1}{2N}\frac{\partial^2 C_i}{\partial x^2} \qquad (26)$$

This equation indicates that the reduced band profiles calculated by solving the equation system of the equilibrium-dispersive model do not depend on the column length or the mobile phase flow velocity, but only on the parameters that appear in Eq. (26) or in the initial and boundary conditions, that is, on the column efficiency and the loading factor. This is true as long as the boundary and initial conditions are kept constant on the reduced scale. This means that in changing the column length from L_1 to L_2 and the linear velocity u_1 to u_2 while keeping the column efficiency constant, we have to keep the reduced injection time unchanged to achieve identical band profiles at the column outlet ($z = L$ or $x = 1$) on the reduced scale:

$$\frac{t_{s,1}}{t_{0,1}} = \frac{t_{s,2}}{t_{0,2}} \qquad (27)$$

Because the dead time is $t_0 = L/u$ and the feed volume injected is $V_s = t_s u \varepsilon S$, Eq. (27) becomes

$$\frac{V_{s,1}}{L_1} = \frac{V_{s,2}}{L_2} = \frac{L_f(1-\varepsilon)Sq_s}{C_1^0 + C_2^0} \qquad (28)$$

which means that the ratio of the actual volume injected to the column length must be kept constant in order to obtain identical band profiles on the reduced time scale. This quantity is proportional to the loading factor.

Although Eq. (28) contains the value of the injected concentrations too, it was shown that the shape of the band profile does not depend on the concentration of the sample injected as long as we do not reach a significant volume overload [23]. That is why the specific production and the recovery yield depend only on the loading factor and the column efficiency, no matter how this latter was achieved. This is not true for the production rate. The latter is given by Eq. (3), which with the dimensionless parameters becomes

$$\mathrm{Pr}_i = \frac{t_s}{\Delta t_c} C_i^0 Y_i u \qquad (29)$$

Production rate depends on the same parameters as the specific production but also on the mobile phase velocity.

C. The Parameters To Be Optimized

As we have seen, solvent consumption depends only on the column efficiency and the loading factor. The production rate depends on more parameters: It is also influenced by the linear velocity of the mobile phase. It was shown [1] that when all design and operation parameters are optimized together for maximum production rate, the column should always be operated at the flow rate corresponding to the maximum inlet pressure allowed, ΔP_M. Rewriting Eq. (7) for the reduced velocity instead of the linear velocity gives

$$\nu = \frac{\Delta P_M \, k_0 d_p^3}{D_m \eta L} \tag{30}$$

Using a simplified plate height equation ($h = c\nu$) that is valid at high mobile phase velocity, Golshan-Shirazi and Guiochon [24] showed that there is no independent optimum value for the column length and the particle diameter and that only the ratio d_p^2/L should be optimized. They expressed the optimum ratio as

$$\left[\frac{d_p^2}{L}\right]_{opt} = \frac{\alpha - 1}{\alpha} \frac{[(\alpha - 1/2)\sqrt{\alpha^2 - \alpha + 1/36} - (\alpha^2 - \alpha - 1/12)]^{1/2}}{4[(k_2' + 1)/k_2']/(2k_0 \Delta Pc/3D_m \eta)^{1/2}} \tag{31}$$

where D_m is the molecular diffusivity of the sample in the mobile phase. By neglecting the competitive interaction between the sample components, we obtain the simplified result derived by Knox and Pyper [25]:

$$\left[\frac{d_p^2}{L}\right]_{opt} = \frac{(\alpha - 1)/\alpha}{4[(k_2' + 1)/k_2']/(2k_0 \Delta Pc/3D_m \eta)^{1/2}} \tag{32}$$

The existence of d_p^2/L was confirmed by the results of numerical calculations made using the equilibrium-dispersion model in overloaded elution [1] and displacement [2] chromatography. As long as the ratio d_p^2/L is unchanged, the production rate remains almost constant. This is illustrated in Fig. 5, where production rate is maximized for various column lengths and particle diameters. The contour lines indicate that in the case of each component the contour lines for the maximum production rate run nearly parallel with the contour lines for constant d_p^2/L. Thus, it is sufficient to optimize only one column design parameter (either d_p or L).

Because it does not matter through which combination of particle size and column length the column efficiency is achieved, we can carry out a two-parameter optimization for the loading factor and the column efficiency (or the column length or particle size, whichever is most convenient). The loading

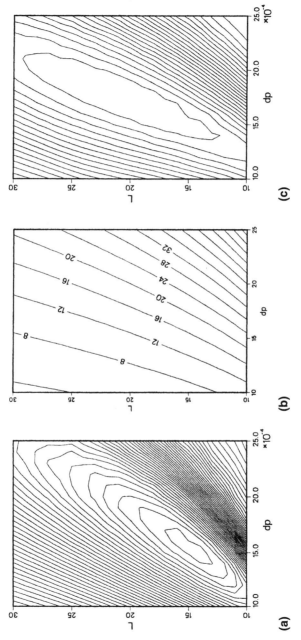

Figure 5 Contour plots of (a) the production rate of the less retained component as a function of the column length and the average particle diameter (mixture composition 3:1, $\alpha = 1.5$, $k'_1 = 6$); (b) constant d_p^2/L; (c) the production rate of the more retained component. (Reprinted with permission from Ref. 2.)

Optimization of Preparative Separations

factor is changed by setting the feed volume injected. As a second parameter, we can optimize the value of the average particle diameter while keeping the column length constant. From the value of the average particle size, we can determine the maximum linear velocity allowed. Then we can calculate the reduced plate height and consequently the column efficiency.

The reduced plate height at this velocity can be calculated by using Eq. (6) for the derivation of the column efficiency $N (= L/hd_p)$:

$$N = L \left[\frac{2D_m \eta L}{\Delta P k_0 d_p^2} + d_p^2 \left(\frac{\Delta P k_0}{D_m \eta L} \right)^{1/3} + \frac{\Delta P k_0 d_p^4}{D_m \eta L} \right]^{-1} \quad (33)$$

We have no practical options to adjust the values of η, D_m, and k_0. Thus, Eq. (33) shows that if the column length is kept constant and the column is operated at the maximum possible inlet pressure, substituting the optimum value of the particle diameter into Eqs. (30) and (33) gives the optimum flow rate and column efficiency for maximum production rate. In this manner, the optimization of the production rate in overloaded elution chromatography can be studied as a function of the same two parameters as that of the specific production. This approach is valid as long as the maximum inlet pressure is constant and the Knox plate height equation [Eq. (6)] remains unchanged. This observation makes easier the comparison of the optimum conditions required for minimum solvent consumption and for maximum production rate.

1. The Effect of Column Length

It is clear from Eq. (26) that band profiles are identical on a normalized time scale at constant values of the loading factor and plate number. However, it is not obvious that the optimum chromatograms obtained under experimental conditions leading to the same values of these parameters should also be identical. To confirm whether there is an influence of the column length, Fig. 6 shows optimization carried out successively with two different column lengths. The optimum chromatograms were obtained in the case of displacement chromatography. In each of the two successive calculations, the loading factor, the plate number, and the displacer concentration were optimized simultaneously. All the other parameters were kept constant and identical for the two column lengths. Figure 6 demonstrates that except for the minor, unavoidable consequences of the uncertainty of numerical origin in finding the global optimum the chromatograms are indeed identical when plotted on the reduced time scale.

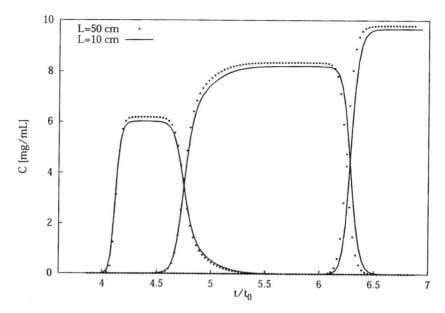

Figure 6 Optimum displacement chromatograms. The loading factor, the plate number, and the displacer concentrations were optimized simultaneously at $k'_1 = 10$ and $\alpha = 1.5$. The column lengths are 10 cm (solid line) and 50 cm (dotted line). The optimum loading factors L_f were 25.1 and 24.1%, respectively; the optimum plate numbers, 1096 and 1271 theoretical plates, respectively; and the displacer concentrations C_d, 9.6 and 9.4 mg/mL, respectively. The optimum conditions for the 10 cm column were Pr $Y_{max} = 0.164$. As a consequence, the other optimum experimental conditions are $d_p = 8.2$ μm, $v = 69.9$, $u = 0.848$ cm/s, $Y = 78.5\%$, and Pr $= 0.209$. For the 50 cm long column, they were Pr $Y_{max} = 0.182$, $d_p = 18.5$ μm, $v = 159$, $u = 0.858$ cm/min, $Y = 84\%$, and Pr $= 0.218$. (Reprinted with permission from Ref. 11.)

Similar results were found in gradient elution chromatography. The loading factor, the plate number, and the gradient steepness were optimized simultaneously. As can be seen in Fig. 7, when the optimum band profiles are plotted on the reduced time scale they match perfectly, regardless of the column length.

On the basis of these results, we can state that when the column length is fixed and the plate number is varied, e.g., by adjusting the average particle diameter, the optimum chromatograms are always identical when plotted on the reduced time scale. These results confirm the assumption that the loading

Optimization of Preparative Separations

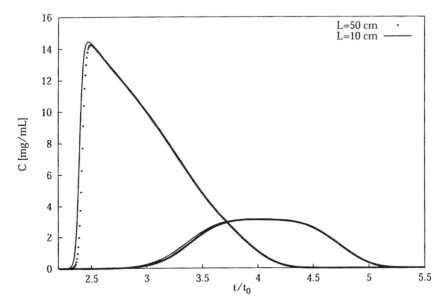

Figure 7 Optimum gradient elution chromatograms. The loading factor, the plate number, and the gradient steepness were optimized simultaneously at $k'_1 = 10$. The column lengths are 10 cm (solid line) and 50 cm (dotted line). The optimum loading factors L_f were 26.5 and 25.8%, respectively; the optimum plate numbers, 162 and 173 theoretical plates, respectively; and the gradient steepness G, 0.50 mg/mL, respectively. The optimum conditions for the 10 cm column were Pr $Y_{max} = 1.34$. As a consequence, the other optimum experimental conditions are $d_p = 14.2$ μm, $v = 361$, $u = 2.53$ cm/min, $Y = 69.5\%$, and Pr = 1.92. For the 50 cm long column, they were Pr $Y_{max} = 1.39$ $d_p = 32$ μm, $v = 812$, $u = 2.54$ cm/s, $Y = 72\%$, and Pr = 1.94. (Reprinted with permission from Ref. 11.)

factor and the plate number are the essential experimental parameters to be optimized in all modes of preparative chromatography.

2. The Effect of Cycle Time

In isocratic elution, the presence of strongly retained impurities—which are not infrequent in binary separations and are not always easy to eliminate in a purification step prior to preparative HPLC—may call for column regeneration. Depending on the specifics of the case, the impurity can be washed off the column more or less rapidly. We can consider two extreme case scenarios. In the first one, there are no impurities and the cycle time is the conventional analysis time under linear conditions. In the second case, a regeneration step of complexity comparable to that of the one required in displacement or gradient elution chromatography is needed and will take about the same time. However, because the displacer must be more strongly retained than all feed impurities, it is only in rare cases that it will take the same time to regenerate the column in elution and displacement chromatography. Figure 8 compares the performance achieved in isocratic elution chromatography with that achieved in displacement chromatography when the definition of the cycle time is altered in the elution mode. The solid line gives the product of the production rate and the recovery yield in displacement chromatography as a function of the retention

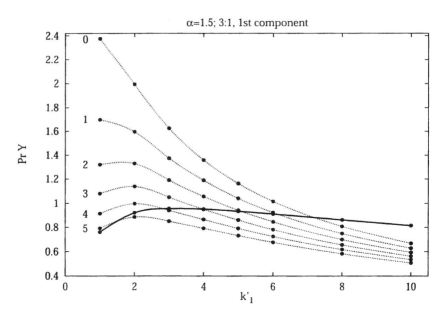

Figure 8 Plot of the optimum value of the objective function (Pr Y) against the retention factor of the less retained component. The solid line gives the value of Pr Y in displacement chromatography when six column volumes of solvent is needed for column regeneration. The dotted lines give the values of Pr Y in isocratic elution chromatography when the amount of solvent needed for column regeneration changes between zero and five column volumes. (Reprinted with permission from Ref. 11.)

Optimization of Preparative Separations

factor of the less retained component. In this case, for the definition of the cycle time we assume that the column regeneration requires six column volumes of solvent. The same Pr Y objective function, calculated for isocratic elution, is plotted for different volumes of solvent, assuming that zero to five column volumes is needed for column regeneration. Figure 8 clearly shows that the definition of the cycle time has a considerable influence on the product rate in isocratic elution.

The production achieved is greatest when there is no need for column regeneration—when the cycle time is simply determined by the retention time of the more retained component. In this particular case, displacement chromatography performs better than isocratic elution only if the retention factor k'_1 is greater than 7. However, as the regeneration need of isocratic elution approaches that of displacement chromatography, the displacement mode begins to outperform isocratic elution for decreasing values of the retention factors. Finally, displacement chromatography performs better than isocratic elution at any retention factor when identical cycle time definitions are used for the two modes of operation.

It is important to note that none of the modifications of the cycle time definition just discussed cause any change in either the optimum experimental conditions or the chromatogram obtained under these conditions. The optimum values of the plate number and loading factor—and accordingly, the throughput, production per cycle, and recovery yield—remain constant, regardless of the definition of cycle time. Only the value of the production rate is affected by the change in the actual value of the cycle time resulting from the use of a different definition. This is illustrated in Fig. 9, in which are plotted the optimum chromatograms calculated for $k'_1 = 4$. The solid line corresponds to displacement chromatography. The broken lines show the elution chromatograms obtained with the different definitions of the cycle time used also in Fig. 8 and just discussed. These optimum chromatograms match quite well, and the figure should be considered an illustration of the uncertainty of the numerical calculations made during the whole optimization routine and their effect on the optimum band profiles.

D. Optimization for Maximum Production Rate and Maximum Pr Y

In the following we compare the fundamental differences in optimum experimental conditions for maximizing the production rate and for maximizing the product of production rate and recovery yield.

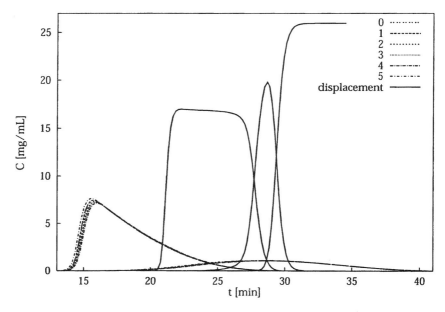

Figure 9 Optimum chromatograms whose data are plotted in Fig. 8 at $k'_1 = 4$ and $\alpha = 1.5$. The solid line is the optimum displacement chromatogram when six column volumes of solvent is needed for column regeneration. The column length is $L = 10$ cm. The optimum conditions are $L_f = 24.5\%$, $N = 747$, $C_d = 26.0$ mg/mL, and Pr $Y_{max} = 0.952$. As a consequence, the other optimum experimental conditions are $d_p = 9.2$ μm, $v = 98.9$, $u = 1.07$ cm/min, $Y = 85\%$, and Pr = 1.12. The other lines show the optimum chromatograms calculated in isocratic elution when the amount of solvent needed for column regeneration changes between zero and five column volumes. The optimum conditions are $L_f = 16.3\%$, $N = 273$, and Pr $Y_{max} = 0.732$. As a consequence, the other optimum experimental conditions are $d_p = 12.4$ μm, $v = 240$, $u = 1.93$ cm/min, $Y = 80\%$, and Pr = 0.91 with six column volumes for regeneration. (Reprinted with permission from Ref. 11.)

1. Overloaded Isocratic Elution

The chromatograms on the left-hand side of Figs. 10 and 11 show the band profiles obtained under the optimum conditions for maximum production rate of the less retained and the more retained component, respectively. In both cases, there is a considerable degree of overlap of the two bands, the mixed zone is very important, and the recovery yield is rather poor. This is especially true for the more retained component because of the important tag-along effect. The displacement effect allows a higher loading factor for the purification

Optimization of Preparative Separations

of the less retained component and leads to a higher recovery yield. The chromatograms on the right-hand side of Figs. 10 and 11 were obtained for the optimum separation when the product of the production rate and recovery yield was maximized instead of the production rate. Obviously, because this objective function is weighed equally by production rate and recovery yield, the optimum loading factor is smaller for both components than in the previous case and the recovery yield is significantly higher. The improvement is particularly spectacular in the case of the more retained component, where the loading factor is almost three times lower and the yield, at 90%, is more than three times higher than it was previously. A recovery yield above 90% is usually considered satisfactory. The considerable improvement in the recovery yield is achieved at the cost of a 20% reduction in the production rate.

We know that columns should be operated at the highest possible flow rate in order to achieve the maximum production rate [1,2,4]. This is a serious inconvenience, because high flow rates mean high pressure drops and, to a

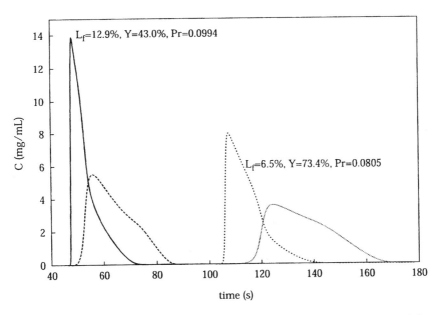

Figure 10 Optimum separations calculated by the equilibrium-dispersive model for isocratic overloaded elution for the purification of the less retained component. The production rate and the product of the production rate and recovery yield were maximized at the left and right, respectively. $\alpha = 1.2$, $k'_1 = 2$, $C^0_1 = 100$ mg/mL, $C^0_2 = 100$ mg/mL. (Reprinted with permission from Ref. 8.)

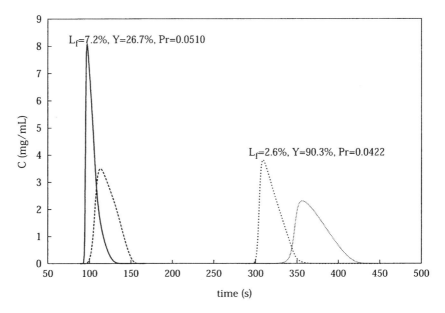

Figure 11 Same as Fig. 10, except that calculations were made for the more retained component. (Reprinted with permission from Ref. 8.)

large extent, significant dilution. When the Pr Y objective function is used, the required column efficiency is slightly higher than when the production rate is being maximized. This demands the use of smaller particles at constant column length. Consequently, the linear velocity of the mobile phase will be lower at the same pressure drop. For this reason, the cycle times are longer when the Pr Y objective function is used. This leads to smaller values of the production rate. However, the decrease in the production rate remains modest due to the improvement of the recovery yield.

The upper part of Fig. 12 illustrates the shift in the position of the optimum experimental conditions for each component when the production rate is replaced by the Pr Y objective function for the optimization. The maximum production rate is at point A, whereas the product of the production rate and the recovery yield reaches its maximum at point B. The contour lines clearly show that although the optimum experimental conditions are markedly shifted, the production rate is only slightly lower at the new optimum. On the other hand, in the middle of Fig. 12 one can see how considerably the recovery yield is improved when the experimental conditions are shifted from point A

Optimization of Preparative Separations

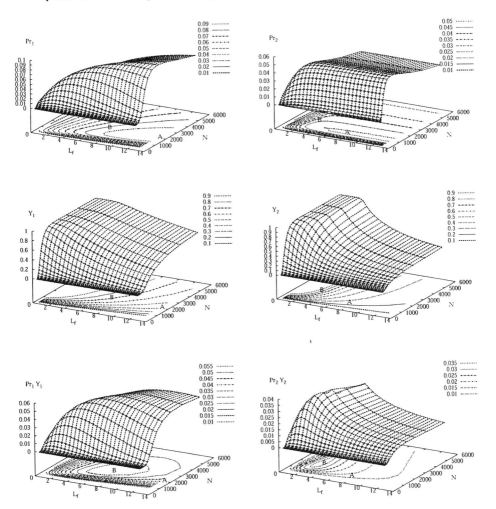

Figure 12 Left-hand column, top to bottom: Plot of the production rate of the less retained component against the loading factor and plate number ($\alpha = 1.2$, $k'_1 = 2$, $C_1^0 = 100$ mg/mL, $C_2^0 = 100$ mg/mL), plot of the recovery yield, and plot of $Pr_1 Y_1$. Right-hand side: Same plots for the more retained component. (Reprinted with permission from 8.)

to point B. Finally, the lower part of Fig. 12 shows that the surface determined by the Pr Y objective function exhibits a well-defined maximum, which makes the numerical optimization stable.

This property of the two surfaces and the corresponding functions, Pr and Pr Y, is more important for the optimization of the production of the more retained component. The production rate of the more retained component reaches a flat plateau at high loading factors, as forecasted by the ideal model. This feature of the surface makes the numerical optimization very problematic. It causes the optimization program to locate the optimum conditions at too high values of the loading factor, conditions for which the recovery yield is unsatisfactory and the result of the optimization somewhat meaningless.

As illustrated in Fig. 12, there are major differences between the behavior of the two components. One of the most noteworthy is in the nature of the shift of the experimental conditions from one objective function to the other.

The loss of production rate is important only at very small separation factors. Equation (12) indicates that the maximum of the Pr Y objective functions is approximately proportional to the fourth power of $(\alpha - 1)/\alpha$, whereas the production rate is proportional to the second power of that expression. This, however, is the price to pay to achieve high values of the recovery yield at small separation factors. Calculations based on the equilibrium-dispersive model in gradient elution chromatography confirm this finding [9].

In Fig. 13, the optimum achieved by maximizing Pr Y is plotted versus $(\alpha - 1)/\alpha$ on a log-log scale; in Fig. 14, the optimum production rate achieved by maximizing the production rate is plotted. In all three cases, as shown in Figs. 13 and 14, the results were well approximated by a linear dependence. In the case of maximizing the production rate, the slopes of the lines are between 2.5 and 2.8. In contrast, the slopes of the straight lines for Pr Y are 3.8 and 3.6 for the more retained component. For the less retained component, the values of the slope are smaller, 2.8 and 2.6. These results confirm the validity of the predictions derived from the ideal model. The optimum production rate of the more retained component increases slightly more slowly than the fourth power of $(\alpha - 1)/\alpha$. If the production rate Pr_2 were optimized instead of $Pr_2 Y_2$, the maximum production rate would increase in proportion to the square of $(\alpha - 1)/\alpha$ [1,12]. This shows that the hybrid Pr Y objective function leads to optimum experimental conditions under which the production rate is far from the maximum possible at small values of the separation factor. This was expected because the Pr Y objective function was designed to give high recovery yields, even in unfavorable conditions, which includes especially low separation factors. A reduction in the production rate is obviously the price to pay for an increase in recovery yield. This permits a quantita-

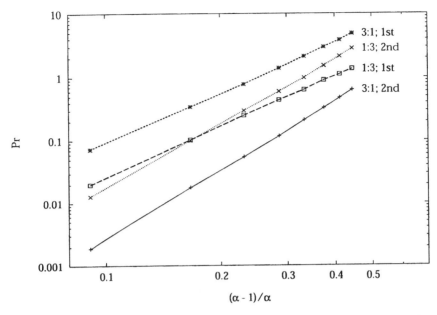

Figure 13 Plot of the production rate under optimum conditions for maximum Pr Y against the separation factor in case of 3:1 and 1:3 relative concentrations and for both the 1st and the 2nd eluted components. (Reprinted with permission from Ref. 9.)

tive estimate of the trade-off between production rate and recovery yield, a difficult compromise.

Consequently, the design of separation in respect to the maximum achievable separation factor is still more critical than predicted by some studies.

2. Overloaded Gradient Elution

We recall that the optimum experimental conditions in isocratic elution depend only on the loading factor and on the column of efficiency, regardless of how the column efficiency is achieved, because these two parameters determine the shape of the elution band profiles. This is also true in gradient elution. The concentration profiles relative to each other do not depend on the gradient steepness [26]. If the gradient steepness is increased, the band profiles are compressed but the relative concentrations, and accordingly the recovery yield, remain constant. If the degree of separation, i.e., the relative band spac-

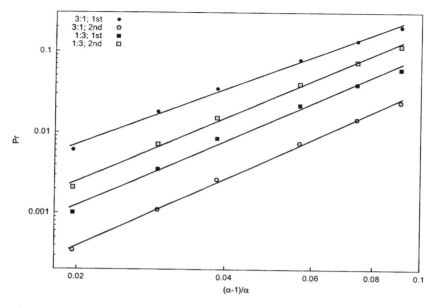

Figure 14 Plot of the production rate under optimum conditions for maximum production rate against the separation factor.

ing, is independent of the gradient steepness, the cycle time depends on it, so there remain three parameters to optimize: the loading factor, the column efficiency, and the gradient steepness.

The Pr Y objective function can be applied successfully to the optimization of overloaded gradient elution chromatography [9,27]. Figure 15 compares the chromatograms obtained under the optimum conditions given by the two objective functions (Pr and Pr Y) for the purification of the more retained component of a binary mixture. The chromatogram on the left-hand side of the figure corresponds to maximum production rate, the one on the right-hand side to maximum Pr Y.

In spite of the strong gradient steepness, both objective functions give values of the optimum loading factor and of the recovery yield achieved under optimum conditions, which are very similar in isocratic and gradient elution. At the maximum production rate, the recovery yield is rather poor. With the Pr Y objective function, the optimum loading factor is smaller but the recovery yield is almost 90%. The optimum experimental conditions found by the two methods are quite different, but the production rate is only 15% lower with

Optimization of Preparative Separations

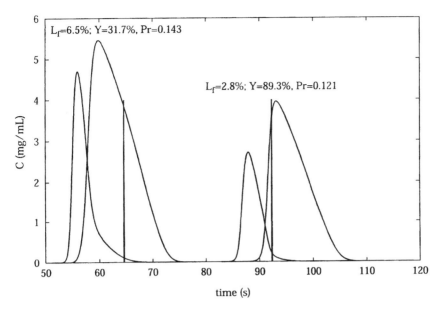

Figure 15 Optimum separations calculated by the equilibrium-dispersive model for gradient overloaded elution for the purification of the more retained component. The production rate (left) and the product of the production rate and recovery yield (right), were maximized. $\alpha = 1.2$, $k'_{0,1} = 10$, $G = 0.5$, $C_1^0 = 100$ mg/mL, $C_2^0 = 300$ mg/mL. (Reprinted with permission from Ref. 8.)

the Pr Y objective function. The amount of pure fraction collected during each run is much greater with the Pr Y objective function owing to the improved yield. The production rate is lower because the new optimum is achieved at a higher column efficiency, i.e., at smaller linear velocity, which increases the cycle time.

The optimum column efficiency is low, particularly in the case of high separation factors. This is due to the fact that the retention factor is very high during most of the elution. Even though it gradually decreases, its average or effective value is much higher than the optimum retention factor found in isocratic elution. This result is similar to one obtained under isocratic conditions: The optimum plate number decreases rapidly when the retention factor increases above its optimum [1].

The optimum gradient steepness is quite different for the first than for the more retained component. It varies with the separation factor. For the less retained component, the production rate usually reaches a plateau without a maximum, and in most cases there is no production rate gain above $G = 0.4–0.6$. Although the cycle time should decrease when the gradient steepness is increased, the optimum column efficiency increases with increasing gradient steepness, and these two effects compensate for each other, resulting in a nearly constant cycle time.

For convergent solutes, flat gradient profiles should be used, flat enough to still provide a separation factor not too close to unity and avoid reversal of the elution order of the two components at the end of the elution. This leads to the use of gradient profiles that are much less steep than those used to separate the two components under analytical conditions.

3. Displacement Chromatography

Displacement chromatography has become very popular recently because it delivers purified fractions that are much more concentrated than those of elution chromatography, even if it is a much more complicated separation process. Currently there is great interest in the application and optimization of displacement chromatography in biotechnology [28–30].

Although the concentration of the displacer is a factor that one may try to optimize, it is very difficult to find an optimum displacer concentration. As a general rule, we can state that the higher the displacer concentration, the higher the production rate. Furthermore, higher displacer concentration will result in higher solute concentration in the collection fraction [2]. Usually the solubility of the displacer will set a constraint on the concentration of the displacer.

The retention factor of the displacer is not a major issue with respect to the optimum performance of displacement chromatography. The displacer, of course, should be more retained than any sample component, but the value of the separation factor between the more retained sample component and the displacer does not really affect the production rate.

In Fig. 16 we compare the concentrations of the collected fractions at maximum production rate for overloaded elution and displacement chromatography. Regardless of the separation factor and retention factor, the concentration of the collected fraction is always higher in displacement chromatography.

Figure 17 compares the displacement chromatograms obtained under the optimum conditions by the two objective functions, maximum production rate and maximum Pr Y. These chromatograms correspond to an optimized purification of the more retained component. The shift of the optimum conditions is quite similar to what we observed earlier in both isocratic and gradient

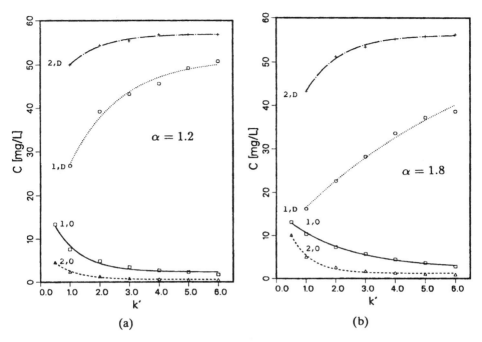

Figure 16 The concentration of the collected fraction versus the retention factor of the less retained component. (a) $\alpha = 1.2$, first ($\square$) and second ($\triangle$) components in overloaded (O) elution mode; first ($\bigcirc$) and second (+) components in displacement (D) mode. (b) $\alpha = 1.8$, first ($\square$) and second ($\triangle$) components in overloaded elution mode; first ($\bigcirc$) and second (+) components in displacement mode. (Reprinted with permission from Ref. 10.)

overloaded elution. At the maximum production rate, the recovery yield is rather poor. When Pr Y is maximized instead, the optimum loading factor is 2.5 times smaller, but the recovery yield is 2.6 times higher. Although the production rate is decreased by 35% in switching to the Pr Y objective function, the amount of pure product obtained during one run is higher, as can be seen in Fig. 17 from the area between the two vertical lines indicating the cut points.

E. Optimization for Maximum Specific Production

Because the specific production depends on two parameters, the column efficiency and the loading factor, a grid search can easily be performed in order

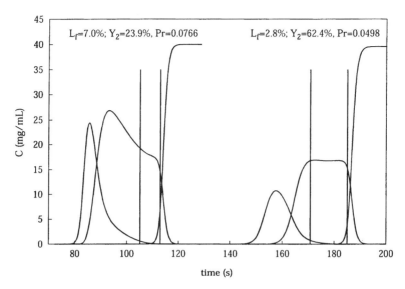

Figure 17 Optimum separations calculated by the equilibrium-dispersive model for displacement chromatography for the purification of the more retained component. The production rate (left) and the product of the production rate and recovery yield (right) were maximized. $\alpha = 1.2$, $k'_1 = 2$, $C^0_1 = 100$ mg/mL, C^0_2 300 mg/mL. (Reprinted with permission from Ref. 8.)

to visualize its dependence on the two parameters. The results of this research are illustrated in Fig. 18, which shows the surface plots for both components for $w = 1$. Figure 18 shows that at each column efficiency there is an optimum loading factor that gives a maximum specific production. This is true for both the less retained and more retained components. However, the higher the column efficiency, the higher the specific production. Owing to the decreasing apparent band dispersion, the optimum loading factor increases with increasing column efficiency, and so does the recovery yield at a given loading factor. After a steep increase at low column efficiency, a plateau is reached where there is not much improvement in the specific production. The higher the separation factor, the lower the column efficiency for which this plateau is reached.

Figure 18 Three-dimensional plot of the hybrid objective function [defined in Eq. (36)] for the (a) less and (b) more retained components of a 3:1 mixture at a separation factor $\alpha = 1.5$ and retention factor $k'_1 = 6$. (Reprinted with permission from Ref. 4.)

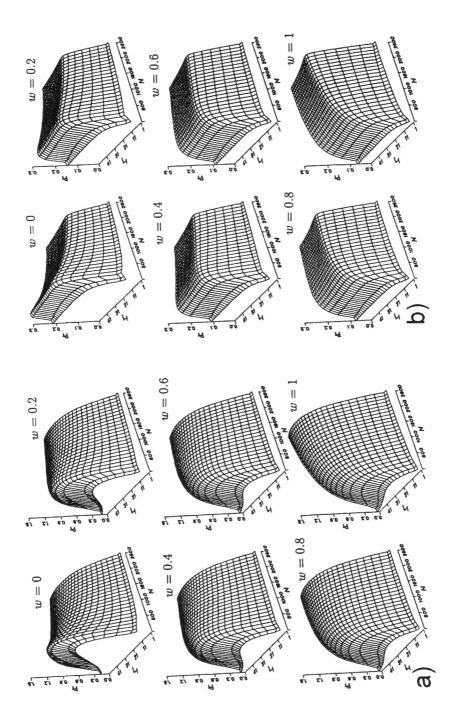

1. Weighted Objective Function

As we have seen, the optimum conditions for minimum solvent consumption and those for maximum production rate are very different. The high efficiency required for low solvent consumption is optimal for the maximum production rate only in a few cases, when both the separation factor and the retention factor are small.

Separations are usually carried out at high values of the retention factor, for practical reasons and because few users realize yet the potential savings afforded by making the separations at low retention factors. This situation requires the use of less efficient columns to achieve the same separation. For this reason, the production rate becomes rather low at minimum solvent consumption. As a trade-off between these two factors, a hybrid objective function can be constructed that considers the importance of the solvent consumption and that of the production rate with a given weight.

The hybrid objective function is based on the optimization of the unit cost of the purification, which is the ratio of the cost of the purification to the production rate. The two factors of the purification cost are the fixed cost F_c and the operating cost, whereas the cost of the wasted feed is usually negligible at high recovery yield. Because the operating cost is attributed to the amount of solvent used, this factor is expressed by the cost of solvent (S_u), the solvent consumption (the reciprocal of specific production), and the production rate, which is therefore, $\Pr_i S_u/\mathrm{SP}_i$. The unit cost of the purification can be written as

$$\frac{P_c}{\Pr_i} = \frac{F_c}{\Pr_i} + \frac{S_u}{\mathrm{SP}_i} \tag{34}$$

In this manner, the hybrid objective function can be transformed into

$$\frac{1}{\Pr_i^*} = \frac{1-w}{\Pr_i} + \frac{w}{\mathrm{SP}_i} \tag{35}$$

Combining Eqs. (12) and (35) gives

$$\Pr_i^* = \frac{\Pr_i}{1 - w + wu} \tag{36}$$

where parameter w ($0 \leq w \leq 1$) reflects the significance of the production rate or solvent consumption. If $w = 0$, the production rate is maximized regardless of the solvent consumption. If $w = 1$, the objective function is simply the specific production, used in the previous section. Intermediate values of w involve both factors and result in a trade-off between production rate and

Optimization of Preparative Separations

solvent consumption. By choosing a weight w that is inversely proportional to the ratio of the capital costs (fixed) and the operating (solvent) costs, we can use this objective function to reflect approximately the total production cost.

Figure 18 show two series of surfaces that illustrate the transition of the hybrid objective function from the maximization of the production rate to the maximization of specific production.

The transition of the location of the optimum on these surfaces is gradual. In the case of the optimization of the production of the less retained component, when w increases from 0 to 0.5 the optimum loading factor increases by nearly 30% and the optimum column efficiency increases by 200%, but the production rate at $w = 0.5$ is still 84% of that at $w = 0$. However, the maximum has nearly vanished for $w = 0.4$, and we can observe only an inconspicuous mound that has seemingly disappeared at $w = 0.6$. At this stage the dependence of the production rate on N becomes completely flat, and any column that has more than 500 theoretical plates will given nearly identical results, but a very flat optimum can always be identified.

Similar results are observed when the purification of the more retained component is optimized. The optimum efficiency and maximum production rate change in nearly the same proportion as for the optimization of the production of the less retained component. The only noticeable differences are the steady improvement in the recovery yield and the decrease in the optimum loading factor.

For both components, the specific production is about half as great at the maximum production rate ($w = 0$) as at $N = 5000$. On the other hand, at the high efficiency where the solvent consumption is minimum, the production rate is reduced to nearly one-third of its maximum. When a compromise is made ($w = 0.5$), the production rate loss is only 15%, whereas around 30–40% more solvent is needed than at $N = 5000$.

Remember that solvent consumption depends only on the loading factor and the column efficiency, whereas the maximum production rate depends also on the ratio d_p^2/L. Furthermore, the more efficient the column, the lower the solvent consumption, and the optimum column efficiency for maximum production rate is not the maximum possible efficiency but a lower value achieved at a high flow velocity. Thus, a compromise between high production rate and low solvent consumption is needed. The compromise will be dictated by the relative importance of fixed and proportional costs.

The maximum production rate and the minimum solvent consumption require quite different operational parameters at high retention factors, whereas when the retention factor is also optimized, all the values of the optimum parameters for maximum production rate or minimum solvent consumption are closer.

As a first approximation, we can separate the production costs into two contributions: a fixed cost contribution and a solvent cost contribution. The former includes all capital costs and labor, because the personnel requirements are determined more by regulations and the need for a minimum size crew and will be the same whether the plant is operated under conditions of maximum production rate or minimum solvent consumption. The latter cost contribution includes solvent and energy costs, because most of the energy is needed to evaporate the solvent, concentrate the purified product from the collected fractions, and purify the recycled solvent by distillation. When a given relative importance is attributed to these two costs, a weighted objective function can be constructed to find a compromise between the requirements of minimum solvent consumption and maximum production rate. In this case we found that there are two regions. When the capital costs are greater than the solvent costs, the optimum column efficiency remains close to the value that is optimal for maximum production rate, though it gradually increases. When the solvent costs are greater than the fixed costs, the optimum column efficiency is high, as for minimum solvent consumption. In the intermediate cases, when the two cost contributions are similar, the optimization problem is ill-posed, and the optimum conditions depend only slightly on the column efficiency in a wide range of this parameter.

VI. COMPARING THE PERFORMANCE OF THE DIFFERENT MODES OF PREPARATIVE CHROMATOGRAPHY

We recall that the band profiles shown in Fig. 9 are plotted against absolute time and not reduced time. This illustrates the fact that a comparison of the performance of the two modes of preparative separation is not straightforward because the actual optimum experimental conditions are quite different: Although the column length is the same in both cases, the holdup times are different, leading to quite different values of reduced time for the same value of absolute time. Although the retention times of the elution and displacement chromatograms in Fig. 9 are not very different, there is a major difference between the two chromatograms. It arises from the higher plate number that displacement chromatography requires compared to elution chromatography. Accordingly, because the column is operated at the maximum possible flow rate and the plate number can be altered by changing either the particle size or the column length, the flow rate is almost twice as high in elution chroma-

Optimization of Preparative Separations

tography as in displacement chromatography. In turn, this difference between the flow rates strongly influences the cycle time, because during regeneration the mobile phase flow rate cannot be higher than the flow rate determined by the maximum pressure drop (the effect of the feed component concentration on the viscosity [10] is neglected). Therefore, in this case, column regeneration will take a much longer time in the displacement mode than in isocratic elution.

Conversely, however, there are other phenomena that have more beneficial effects in displacement chromatography than in elution chromatography: The former mode is operated with a higher loading factor and shorter retention times. For example, in the case illustrated in Fig. 9, the loading factor is approximately 50% higher in the displacement mode than in isocratic elution, whereas the breakthrough time of the displacer is about 66% of the analytical retention time of the more retained component in isocratic elution. Another well-known advantage of displacement chromatography can also be seen in Fig. 9. The band profiles are more compact, resulting in a higher concentration of the collected fractions.

One of the important findings of the optimization studies is the existence of an optimum value of the retention factor for which the objective function is maximum. Although the discussion of the dependence of the objective function on the retention factor is simple in elution and displacement chromatography, it is not straightforward in gradient elution chromatography, because the retention factor changes continuously during a gradient elution separation. The following transformation is proposed to include gradient elution into a general comparison of the performance of the various modes of chromatography. In gradient elution, the retention time of a band under linear conditions (i.e., assuming so-called analytical injection) is given by [16]

$$t_R = t_0 \left(1 + \frac{1}{G} \ln[1 + k'_0 G]\right) \tag{37}$$

In the case of a Langmuir-type isotherm, the analytical retention time coincides with the elution time of the diffuse rear of nonlinear band profiles. Therefore, Eq. (37) can be used to estimate the time required to completely elute the material from the column. We can reformulate Eq. (37) as

$$t_R = t_0 (1 + k'_g) \tag{38}$$

where $k'_g = (1/G) \ln (1 + k'_0 G)$ is the *gradient retention factor* that can be used for the characterization of solute retention in gradient elution in much the same way as the classical retention factor k' is used in isocratic separations.

To compare the performance of gradient elution with those of isocratic elution and displacement chromatography, k'_g was calculated from the initial retention factor and the gradient steepness. The maximum value obtained for the objective function was then plotted against k'_g. The applicability of the gradient retention factor k'_g for performance comparison is demonstrated in Fig. 19. The solid lines are the profiles of the two component bands in an optimized isocratic elution. Two quite different gradient separations are also shown.

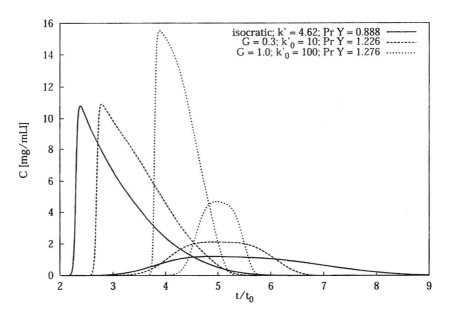

Figure 19 Comparison of the band profiles for optimum separations carried out by isocratic elution (solid line) and gradient elution (dashed and dotted lines). The gradient retention factor in each of the gradient elution separations is equal to the retention factor in isocratic elution (4.62). For $G = 0$ (isocratic elution), $L_f = 25.0\%$ and $N = 209$. The maximum value of the objective function was Pr $Y_{max} = 0.888$. The other optimum experimental conditions were $d_p = 13.3$ µm, $v = 294$, $u = 2.21$ cm/min, $Y = 77\%$, and Pr $= 1.155$. For $G = 0.3$ (isocratic elution), $L_f = 26.1\%$ and $N = 135$. The maximum value of the objective function was Pr $Y_{max} = 1.226$. The other optimum experimental conditions were $d_p = 15$ µm, $v = 420$, $u = 2.80$ cm/min, $Y = 68\%$, and Pr $= 1.79$. For $G = 1.0$ (isocratic elution), $L_f = 27.6\%$ and $N = 124$. The maximum value of the objective function was Pr $Y_{max} = 1.276$. The other optimum experimental conditions were $d_p = 15.3$ µm, $v = 450$, $u = 2.93$ cm/min, $Y = 63\%$, and Pr $= 2.03$. (Reprinted with permission from Ref. 11.)

Optimization of Preparative Separations

Their common feature is that they have the same gradient retention factor. Because of the rather different values of the initial retention factors and the gradient steepness of these two gradient elution experiments, the band profiles obtained under optimum conditions (dotted and dashed lines, respectively) are quite different. However, very similar optimum values of the objective function are obtained for the two separations. Both values are about 40% higher than the maximum Pr Y obtained for isocratic elution with a retention factor equal to the two gradient retention factors. This result demonstrates that the gradient retention factor characterizes the retention in preparative gradient function chromatography sufficiently well to allow meaningful comparison of the performance achieved with the different modes.

Figure 20 summarizes the results of a comparison of the maximum performance of the three modes of chromatography for a separation factor $\alpha = 1.5$ for different mixture compositions and elution orders. One can conclude

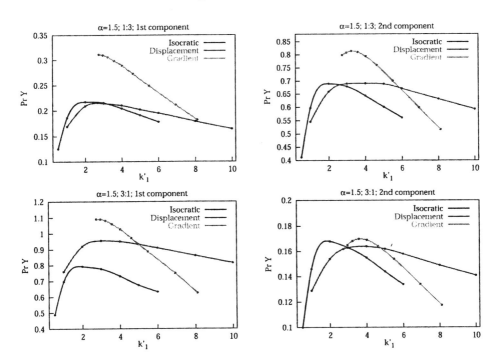

Figure 20 Comparison of the performance of isocratic elution, gradient elution, and displacement chromatography for various mixture compositions and elution orders at $\alpha = 1.5$. (Reprinted with permission from Ref. 11.)

that the dependence of the separation performance on the retention factor is most significant in gradient elution. Displacement chromatography is the mode that is the least sensitive to changes in the retention factor. Although the particular experimental setup (mixture composition, elution order, separation factor, etc.) selected arbitrarily may vastly influence the separation performance, we can conclude that both gradient elution and displacement chromatography generally outperform isocratic elution in all cases in which this last method requires extensive column regeneration (with five column volumes or more). On the other hand, when the isocratic separation does not require column washing, all three modes give similar results. In the specific calculations carried out for this work, a gradient steepness that results in a gradient retention factor no greater than $k'_g = 5$ is advantageous. The maximum performance of gradient elution was found for values of k'_g slightly below 3 for the purification of the less retained component and slightly above 3 for the purification of the more retained component. Displacement chromatography appears to be the method of choice if the retention factor (or the gradient retention factor) is greater than 5. Displacement chromatography is more attractive than isocratic elution for retention factors in excess of 2 when extensive washing is needed for column regeneration. If a cleaner feed can be used, elution becomes more attractive in a much wider range of retention factors.

Jandera et al. [15] optimized the gradient steepness in the preparative reversed-phase separation of phenol and o-cresol and concluded that even if six column volumes of solvent is used to regenerate the column, gradient elution is still more beneficial for that separation than isocratic elution. They found that the initial mobile phase composition had a strong influence on the production rate [31].

The choice of a uniform cycle time definition—with the requirement of six column volumes of solvent for the regeneration of the column in all cases—is arbitrary and might seem pessimistic in the case of isocratic elution. To study the consequences of this choice, a comparison of the performance of the three modes of preparative chromatography shown in Fig. 20 for the case of $\alpha = 1.5$ was also carried out with different values for the volume of solvent required for column regeneration. The results are summarized in Fig. 21. The narrow broken lines correspond to different volumes required for the regeneration, from zero to six column volumes. The results show that isocratic elution outperforms the other two modes when there is no need for column regeneration, for instance in the case of an impurity-free isocratic separation. As the time needed for column regeneration increases in isocratic elution, the other two modes of separation become more and more attractive.

Optimization of Preparative Separations

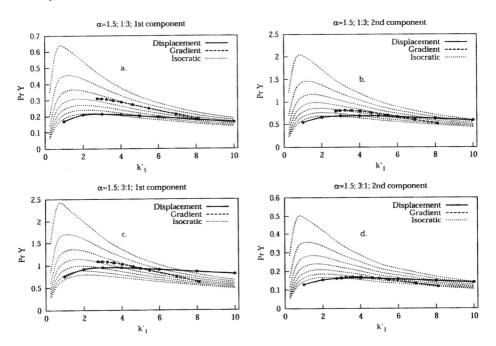

Figure 21 Comparison of the performance of isocratic elution, gradient elution, and displacement chromatography for various mixture compositions and elution orders at $\alpha = 1.5$. The narrow broken lines give the performance of isocratic elution (downward) when zero to six column volumes of solvent is required for column regeneration. (Reprinted with permission from Ref. 11.)

These results illustrate the considerable influence of the solvent volume needed for column regeneration on the performance achieved at small retention factors. At $k'_1 = 1$, for instance, the performance (i.e., the value of Pr Y) of isocratic elution is about 3.5 times higher if there is no need for column regeneration than if six column volumes is required. The gain in performance is a factor of approximately 2.5 at $k'_1 = 2$ and only 1.4 at $k' = 10$. This phenomenon arises from the greater contribution of the retention time to the cycle time observed at higher retention factors. Correspondingly, the contribution of the regeneration time to the cycle time decreases. Accordingly, the solvent volume needed for column regeneration must be determined very carefully when working at small retention factors.

The optimum retention factor afforded by the Pr Y objective function is somewhat higher than the one obtained when the production rate itself is optimized. The definition of the cycle time also influences the optimum retention factor. Figure 8 illustrates how, in the case of isocratic elution, the optimum value of k'_1 shifts toward higher retention factors as the solvent volume needed for column regeneration increases. The optimum retention factor is always higher in displacement chromatography than in the other two modes. It was found to be $k'_1 = 3$–4 for $\alpha = 1.5$ and $k'_1 = 4.5$ for $\alpha = 1.2$. The strong dependence and the value of the optimum retention factor on the separation factor is illustrated in Fig. 22. The optimum value decreases from $k'_1 = 4.5$ to $k'_1 = 1.7$ when the separation factor improves from $\alpha = 1.1$ to $\alpha = 1.8$, regardless of the elution order or the feed composition. It has always been known in chromatography that separation requires retention. It is not surprising that, whatever the criterion, easy separations require less retention than difficult ones.

Seidel-Morgenstern [32] compared the performance of isocratic elution, recycling, simulated moving bed, and annular chromatography for the separation of the two isomers of a steroid compound. That optimization was restricted, because it was carried out for a given chromatographic system; there-

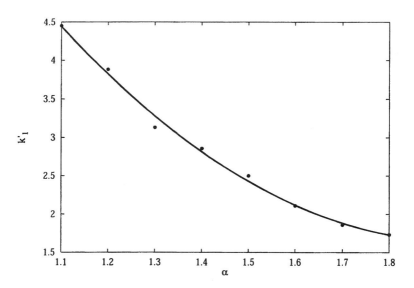

Figure 22 Optimum value of the retention factor plotted against the separation factor in isocratic elution chromatography. (Reprinted with permission from Ref. 11.)

fore, only the experimental conditions were optimized. For that application the simulated moving bed separation was the superior method. It offered an approximately 2.5-fold higher production rate than any batch process. The three batch separations (isocratic elution, recycling, and annular chromatography) showed very similar performance with respect to both solvent consumption and production rate.

The choice of the most advantageous mode to perform a given separation relies mostly on the actual values of the retention factors of the two components to be separated. There is probably much work to pursue along this line. Optimizing the retention factor, when possible, would bring important improvements in the degree of performance achieved. However, this work neglects the actual economy of separation processes. Even if elution gives a lesser degree of performance and a lower production rate than gradient elution and displacement, it may turn out to be the less costly solution because it does not cause the obvious additional problems of solvent recovery.

REFERENCES

1. A Felinger, G Guiochon. Optimization of the experimental conditions and the column design parameters in overloaded elution chromatography. J Chromatogr 591:31–45, 1992.
2. A Felinger, G Guiochon. Optimization of the experimental conditions and the column design parameters in displacement chromatography. J Chromatogr 609:35–47, 1992.
3. H Colin. Large-scale high-performance preparative chromatography. In: G Ganetsos, PE Barker, eds. Preparative and Production Scale Chromatography. New York: Marcel Dekker, 1993, pp 11–45.
4. A Felinger, G Guiochon. Optimizing experimental conditions for minimum production cost in preparative chromatography. AIChE J 40(4):594–605, 1994.
5. P Jageland, J Magnusson, M Bryntesson. Optimization of industrial-scale high-performance liquid chromatography applications using a newly developed software. J Chromatogr A 658:497–504, 1994.
6. AM Katti, P Jageland. Development and optimization of industrial scale chromatography for use in manufacturing. Analusis 26:M38–M46, 1998.
7. AM Katti, EV Dose, G Guiochon. Comparison of the performances of overloaded elution and displacement chromatography for a given column. J Chromatogr 540:1–20, 1991.
8. A Felinger, G Guiochon. Optimizing preparative separations at high recovery yield. J Chromatogr A 752:31–40, 1996.
9. A Felinger, G Guiochon. Optimizing experimental conditions in overloaded gradient elution chromatography. Biotechnol Prog 12(5):638–644, 1996.

10. A Felinger, G Guiochon. Comparison of maximum production rates and optimum operating/design parameters in overloaded elution and displacement chromatography. Biotechnol Bioeng 41(1):134–147, 1993.
11. A Felinger, G Guiochon. Comparing the optimum performance of the different modes of preparative liquid chromatography. J Chromatogr A 796:59–74, 1998.
12. G Guiochon, S Golshan-Shirazi, A Katti. Fundamentals of Preparative and Nonlinear chromatography. Boston: Academic Press, 1994.
13. AM Katti. Strategies for the development of process chromatography as a unit operation for the pharmaceutical industry. In: K Valkó, ed. Handbook of Analytical Separations: Separation Methods in Drug Synthesis and Purification. Amsterdam: Elsevier, 2000, pp 213–291.
14. SR Gallant, S Vunnum, SM Cramer. Optimization of preparative ion-exchange chromatography of proteins: linear gradient separations. J Chromatogr A 725: 295–314, 1996.
15. P Jandera, D Kromers, G Guiochon. Effects of the gradient profile on the production rate in reversed-phase gradient elution overloaded chromatography. J Chromatogr A 760:25–39, 1997.
16. LR Snyder. Gradient elution. In: CS Horváth, ed. High-Performance Liquid Chromatography. Advances and Perspectives, Vol 1. New York: Academic Press, 1980, pp 207–316.
17. JH Knox. Practical aspects of LC theory. J Chromatogr Sci 15:352–364, 1977.
18. MD LeVan, T Vermeulen. Binary Langmuir and Freundlich isotherms for ideal absorbed solutions. J Phys Chem 85:3247–3250, 1981.
19. SC Jacobson, S Golshan-Shirazi, G Guiochon. Isotherm selection for band profile simulation in preparative chromatography. AIChE J 37:836–844, 1991.
20. S Golshan-Shirazi, JX Huang, G Guiochon. Comparison of an experimental competitive isotherm and the LeVan–Vermeulen model and prediction of band profiles in a case of selectivity reversal. Anal Chem 63:1147–1154, 1991.
21. MZ El Fallah, G Guiochon. Comparison of experimental and calculated results in overloaded gradient elution chromatography for a single-component band. Anal Chem 63(9):859–867, 1991.
22. S Golshan-Shirazi, G Guiochon. Theory of optimization of the experimental conditions of preparative elution using the ideal model of liquid chromatography. Anal Chem 61(11):1276–1287, 1989.
23. A Katti, G Guiochon. Optimization of sample size and sample volume in preparative liquid chromatography. Anal Chem 61:982–990, 1989.
24. S Golshan-Shirazi, G Guiochon. Theory of optimization of the experimental conditions of preparative elution chromatography: optimization of the column efficiency. Anal Chem 61(13):1368–1382, 1989.
25. JH Knox, M Pyper. Framework for maximizing throughput in preparative liquid chromatography. J Chromatogr 363:1–30, 1986.
26. A Felinger, G Guiochon. Multicomponent interferences in overloaded gradient elution chromatography. J Chromatogr A 724:27–37, 1996.

27. H Schramm, H Kniep, A Seidel-Morgenstern. Optimization of solvent gradients for chromatographic separations. Chem Eng Technol 24:133–138, 2001.
28. SR Gallant, SM Cramer. Productivity and operating regimes in protein chromatography using low-molecular-mass displacers. J Chromatogr A 771:9–22, 1997.
29. V Natarajan, BW Bequette, SM Cramer. Optimization of ion-exchange displacement separations. I. Validation of an iterative scheme and its use as a methods development tool. J Chromatogr A 876:51–62, 2000.
30. V Natarajan, SM Cramer. Optimization of ion-exchange displacement separations. II. Comparison of displacement separations on various ion-exchange resins. J Chromatogr A 876:63–73, 2000.
31. P Jandera, D Komers, G Guiochon. Optimization of the recovery yield and of the production rate in overloaded gradient-elution reversed-phase chromatography. J Chromatogr A 796:115–127, 1998.
32. A Seidel-Morgenstern. Optimization and comparison of different modes of preparative chromatography. Analysis 26:M46–M55, 1998.

4
Engineering Aspects of Ion-Exchange Chromatography

Peter Watler, Oliver Kaltenbrunner, and Daphne Feng
Amgen Inc., Thousand Oaks, California, U.S.A.

Shuichi Yamamoto
Yamaguchi University, Ube, Japan

I. MECHANISM OF ION-EXCHANGE CHROMATOGRAPHY

Ion-exchange chromatography (IEC) is an adsorptive technique by which molecules are separated on the basis of differences in their surface charge distribution (Yamamoto et al., 1988; Sofer, 1995). The molecular forces involved in chromatography have been studied in detail by Forgacs and Cserhati (1997) and include Coulombic (electrostatic) interaction between oppositely charged ions, ion–dipole interactions, hydrogen bonds, and hydrophobic forces such as van der Waals and London dispersion forces. The mechanism of IEC and its adsorption isotherms is based on co-ion adsorption as described by Kokotov (2000). In IEC the binding is generally enthalpy-driven with $\Delta G < 0$ and $\Delta H < 0$ and is primarily due to the interaction energy of electrostatic and van der Waals forces. Binding energy is typically higher than for entropy-driven processes, and for proteins, ΔG values are typically < -10 kJ/mol. ΔH values of -6 to -13 kJ/mol have been reported for electrostatic binding of β-lactoglobulin A (Chen, 2000; Lin, 2001). Proteins possess either a net positive or net negative charge at pH values away from their pI and will bind to an oppositely charged ion exchanger. As such, IEC is one of the most widely used techniques for protein separation and is typically found in all bioprocesses. Proteins are generally adsorbed at low salt concentrations (0.01–0.05 M) and eluted with a step or gradient increase in salt concentration, typically up to 1.0 M (Yamamoto et al., 1987b). Typical ligand chemistries and their charged pH range include, for cation exchange, carboxymethyl (CM, pH 6–10) and sulfyl-propyl (SP, pH 2–10), and for anion exchange, diethylaminoethyl (DEAE, pH 2–9) and quaternary amine (Q, pH 2–10) (Scopes, 1994).

A. Distribution Coefficient

The distribution coefficient K describes the distribution of a solute between the stationary and mobile phases. It is not dependent upon flow rate but is a measure of the affinity of the protein for the stationary phase matrix. This parameter is important for understanding, optimizing, controlling, and troubleshooting separations (Snyder and Kirkland, 1974; Schoenmakers, 1986; Horváth and Lin, 1976). K can be used to determine and compare when species will elute from a given chromatographic medium (Fig. 1). K can be measured by using a small pulse injection of the pure protein. Alternatively, if a small injection of the sample gives baseline separation of the species of interest, the distribution coefficient can be measured from the respective peaks. K is calculated as

$$K = \frac{V_R - V_0}{V_t - V_0} \qquad (1)$$

The greater the difference in the distribution coefficients of species, the greater will be the difference in retention times. A protein with $K = 0$ is completely excluded from the resin pore structure and will elute in the column void volume V_0. A protein with $K = 1$ will be equally distributed between

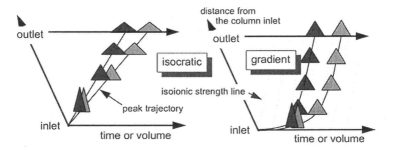

Figure 1 Zone movement during isocratic and gradient elution.

Ion-Exchange Chromatography

the stationary and mobile phases and will elute at V_t. In preparative IEC, K is typically $\gg 1$ during loading and $K \approx 1$ during elution. In isocratic and gradient chromatography, the protein zone moves down the column according to its distribution coefficient as shown in Fig. 1.

II. DEVELOPMENT OF AN IMPROVED SEPARATION

It has been shown that the separation of a particular protein is proportional to a single dimensionless parameter (Yamamoto, 1995; Snyder and Stadalius, 1986). This equation has been verified experimentally for ion-exchange chro-

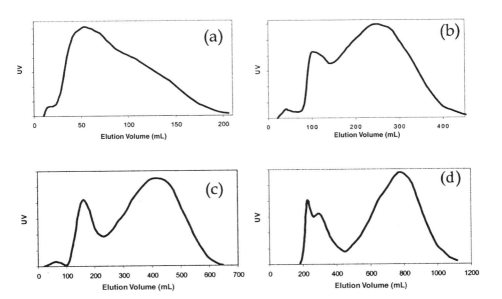

Figure 2 Improved separation due to increasing bed height and reducing protein loading. (a) Chromatogram obtained with initial, unoptimized conditions showing little resolution of species; 1.0 cm diameter column × 6.1 cm bed height, 42 mg/mL loading, $O' = 1102$. (b) Improved resolution of pre-peaks and product peak; 1.6 cm diameter column × 6.1 cm bed height, 16 mg/mL loading, $R_s = 1.25$, $O' = 420$. (c) Further resolution of pre-peaks and product peak; 1.6 cm diameter column × 9.6 cm bed height, 10 mg/mL loading, $R_s = 1.6$, $O' = 167$. (d) Separation into three peaks; 1.6 cm diameter column × 16 cm bed height, 6.2 mg/mL loading, $R_s = 2.2$, $O' = 62$. All at $u_0 = 160$ cm/h.

matography of proteins and has been used for scale-up in an industrial application (Yamamoto et al., 1987c, 1988; Slaff, 1993). This proportionality was further modified to include the protein loading on the resin, and for isocratic elution,

$$R_s \propto \frac{1}{O'} \tag{2}$$

where

$$O' = \frac{M \, \text{HETP}}{ZI} \tag{3}$$

where M is the protein mass loading on the resin (grams of protein per liter of resin), and I represents a dummy variable having a numerical value of 1 so that O' becomes dimensionless. If the separation is limited to the range of linear velocities such that height equivalent to a theoretical plate (HETP) is linearly proportional to the superficial velocity (i.e., HETP = ku_0, intercept $b \approx 0$), then the number of plates is proportional to the residence time and Eq. (3) becomes

$$O' = \frac{Mu_0}{ZI'} \tag{4}$$

The resolution of a separation can be improved by choosing conditions that give a lower value of O' as shown in Fig. 2.

III. SCALE-UP OF AN OPTIMIZED SEPARATION

As shown in Fig. 3, O' can be used to specify various operating conditions that provide the desired resolution. An iterative approach can be used to identify flow rates and bed heights that give acceptable pressure drops for the production scale column. After an O' value that gives the desired resolution is identified, alternative operating conditions giving the same O' can be calculated. In this way, O' is used to specify various column geometries and flow rates that provide the required purity, cycle time, and product volume. For example, Fig. 3 with $O' = 63$ gives similar resolution to Fig. 2d with $O' = 62$, but at a bed height 25% lower and a much higher resin loading. Using the Fig. 3 O', the superficial velocity which yields similar separation but with a smaller column volume and lower elution volume can be calculated.

This separation finally scaled up to purify ~100 g of protein with a 25 cm diameter column that was limited to an 8.2 cm bed height. To ensure

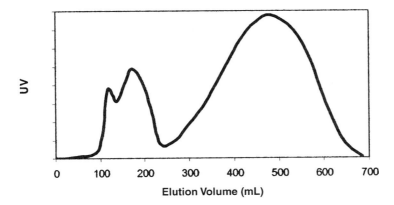

Figure 3 Chromatogram showing resolution similar to that of Fig. 2d. 1.6 cm diameter column × 12 cm bed height, operated with increased protein loading to 14 mg/mL and lower elution velocity of 54 cm/h; $O' = 63$.

separation of the pre-peaks from the product peak in plant operation, a slightly greater resolution than that of Fig. 2c ($O' = 167$) but slightly less than that of Fig. 2d was desired. For this operation, a value of $O' = 128$ was chosen, and the elution velocity was calculated. Upon scale-up, the enhanced resolution was achieved as shown in Fig. 4.

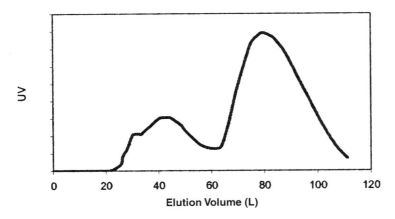

Figure 4 Chromatogram showing resolution at scaled-up conditions, 25 cm diameter column × 8.2 cm bed height, 49 cm/h, $O' = 128$.

IV. MODELING AND CHARACTERIZATION OF A CHROMATOGRAPHIC SEPARATION

Ion-exchange chromatography separations can be sensitive to small changes in operating parameters such as pH, ionic strength, and ion-exchange capacity (Yamamoto and Ishihara, 1999; Yamamoto, et al., 1999). Chromatography modeling is a useful engineering design tool to scale up, characterize, troubleshoot, and optimize productivity. Early models, including that by Martin and Synge, appeared in the 1940s (Wicke, 1939; DeVault, 1943; Wilson, 1940; Martin and Synge, 1941). Several solutions to the plate model and other approaches appeared in the 1950s–1970s (Lapidus and Amundson, 1952; van Deemter et al., 1956; Kucera, 1965; Kennedy and Knox, 1972; Crank, 1975; Yamamoto et al., 1979). With the advent of the personal computer, numerous solutions have been offered in the past two decades (Knox et al., 1983; Yamamoto and Scott, 1983; Arnold et al., 1985b; Carta, 1995; Lightfoot et al., 1997; Kaltenbrunner et al., 1997; Felinger and Guiochon, 1998; Natarajan and Cramer, 2000). However, most models require complex analytical measurements, and hence few have met practical application in industrial settings. In addition, such models have rarely been employed for preparative scale-up to large columns and high protein loadings (Yamamoto et al., 1988).

The model described here is based on the work of Yamamoto and coworkers (Yamamoto, 1995; Yamamoto and Ishihara, 2000; Watler et al., 1999) and requires only the determination of the distribution coefficient K and the number of theoretical plates N. K is influenced by both protein concentration and ionic strength and is needed for predicting the retention of the various species. The distribution coefficient is related to the effective ionic capacity and the number of adsorption sites and provides an understanding of the separation mechanism. N is a measure of column performance and is defined as the degree of peak broadening by relating peak width to retention volume. N is influenced by the structure of the chromatographic medium, the protein, the column packing quality, and operating conditions. Data for the model can be obtained from as few as 16 linear gradient elution (LGE) experiments (Yamamoto et al., 1988).

A. Model Derivation

The LGE model is based on the ion-exchange equilibrium model (Boardman and Partridge 1955; Haff et al., 1983; Kopaciewicz et al., 1983; Yamamoto et al., 1987a; Yamamoto, 1995; Roth et al., 1996; Gallant et al., 1996) and requires a term for determining the peak position (Yamamoto and Ishihara

Ion-Exchange Chromatography

1999). For a protein pulse traveling through the void volume and part of the gel volume, the elution volume is

$$V_e = V_0 + KV_g = V_0 + K(V_t - V_0) \tag{5}$$

K implies the fraction of the gel volume through which the protein travels; K by definition is the distribution coefficient,

$$K = \frac{\overline{C}}{C} \tag{6}$$

Defining H as the ratio of the stationary phase volume to the mobile phase volume,

$$H = \frac{V_t - V_0}{V_0} \tag{7}$$

the elution volume becomes

$$V_e = V_0 + KV_0H = V_0(1 + HK) \tag{8}$$

By definition, the peak retention time is

$$t_r = \frac{V_e}{u_0 A_c} \tag{9}$$

where u_0 is the superficial velocity.

The protein velocity is

$$\frac{dz_p}{dt} = \frac{z}{t_r} = z\frac{u_0 A_c}{V_e} \tag{10}$$

Substituting Eq. (9) into Eq. (10) gives

$$\frac{dz_p}{dt} = \frac{zu_0 A_c}{V_0(1 + HK)} \tag{11}$$

Because $zA_c = V_t$,

$$\frac{dz_p}{dt} = zu_0\left(\frac{1}{V_0}\right)A_c\left(\frac{1}{1 + HK}\right) \tag{12a}$$

$$\frac{dz_p}{dt} = u_0 V_t\left(\frac{1}{V_0}\right)\left(\frac{1}{1 + HK}\right) \tag{12b}$$

Because $u = u_0\left(\frac{V_t}{V_0}\right)$

$$\frac{dz_p}{dt} = \frac{u}{1 + HK} \tag{13}$$

Equation (13) describes the migration velocity of the protein peak and shows that it decreases with increasing K. An equation is required for relating K to the ion-exchange system. According to Yamamoto et al. (1988), the distribution coefficient K can be derived as follows. The ion-exchange reaction can be expressed from the law of mass action as

$$P + \frac{Z_P}{Z_B}\overline{B} \xrightleftharpoons{K_e} \overline{P} + \frac{Z_P}{Z_B}B \tag{14}$$

where the overbar indicates the bound state and P = protein in solution, B = counterion in solution, Z_P = protein valence, Z_B = counterion valence, and K_e is the equilibrium constant. By definition,

$$K_e = \frac{\overline{a_P}^{|Z_B|} a_B^{|Z_P|}}{a_P^{|Z_B|} \overline{a_B}^{|Z_P|}} \tag{15}$$

where a = activity of the species. If $Z_B = 1$, then

$$K_e = \frac{\overline{a_P}}{a_P}\left(\frac{a_B}{\overline{a_B}}\right)^{|Z_P|} \tag{16}$$

The activity coefficient is defined as

$$\gamma_i = \frac{a_i}{m_i} \tag{17}$$

where m_i = molarity of species i. Substituting Eq. (17) into Eq. (16) gives

$$K_e = \frac{\overline{m_P}\,\overline{\gamma_P}}{m_P \gamma_P}\left(\frac{m_B \gamma_B}{\overline{m_B}\,\overline{\gamma_B}}\right)^{|Z_P|} \tag{18a}$$

$$K_e = \frac{\overline{m_P}}{m_P}\left(\frac{m_B}{\overline{m_B}}\right)^{|Z_P|}\left[\frac{\overline{\gamma_P}}{\gamma_P}\left(\frac{\gamma_B}{\overline{\gamma_B}}\right)^{|Z_P|}\right] \tag{18b}$$

For electroneutrality,

$$m_R = |Z_P|\overline{m_P} + \overline{m_B} \tag{19a}$$

or

$$\overline{m_B} = m_R - |Z_P|\overline{m_P} \tag{19b}$$

where m_R is the ion exchanger molarity and is a fixed value. The distribution coefficient, Eq. (6), can also be written as

Ion-Exchange Chromatography

$$K = \frac{\overline{m_P}}{m_P} \tag{20}$$

Rearranging Eq. (18b) gives

$$\frac{\overline{m_P}}{m_P} = K_e \left(\frac{\overline{m_B}}{m_B}\right)^{|Z_P|} \left[\frac{\gamma_P}{\overline{\gamma_P}}\left(\frac{\overline{\gamma_B}}{\gamma_B}\right)\right]^{|Z_P|} \tag{21}$$

Substituting Eq. (19b) into Eq. (21) gives

$$\frac{\overline{m_P}}{m_P} = K_e \left(\frac{m_R - |Z_P|\overline{m_P}}{m_B}\right)^{|Z_P|} \left[\frac{\gamma_P}{\overline{\gamma_P}}\left(\frac{\overline{\gamma_B}}{\gamma_B}\right)\right]^{|Z_P|} \tag{22}$$

$$K = K_e \left[\frac{\gamma_P}{\overline{\gamma_P}}\left(\frac{\overline{\gamma_B}}{\gamma_B}\right)\right]^{|Z_P|} \left(\frac{m_R - |Z_P|\overline{m_P}}{m_B}\right)^{|Z_P|} \tag{23}$$

Hence, K depends on the molarity of the ion-exchange groups, the molarity of the ions in the mobile phase, the molarity of the bound protein, and the number of charges on the protein. If the concentration of the protein in solution, m_P, is very small compared to the concentration of the groups on the ion exchanger, m_R, then the concentration of ions in solution will change very little as the protein is adsorbed:

If $m_P \ll m_R$, then $m_R - |Z_P|\overline{m_P} \approx m_R$.

Equation (23) becomes

$$K = K_e \left[\frac{\gamma_P}{\overline{\gamma_P}}\left(\frac{\overline{\gamma_B}}{\gamma_B}\right)\right]^{|Z_P|} \left(\frac{m_R}{m_B}\right)^{|Z_P|} \tag{24}$$

because for a given ion exchanger and protein system, $|Z_P|$, m_R, and K_e are constants.

Defining a new constant,

$$\Lambda = \frac{\gamma_P}{\overline{\gamma_P}}\left(\frac{\overline{\gamma_B}}{\gamma_B}\right) m_R \tag{25}$$

where $m_B = I$, the ionic strength of the buffer, and $|Z_P| = B$, representing the valence or number of charges on the protein, then Eq. (24) can be written as

$$K = K_e \Lambda^{|Z_P|} I^{-B} \tag{26}$$

For this separation system, Λ is the effective total ion-exchange capacity, I is the ionic strength (e.g., NaCl concentration), and B is the number of sites (charges) involved in protein adsorption, which is basically the same as the Z number in the literature (Yamamoto, 1995; Kopaciewicz et al., 1983; Karlson et al., 1998). If A is a constant implying the ionic capacity of the resin, defined as

$$A = K_e \Lambda^{|Z_p|} \tag{27}$$

then

$$K = AI^{-B} \tag{28}$$

accounting for the distribution coefficient of the ion species,

$$K = AI^{-B} + K' \tag{29}$$

The ionic strength at position z_p in the column is given by

$$I = I_0 + G\frac{ut - z_p(1 + HK')}{z} \tag{30}$$

Differentiating with respect to z_p,

$$\frac{dI}{dz_p} = \frac{dI_0}{dz_p} + \frac{G}{z}\left(u\frac{dt}{dz_p} - \frac{dz_p}{dz_p}(1 + HK')\right) \tag{31}$$

$$= G\frac{u(dt/dz_p) - (1 + HK')}{z}$$

Rearranging Eq. (13) gives

$$\frac{dt}{dz_p} = \frac{1 + HK}{u} \tag{32}$$

Inserting Eq. (32) into Eq. (31) gives

$$\frac{dI}{dz_p} = \frac{G}{z}\left[u\frac{1 + HK}{u} - (1 + HK')\right]$$

$$= \frac{G}{z}(HK - HK') \tag{33}$$

$$= \frac{GH(K - K')}{z}$$

Ion-Exchange Chromatography

Rearranging,

$$\frac{dI}{K - K'} = GH \frac{1}{z} dz_p \tag{34}$$

Integrating,

$$\int_{I_0}^{I_R} \frac{dI}{K - K'} = GH \int_0^{I_z} \frac{1}{z} dz_p = GH \tag{35}$$

Rearranging gives

$$\int_{I_0}^{I_R} \frac{dI}{K - K'} = GH \tag{36}$$

Inserting Eq. (29) into Eq. (36) gives

$$\int_{I_0}^{I_R} \frac{dI}{AI^{-B} + K' - K'} = GH \tag{37}$$

$$\int_{I_0}^{I_R} \frac{dI}{AI^{-B}} = GH \tag{38}$$

Integrating Eq. (38) gives

$$GH = \frac{I_R^{B+1}}{A(B + 1)} \tag{39}$$

which can be written as

$$GH = \frac{1}{A(B + 1)} I^{B+1} \tag{40}$$

This is an equation of the power law form where

$$\text{constant} = \frac{1}{A(B + 1)} \quad \text{and} \quad \text{exponent} = B + 1 \tag{41}$$

Using a graph of GH vs. I_R with a power fit will give the values of A and B; then K can be calculated from Eq. (29). Peak retention time can be calculated from the I_R value and with the aid of Eq. (30) as,

$$t_{R,\,LGE} = \frac{z}{u} \left(\frac{I_R - I_O}{G} + (1 + HK') \right) \tag{42}$$

Hence, this gives a method for predicting the peak retention time from linear gradient elution experiments. This method is based on the law of mass action and the velocity of the protein zone.

B. GH–I_R Curves

Curves of GH vs. I_R indicate the ionic retention strength at different gradient elution conditions. For a given separation, GH–I_R curves are prepared by conducting the separation at different normalized gradient slopes (GH) and measuring the ionic strength of each peak maximum. The data are then fit with power law regression curves. The shape of the curve reflects the nature and characteristics of the protein on the medium. As the net charge on the protein increases, the dependence of I_R on GH becomes stronger and the curves shift to higher I_R values with steeper slopes. The physical meaning of A in Eq. (29) is a combination of influences from the resin ligand density, the equilibrium constant between resin and protein, and their activity coefficients in the system (Yamamoto et al., 1999). The B term in Eq. (29) implies the number of protein charges involved in the adsorption (Yamamoto et al., 1988).

The GH–I_R curves provide information about the nature of the separation. For the separation described above, the slopes of the GH–I_R curves for the three species are almost equal (Fig. 5). B values for pre-peak 1, pre-peak 2, and the product peak are 1.5, 2.0, and 2.7, respectively. Hence, the very fine separation of these species was achieved under conditions where the number of charges in the adsorption equilibria was low but there was a relatively large (~35%) difference in charge between species. The GH–I_R curves also show that at all gradient slopes the product will elute last. At shallow gradients, the separation of the pre-peaks from the product peak is greatest, whereas at steep gradients the pre-peaks elute closer to the product.

1. Effect of pH on GH–I_R Curves

To study the effect of pH on the separation, gradients at different pH values were conducted and the ionic strength at peak maximum was measured. For this separation, the curves become shallower and shift to lower I_R values with increasing pH (Fig. 6). This is a result of decreasing charge as the pH moves closer to the protein pI (p$I \approx 6.3$). For this recombinant protein separation, this means that all species will elute earlier with increasing pH. With increasing pH, the GH–I_R regression lines between the different species also move closer together, indicating poorer separation of the species. Smaller differences between the GH–I_R curves indicate smaller differences between the distribu-

Ion-Exchange Chromatography

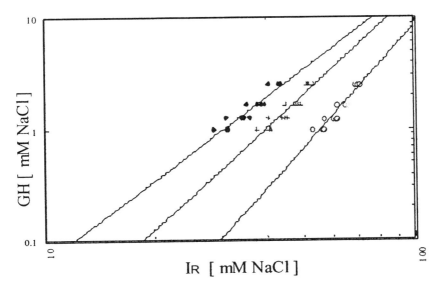

Figure 5 $GH-I_R$ curves for (●) pre-peak 1 ($y = 2.082 \times 10^{-4} X^{2.477}$), (+) pre-peak 2 ($y = 1.466 \times 10^{-5} X^{3.020}$), and (○) product peak ($y = 4.071 \times 10^{-7} X^{3.657}$) on CM Sepharose FF at pH 5.4.

tion coefficients of the species and consequently smaller differences in retention times of the proteins. Hence, with increasing pH, resolution will decrease as the peaks elute close together.

The separation characteristics can be explained from the model of the distribution coefficient, $K = AI^{-B}$ [Eq. (29)]. For this separation, the cationic carboxymethyl (CM) ligand is totally ionized above pH 6, whereas only a few groups remain charged below pH 4. Figure 7a shows that the A value, which implies the charge on the ion exchanger, decreases significantly with decreasing pH. In addition, Fig. 7b shows that the B value, which implies the charge on the recombinant protein, decreases with increasing pH. This is due to the decreasing total charge on the protein as it nears its pI. Hence, there is an inverse relationship between the number of charges involved in the separation, which decreases with pH, and the charge of the ion exchanger, which increases with pH. This creates a complex separation system that is sensitive to pH but where a very fine separation of the species is possible. Hence, preparation of the $GH-I_R$ curves provides information that can be used to optimize the separation and aids in understanding the mechanism of the ion-exchange separation.

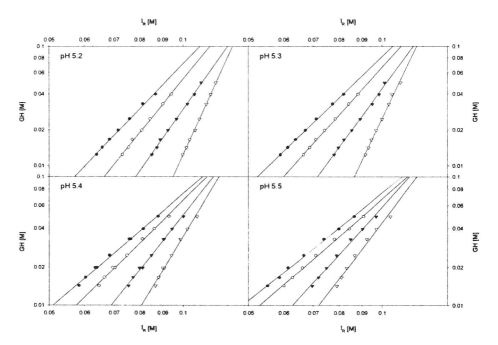

Figure 6 Linearized $GH-I_R$ curves for pH values of 5.2, 5.3, 5.4, and 5.5. (●) Pre-peak 1; (○) pre-peak 2; (▼) product peak; (▽) post-peak. CM Sepharose FF column.

C. Determination of the *K–I* Curve

Stepwise elution of proteins falls into two categories. Type I elution occurs when a protein is desorbed completely in the elution buffer and elutes as a sharp peak at or near the front boundary of the elution buffer. In this case, K_{protein} at $I = I_e$ is less than or nearly equal to that of the salt or displacer K'. In type I elution, the shape of the front boundary of the elution buffer from I_0 to I_e greatly influences peak sharpness and plays a role analogous to that of the steepness of the gradient slope in gradient elution chromatography. With type II elution, the protein peak emerges from the column well after the front boundary of the elution buffer emerges from the column. In type II elution, the peak is considerably more dilute than peaks in type I elution. K_e in this case is greater than K'. Type I elution is applicable only when the properties of the proteins to be separated are sufficiently different. If a suitable condition is not found, type II elution must be employed.

For the recombinant protein separation shown in Fig. 3, the *K–I* curve

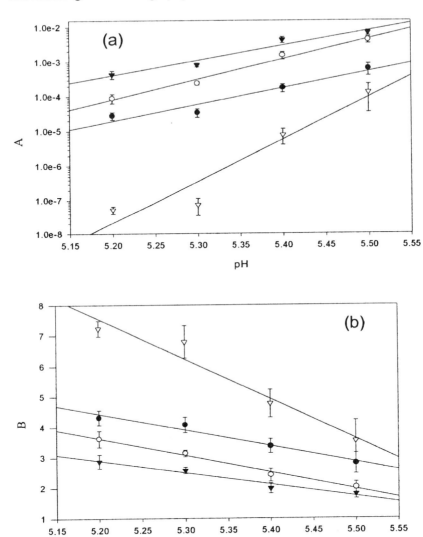

Figure 7 *A* and *B* values for the four different species as a function of the pH. (▼) Pre-peak 1; (○) pre-peak 2; (●) product peak; (▽) post-peak. CM Sepharose FF column.

(Fig. 8) shows that the elution order of the peaks is $K_{product} > K_{pre-peak2} > K_{pre-peak1}$. The K–I curve provides a practical interpretation of the separation by showing the relative retention of the peaks at different ionic strengths. The K–I curve is a practical tool that can be used to select operating conditions and troubleshoot the separation. In this separation, the K–I curve shows that all peaks will co-elute at ionic strengths ≥70 mM. In this region, type I elution occurs for all species and no separation is possible because the species have very similar properties. At ionic strengths below 35 mM, the distribution coefficient of the product is significantly higher than that of the pre-peaks. Here, the product peak is significantly retained and a very fine separation of the species is possible. However, in this region the K–I curve shows that $K_{product}$ changes very rapidly with slight changes in eluent ionic strength. Thus, small buffer fluctuations can result in significantly delayed elution of the product, requiring a large elution volume, or the species will elute at similar times, resulting in loss of resolution.

D. Use the Model for Scale-Up to an Industrial Size Column

The model parameters K and HETP are determined from a series of small-scale linear gradient elution (LGE) experiments. The model can also be adapted for isocratic elution. As shown in Fig. 9, the model can be used to predict isocratic elution behavior of a large-scale industrial column. The predicted chromato-

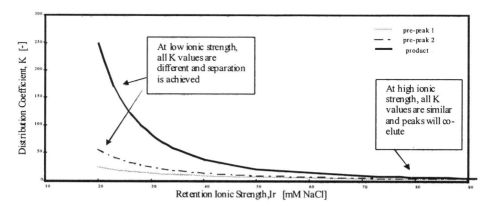

Figure 8 K–I curves for pre-peak 1, pre-peak 2, and product peaks of the recombinant protein: CM Sepharose FF, pH 5.4.

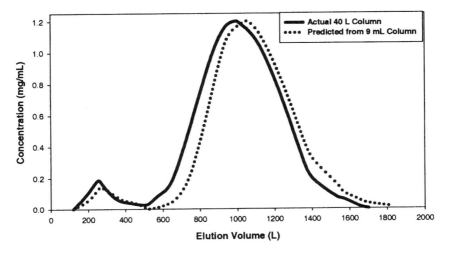

Figure 9 Comparison of isocratic elution curve predicted from model using 9 mL column and linear gradient elution data with actual elution curve of a 40 L column, CM Sepharose FF, pH 5.4, 40 mM NaCl.

gram closely approximated the actual chromatogram from a 40 L column, with predicted retention volumes within ~15% of the actual retention volume (see Table 1).

The model also predicted peak widths of the larger column, although that of pre-peak 1 was significantly overestimated (Table 1). This may be due

Table 1 Comparison of Predicted Retention Volumes and Peak Widths for Isocratic Elution of the Separations Shown in Fig. 9

Peak	Retention volume (CV)		Peak width (CV)	
	Predicted	Actual	Predicted	Actual
Pre-peak 1	6.20	6.26 ± 0.09	4.4	1.8 ± 0.1
Pre-peak 2	9.61	8.3 ± 1.0	7.0	6.0 ± 1.5
Product peak	26.60	23.2 ± 3.1	19.9	20.8 ± 2.9

Predicted values from model using 9 mL column and linear gradient elution data. Actual values measured from chromatogram of a 40 L column ($n = 4$). CM Sepharose FF, pH 5.4. CV = column volume.

to the difficulty in measuring the widths of very small peaks. Finally, resolution can be predicted by the model and was in very close agreement with predicted values (Table 2). Prediction of such measures can be used to verify that column operation and performance have been accurately translated across different size columns during chromatography scale-up.

E. Use of the Model for Linear Gradient Elution Optimization

One of the most useful applications of a chromatography model is for screening operating conditions to characterize and optimize the separation. To assess the separation and challenge the model, extremes of column length, gradient length, and initial ionic concentration were evaluated.

A tall bed height and long gradient typically maximize peak separation but also increase band spreading and peak width, which reduces resolution. Such conditions were screened with the aid of the model by adjusting the operating conditions to a 20 cm bed height and 30 column volume (CV) gradient. Under these conditions the model predicts relatively broad and late-eluting peaks with baseline separation (Fig. 10a). To verify the model, the actual chromatogram obtained under the identical operating conditions is shown in Fig. 10b. The predicted and actual chromatograms show similar peak shapes, peak widths, retention times, and baseline separation, with the main peak eluting at ~70% of the gradient. The model provided an accurate prediction of the separation under conditions for high resolution.

After identifying and verifying a high resolution separation, the operating conditions were modified to search for conditions that gave good resolution but were more scalable and offered higher productivity. A shorter bed height will have lower pressure drop upon scale-up, and a more moderate gradient will reduce buffer and tank requirements. To compensate for the

Table 2 Comparison of Predicted Resolution for Isocratic Elution from a 9 mL Column and Linear Gradient Elution Data and Actual Resolution for a 40 L Column ($n = 4$), CM Sepharose FF.

Separated peaks	Resolution	
	Predicted	Actual
Pre-peak 1 and pre-peak 2	0.5	0.52
Pre-peak 2 and product	1.27	1.18 ± 0.08

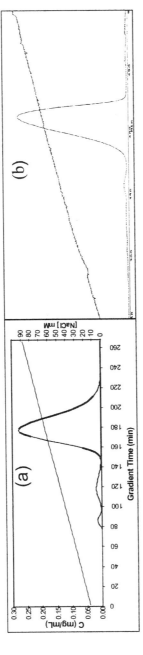

Figure 10 Comparison of predicted chromatogram (a) and actual chromatogram (b) for CM Sepharose FF, pH 5.4. Gradient = 30 CV, diameter = 1.6 cm, height = 20 cm, I_0 = 16 mM NaCL, superficial velocity = 120 cm/h.

shorter gradient length, the gradient slope was reduced by increasing the initial ionic concentration I_0. The predicted chromatogram for the more moderate operating conditions of 12 cm bed height, 20 CV gradient, and $I_0 = 28$ mM is shown in Fig. 11a. The chromatogram shows near baseline separation of the peaks, with the product peak eluting earlier at ~55% of the gradient and with a narrower peak width. This separation was verified as shown in Fig. 11b, demonstrating that the LGE model provided an accurate prediction under moderate operating conditions.

To further characterize the separation, behavior under extremely low resolution conditions was studied. A very low bed height of 5 cm and a very short gradient of 7 CV will require minimal resin, buffer, and tanks. However, the model predicts that the pre-peaks will merge with the product peak, appearing as a small shoulder on the product peak and resulting in very poor resolution (Fig. 12a). In addition, the model predicts that the product peak will now elute toward the end of the gradient (~90%). The actual chromatogram obtained at these operating conditions is very similar to the predicted chromatogram, showing a very sharp, late-eluting peak with little resolution (Fig. 12b).

As discussed above, chromatography models can be used to screen a wide range of conditions in order to optimize column operation and characterize the separation. It is important to empirically verify the proposed conditions prior to specifying them for scale-up. Table 1 shows that the model predicted how the elution volume and peak width changed at various column and gradient conditions. Predicted retention volume was within 3% of the actual volume for the product peak, demonstrating that the LGE model accurately estimated the distribution coefficient K of the product peak. The model showed that peak width decreased with sharper gradients and lower bed heights but underpredicted values by 21–41%.

Ion-Exchange Chromatography

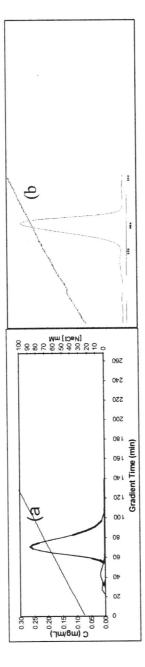

Figure 11 Comparison of predicted chromatogram (a) and actual chromatogram (b), for CM Sepharose FF, pH 5.4, gradient = 20 CV, diameter = 1.6 cm, height = 12 cm, I_0 = 28 mM NaCL, superficial velocity = 120 cm/h.

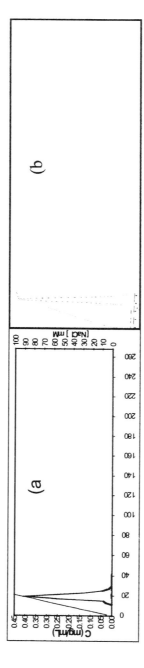

Figure 12 Comparison of predicted chromatogram (a) and actual chromatogram (b), for CM Sepharose FF, pH 5.4, gradient = 7 CV, diameter = 1.6 cm, height = 5 cm, I_0 = 16 mM NaCL, superficial velocity = 120 cm/h.

Table 3 Comparison of Predicted and Actual Experimental Retention Volume and Peak Width for the Product Peak Under Various Gradient Elution Conditions and Column Geometries, CM-Sepharose FF, pH 5.4, Superficial Velocity = 120 cm/h.

Operating conditions			Retention volume (mL)		Product peak width (mL)	
Bed height (cm)	Gradient (CV)	Initial salt concn (mM)	Predicted	Actual	Predicted	Actual
20	30	16	164	162	71	44
12	20	28	110	110	56	42
12	50	16	230	228	—	—
5	7	16	66	68	42	33

F. Use of the Model to Improve Productivity

Productivity is defined as the amount of protein of a given purity produced per unit time per liter of chromatographic resin (Gallant et al., 1995; Yamamoto and Sano, 1992).

$$P = \frac{V_f c_o R}{t_c V} = \frac{V_f c_o R}{t_c \pi r^2 Z(1 - \varepsilon)} \tag{43}$$

Models can be employed to rapidly scout for conditions that give the highest productivity for a given purity requirement. For the separation shown in Fig. 3, the optimized isocratic operating condition gave a productivity of 0.16 g L^{-1} h^{-1} (Table 4). This is largely because of the long cycle time due to the high distribution coefficient of the product peak under isocratic elution at 40 mM NaCl. One method of addressing a high K value is to use gradient elution. The model showed that a very long gradient (60 CV) gave comparable resolution with a somewhat higher productivity of 0.21 g L^{-1} h^{-1}. From this starting point, operating conditions were varied to search for increased productivity. In this study, a moderate bed height of 10 cm and a moderate gradient of 10 CV gave sufficient resolution to achieve 99% purity and very high (99%) recovery (Fig. 13). One of the features of the model is that the starting ionic strength can be varied to adjust the gradient slope. Increasing the starting salt concentration to 40 mM resulted in almost immediate desorption of the prepeaks, as predicted by the K–I curve. At this gradient slope, the product peak elutes earlier, resulting in a short cycle time and higher productivity. Another feature of this separation is that the steeper gradient results in sharper peaks and a lower product pool volume. Productivity under these conditions was

Table 4 Optimization of Operating Conditions Using Chromatography Model Simulations[a]

Bed height (cm)	Operating conditions			Recovery (%)	Productivity (g L^{-1} h^{-1})
	Gradient (CV)	I_0 (mM)			
12	Isocratic	40		100	0.16
12	60	0		100	0.21
20	30	16		100	0.24
5	7	16		28	0.53
10	10	40		99	0.61

[a] Comparison of recovery and productivity for various operating conditions, CM-FF, pH 5.4, I_f = 100 mM NaCL.

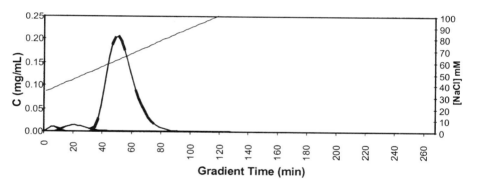

Figure 13 Optimized gradient conditions for the separation, gradient = 10 CV, bed height = 10 cm, I_0 = 40 mM NaCl, superficial velocity = 120 cm/h, CM Sepharose FF, pH 5.4, purity = 99.2%, recovery = 99%.

0.61 g L^{-1}h^{-1} nearly 400% higher than the initial optimized isocratic conditions.

V. CHARACTERIZATION AND TROUBLESHOOTING OF THE SEPARATION

To maintain process consistency, validated large-scale processes typically have a constraint to maintain elution volume within ±5%. This is often re-

Ion-Exchange Chromatography

quired, because tank volumes are fixed and elution must be completed within planned production times. Although column size, flow rate, and protein loading are fixed by the manufacturing procedure, resin ionic capacity, buffer pH, and ionic strength will vary from lot to lot. To assess the consistency of the separation, the model can be used to characterize and troubleshoot the effect of these fluctuations. In this application, the model can be used to elucidate the binding characteristics and to quantify changes in elution volume with variations in ion-exchange capacity, buffer ionic strength, and pH (Yamamoto et al., 1999). For illustration, a model separation system of β-lactoglobulin near its isoelectric point on a weak cation exchanger, CM Sepharose at pH 5.2, was selected (Yamamoto and Ishihara, 2000). Although the isoelectric point of this protein is 5.1–5.2, it is retained on both anion and cation-exchange chromatography columns at pH ~5.2 (Yamamoto et al., 1987a; Haff et al., 1983; Kopaciewicz et al., 1983). The values of $B = 2$ and $K_e \Lambda^B = 0.1179$, where Λ is the effective total ion-exchange capacity, were obtained from linear gradient elution experimental data (Yamamoto et al., 1987a, 1988; Yamamoto, 1995).

For this system, stepwise elution is performed with an elution buffer of $I_E = 0.1$ M, at which $K = 11.8$ from Eq. (20). According to the manufacturer, the ion-exchange capacity Λ of CM Sepharose FF ranges from 90 to 130 µmol per µmol/mL of gel. Such variations can greatly affect retention time, because the distribution coefficient is related to Λ. The impact of such lot-to-lot variations on the relative elution volume V_R/V_t can be calculated as a function of Λ by using Eq. (26) and the equation:

$$\frac{V_R}{V_t} = \varepsilon + (1 - \varepsilon)K \tag{44}$$

The void fraction ε is assumed to be 0.4. V_R/V_t at $\Lambda = 110$ was set to be a reference value where the protein eluted at 7.47 column volumes. The target retention volume and the +5% and −5% retention volumes are shown in Fig. 14. Even a relatively modest 5% variation in Λ (105 to 116 µmol/mL gel) significantly affects the elution volume well beyond the desired ±5%.

For the recombinant protein separation shown in Fig. 3, a 22% increase in the total ionic capacity from 90 to 110 mM µmol/mL resulted in a 47% increase in the distribution coefficient and consequently the relative retention volume (Fig. 15) (Watler et al., 1996).

In a production environment it is impractical to consistently obtain resin of a specific ionic capacity. In order to control the retention volume, it is necessary to adjust the salt concentration I_E of the elution buffer. Figure 16 illustrates the relationship between the relative elution volume and I_E. When

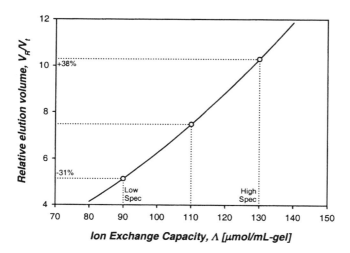

Figure 14 Relative elution volume vs. ion-exchange capacity. The reference value = $V_R/V_t = 7.5$, +5% value = 7.9, −5% value = 7.1.

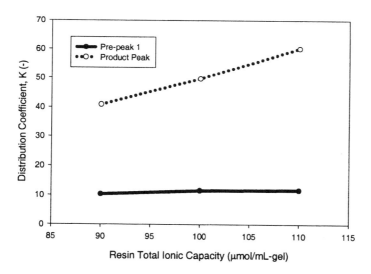

Figure 15 Distribution coefficient K of the recombinant protein pre-peak 1 and product peak with different lots of CM Sepharose FF resin having a total ionic capacity of 90–110 μmol/mL gel.

Ion-Exchange Chromatography

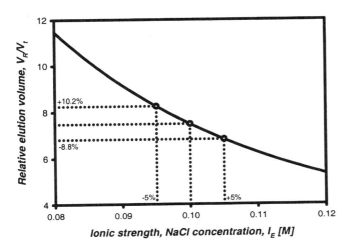

Figure 16 Relative elution volume vs. ionic strength. Curve r = reference value, $\Lambda = 110$; curve *1*: $\Lambda = 90$; curve *2*: $\Lambda = 130$.

the reference value curve r ($\Lambda = 110$) is compared with the low Λ value curve (*1*: $\Lambda = 90$) and with the high Λ value curve (*2*: $\Lambda = 130$), it is seen that the NaCl concentration must be adjusted by ± 0.015 mol/L to elute the protein within 5% of the reference elution volume. While this is typically done by trial and error as discussed above, the distribution coefficient as a function of ionic strength can be obtained from gradient elution experimental data. Once this $K–I$ information is obtained for a given Λ, the elution volume can be predicted and the elution buffer ionic strength can be adjusted. Hence, the chromatography model can serve as a convenient tool for tuning and troubleshooting very sensitive isocratic chromatographic processes. In addition, there is usually a variation in the salt concentration of elution buffers prepared at production scale. Figure 16 indicates that the salt concentration of the buffer must be within ± 0.002 mol/L in order to meet the $\pm 5\%$ elution volume criteria.

Owing to inherent variability during preparation, buffer pH will also vary at production scale. Buffer pH affects the charged state of the ion exchanger, which affects elution volume. The ion-exchange capacity of CM Sepharose decreases with pH below pH 6 (Pharmacia, 1999; Scopes, 1994). The relative change in elution volume resulting from changes in the ion-exchange capacity Λ as a function of operating pH is shown in Fig. 17. Al-

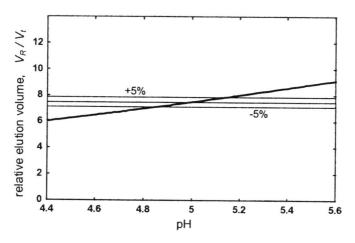

Figure 17 Relative elution volume vs. pH. Reference value = pH 5. (Yamamoto et al. 1999a).

though variations in the relative elution volume are smaller than those shown in Figs. 14 and 16, it is still important that the buffer pH be within ±0.1 pH unit. It is additionally important to control buffer pH, because, as discussed above, the interaction between the protein and the ion exchanger changes with pH, especially near the isoelectric point. In particular, the values of B and $K_e \Lambda^B$ as well as Λ vary with pH in this region. However, controlling the effect of pH is more complex than controlling salt concentration or the ion-exchange capacity.

VI. DISPERSION IN CHROMATOGRAPHY—HETP

Peak sharpness affects resolution because of the overlap of adjacent peaks according to the equation

$$R_s = \frac{t_2 - t_1}{(W_1 + W_2)/2} \tag{45}$$

Peak width is assessed by the height equivalent to a theoretical plate (HETP), which is a measure of column efficiency. Column efficiency denotes peak spreading or the rate of generation of variance with column length. The column performance indicator term N was borrowed from distillation theory. Unlike distillation, it is defined mathematically as the statistical population variance

Ion-Exchange Chromatography

by approximating a peak as a Gaussian distribution. This statistical definition defines a loss of performance or peak broadening by relating the width of a peak, W, to its elution volume V_R. HETP is calculated from N as follows (Bidlingmeyer and Warren, 1984):

$$N = \left(\frac{t_R}{\sigma}\right)^2 \tag{46}$$

$$\text{HETP} = \frac{Z}{N} \tag{47}$$

$$\text{HETP} = Z\left(\frac{\sigma}{t_R}\right)^2 \tag{48}$$

From Fig. 18 the peak width at half the peak height is

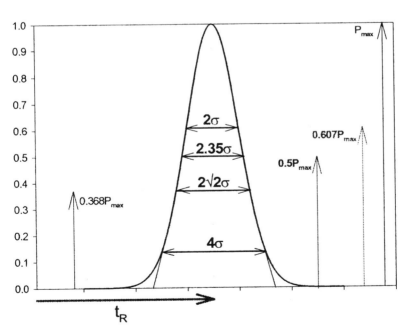

Figure 18 Peak width values at different peak heights for a Gaussian peak.

$$W_{1/2} = 2.35\sigma \tag{49}$$

and

$$\text{HETP} = \frac{Z}{5.54}\left(\frac{W_{1/2}}{t_R}\right)^2 \tag{50}$$

HETP is influenced by both the chromatography media and the column operating conditions. Influential characteristics of the media include particle size, particle type (spherical, irregular), and particle size distribution. Influential operating conditions include mobile phase velocity, mobile phase viscosity, column packing homogeneity, and protein diffusivity. Under nonbinding conditions, the protein's HETP provides critical information on the mass transfer of the protein–column system. Plots of HETP vs. linear velocity u are useful for examining the influence of factors such as d_p, u, and molecular diffusivity and are important for comparing the performance of different media and operating modes. HETP data can also provide information on how to adjust flow rate, particle diameter, and bed height to obtain a desired separation.

HETP depends on the mass transport rate of the protein in the mobile phase and between the mobile and stationary phases. van Deemter (1956) and others (Kennedy and Knox, 1972; Jönsson, 1987; Giddings, 1961, 1965; Horváth and Lin, 1976; Scott, 1992) have quantified the HETP with respect to physical variables such as tortuosity of the intraparticle space, molecular diffusivity, eddy diffusivity, stationary phase diffusion coefficient, particle diameter, and column packing factor. The accumulation of these mass transfer effects are collectively referred to as "axial dispersion" and contribute to band broadening. Axial dispersion is described by the modified van Deemter equation as

$$\text{HETP} = A + \frac{B}{u} + Cu \tag{51}$$

The A term expresses the contribution due to axial mixing or eddy diffusion, the B term expresses molecular diffusion in the mobile phase, and the C term expresses mass transfer resistance between the mobile and stationary phases or pore diffusion (Fig. 19). For fixed bed chromatography, molecular diffusion in the mobile phase is usually negligible, and dispersion is mainly due to mobile phase axial dispersion and stationary phase pore diffusion, allowing

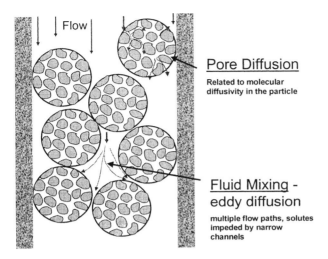

Figure 19 Schematic of eddy diffusion and pore diffusion to axial dispersion in a packed bed.

simplification of the van Deemter equation to (Yamamoto et al., 1988; Mann, 1995; Wang, 1990)

$$\text{HETP} = A + Cu \tag{52}$$

where

$$A = \frac{2D_L}{u} = 2\lambda d_p \tag{53}$$

$$C = \frac{d_p^2}{3^\circ D_s}\left(\frac{HK}{(1 + HK)^2}\right) \tag{54}$$

Because HETP values depend on the media, protein, and flow rate, it is difficult to quote typical ranges. Yamamoto et al. (1986) presented data for protein HETP values in the range of ~0.01–0.1 cm showing that approximately 100 plates are required for good resolution. Hence, once a desirable separation is obtained with a particular resin, the HETP can be calculated, and this or a lower value can be the target HETP for selecting a suitable media and operating conditions. The reduced HETP, h ($h = \text{HETP}/d_p$), and reduced velocity, v ($v = ud_p/D_m$), normalize data against particle diameter and protein diffusivity, enabling comparison of the efficiency of various column systems.

For typical porous media, the h–v data fall within a narrow range. This is because v increases with molecular weight at certain u and d_p values owing to the decrease in D_m with molecular weight. h shows a minimum at $v < 5$. At flow rates typically used in preparative protein chromatography, the minimum band spreading will not be achieved, owing to the low diffusivity of proteins ($D_m < 1 \times 10^{-6}$ cm^2/s).

A. Calculation of HETP from Residence Time Distribution Curves

For Gaussian peaks, HETP measurements can be approximated as in Eq. (50). A more accurate measure involves using the statistical derivation of HETP as calculated from residence time distribution (RTD) curves (Levenspiel, 1972; Swaine and Daugulis, 1988). The pulse response curve method for RTD measurements is more sensitive to axial mixing and is suitable for visual observation of zone movement in the bed. The first absolute moment of the response curve, μ_1', which implies the peak retention time, is given by (Morton and Young, 1995)

$$\mu_1' = \frac{\int_0^\infty Ct\,dt}{\int_0^\infty C\,dt} = \frac{\int_0^\infty Ct\,dt}{\mu_0} \tag{55}$$

where $C = C(t)$ is the solute concentration at the bed exit, t is the time from the start of the pulse injection, and μ_0 is the zeroth moment (area of the curve). The normalized second central moment μ_2, which implies the peak variance, is given by

$$\mu_2 = \frac{\int_0^\infty C(t - \mu_1')^2\,dt}{\mu_0} \tag{56}$$

When the response curve is symmetrical and Gaussian, $\mu_1 = t_R$ and $\mu_2 = \sigma^2$. The observed moments can be corrected for the contribution due to the sample injection $t_F = V_F/Q$ by

$$\mu_1 = \mu_{1,\text{obs}} - \frac{t_F}{2} \tag{57}$$

Ion-Exchange Chromatography

$$\mu_2 = \mu_{2,obs} - \frac{t_F^2}{12} \tag{58}$$

The HETP is calculated from the moments by

$$\text{HETP} = \frac{L}{N} = \frac{L\mu_2}{\mu_1^2} \tag{59}$$

$$\text{HETP} = \frac{L}{N} = \frac{L\sigma^2}{t_R^2} \tag{60}$$

There are several dimensionless groups describing the axial mixing that can be related to N. The Peclet number based on the bed height, Pe_z (also called the Bodenstein number, Bo), is defined as (Thömmes et al., 1995)

$$\text{Pe}_{e_z} = \frac{Zu}{D_L} = 2N \tag{61}$$

The Peclet number based on the particle diameter, Pe_p,

$$\text{Pe}_p = \frac{ud_p}{D_L} \tag{62}$$

is often employed for describing packed bed chromatography (Yamamoto and Nakanishi, 1988). The number of transfer units for axial dispersion is given by (Arnold et al., 1985a; LeVan et al., 1997)

$$\text{NTU}_d = \text{Pe}_z = 2N \tag{63}$$

Figure 20 shows typical RTD pulse response curves for fixed and expanded beds with different media. The plate numbers obtained by Eqs. (55)–(60) are similar to those calculated by assuming a Gaussian peak shape and using Eq. (46) [i.e., $N = 8(t_R/w)^2$].

B. HETP Values in Fixed Bed Chromatography

Measurements of HETP can be used to assess the influence of chromatographic media and operating conditions on the separation. These measurements can be used to select conditions that give an acceptable level of band spreading. This is illustrated with vitamin B_{12}, where the distribution coefficient $K \approx 0.7$ (streamline SP) implies diffusive mass transfer limitations within the bead. According to the van Deemter equation, these mass transfer limitations will increase with velocity. Figure 21 shows that with increasing superficial velocity the HETP values increased linearly from 0.05 to 0.3 cm.

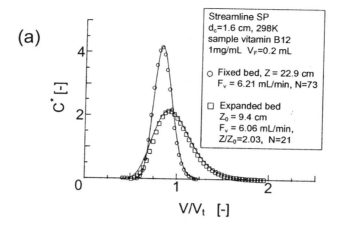

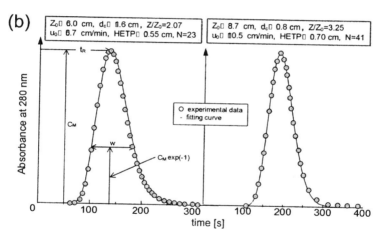

Figure 20 Typical pulse response curves for measuring HETP. (a) Streamline SP expanded bed (EBC) and fixed bed chromatography (FBC) at 298 K. (□) EBC: d_c = 1.6 cm, L_0 = 9.4 cm, L = 19.1 cm, u_0 = 3.0 cm/min. (○) FBC: d_c = 1.6 cm, L_0 = L = 23 cm, u_0 = 3.1 cm/min. The sample was vitamin B_{12}. The data are fit (solid curve) using a curve calculated by the plate model with N = 21 for the EBC and N = 73 for the FBC. (b) Hydroxyapatite expanded bed at 298 K, tracer = vitamin B_{12}. The data were fit with the plate model with (left) N = 23 (u_0 = 6.7 cm/min) and (right) N = 41 (u_0 = 10.5 cm/min). (Yamamoto et al. 2001).

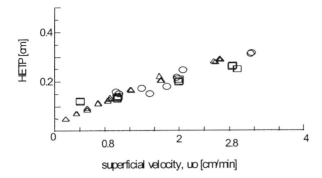

Figure 21 HETP as a function of superficial velocity at 298 K for a packed bed of Streamline SP, 1 mg/mL vitamin B_{12} as tracer, injection volume 0.2 mL. (○) d_c = 1.6 cm, Z = 22.9 cm, ε = 0.39. (□) d_c = 2.6 cm, Z = 8.7 cm, ε = 0.40. (△) d_c = 0.9 cm, Z = 20 cm, ε = 0.37. (Yamamoto et al. 1999b).

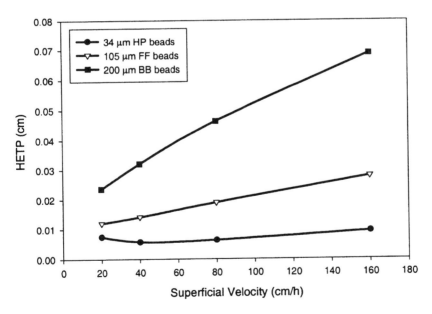

Figure 22 HETP vs. superficial velocity for fixed bed chromatography with different particle diameters, mean particle size SP Sepharose, HP = 40 μm, FF = 90 μm, and BB = 155 μm, I.D. = 2.6 cm. Tracer: 2% column volume of 1 M NaCl.

Such a plot can be used to determine the maximum operating velocity for the separation. This plot also illustrates that HETP is independent of column geometry.

Figure 22 shows HETP for different particle diameters at increasing superficial velocity. As predicted from the van Deemter equation, HETP increases with increasing particle diameter. In the flow range tested, there is little contribution to zone spreading from pore diffusion for the small 40 μm particles. Larger, 90 μm, particles are commonly used in protein production and show increasing HETP owing to diffusion limitations with increasing superficial velocity (Fig. 22). At typical production scale bed heights and flow rates, a column packed with these particles should operate with $N = 500$ plates. Such plate counts are often sufficient for good separation of proteins. If enhanced separation is needed through sharper peaks, smaller diameter (40 μm) beads or lower flow rates could produce plate numbers of ∼1500 for a typical column. However, columns packed with small beads are subject to significant pressure drop restrictions, which determine the maximum operating flow rate and thus limit productivity.

VII. PRESSURE DROP IN FIXED BED CHROMATOGRAPHY

Pressure drop limits the column flow rate and the bed height and can determine the maximum operating flow rate and column geometry (Janson and Hedman, 1982; Kaltenbrunner et al., 2000). Thus by affecting throughput, pressure drop influences the productivity of chromatography systems and can be an important parameter when comparing the performance of various chromatography media. For incompressible particles, fluid flow through a packed bed causes a pressure drop due to friction forces, as described by the Kozeny–Carman equation (Giddings, 1991; Dolejs et al., 1998; Bird et al., 2002),

$$\Delta P = u_0 L \frac{150\mu}{d_p^2} \left(\frac{(1-\varepsilon)^2}{\varepsilon^3} \right) \tag{64}$$

However, most polymeric media used in preparative chromatography will compress to an extent depending on the amount of cross-linking (or gel rigidity), column diameter, bed height, and the frictional properties of the media. The presence of a bounding wall containing the particles presents additional wetted surface area and affects the properties of the packing as well as the viscous and inertial flow through the bed. Wall effects provide support for the particles and reduces bed compression. Wall effects depend on the aspect ratio

Ion-Exchange Chromatography

of the packed bed. Different aspect ratios will result in different values for the critical velocity, which is the velocity beyond which pressure increases asymptotically and no additional flow is achieved. The flow instability and point of criticality are primarily due to the formation of a small, highly compressed region near the bottom of the column (Soriano et al., 1997). As column diameter increases, wall support of the media decreases and vertical compressive forces increase, resulting in a decrease in column porosity and flow instability.

For many preparative media, significant bed compression occurs, resulting in porosity reductions as high as 50% (e.g., from $\varepsilon = 0.38$ to $\varepsilon = 0.2$) have been reported (Colby et al., 1996; Danilov et al., 1997). For such compressible particles, the void fraction must be adjusted to lower values to account for the bed compression which occurs at higher flow rates (Stickel and Fotopoulos, 2001). Compression at a variety of column diameters must be determined in order to estimate production scale pressure drop. For example, pressure drop can increase by as much as 80% in scaling from a 5 cm diameter column to a 20 cm diameter column (Colby et al., 1996). For most preparative media, little wall support is seen above column diameters of 20 cm.

Pressure flow curves for various media can be measured in the production column or may be supplied by the manufacturer. From such curves, it can be seen that the increase in pressure drop is due to the increasing fluid velocity and decreasing height and porosity of the bed (Figs. 23 and 24).

Only recently have methodologies for predicting the pressure drop in fixed beds of compressible chromatographic media been presented. Colby et al. (1996) modified the volume-averaged continuum theory to account for wall effects and changes in bed porosity. The key parameters in this approach are interstitial porosity, specific surface area, internal angle of friction, and the angle of wall friction. Interstitial porosity and specific surface area are empirically related to pressure drop by fitting one set of bed compression data obtained from a small column. The internal angle of friction and the angle of wall friction can be measured by using standard tests on suspensions of the chromatography beads. The equations for frictional support,

$$\frac{\partial \tau_{zz}^s}{\partial z} = \mu u_0 \chi(\tau_{zz}^s) - \frac{4}{d_c} \tan \vartheta \frac{1 + \sin \xi}{1 - \sin \xi} \tau_{zz}^s \tag{65}$$

stress on the particles,

$$\chi(\tau_{zz}^s) = \frac{200 \varepsilon_s^2}{d_p^2 \phi^2 \varepsilon_f^3} \tag{66}$$

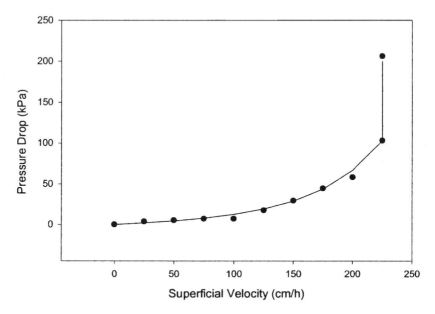

Figure 23 Experimental (●) and predicted (———) pressure drop CM Sepharose FF, $d_c = 10$ cm, $L_0 = 24.8$ cm, mobile phase = H_2O. Predicted using method of Stickel and Fotopoulos (2001) with $m = 1200$ cm^2/h, $b = 2200$ cm^2/h.

and the pressure gradient,

$$\frac{\partial P}{\partial z} = -\mu u_0 \chi(\tau_{zz}^s) \tag{67}$$

can then be used to predict the pressure drop of large-scale columns. These equations can then be solved using fourth-order Runge-Kutta integration and Newton's method for data regression to predict pressure drop at different superficial velocities. This method gave pressure drop predictions within ±20%.

Soriano et al. (1997) employed Janssen's (Ayazi-Shamlou, 1988) one-dimensional elemental slice differential force balance to predict column pressure drop. This approach takes into account gravity and the forces acting on the top and bottom faces and sides of the slice. The force balance on the mass of medium gives an equation for the pressure gradient and employs the Blake–Kozeny equation,

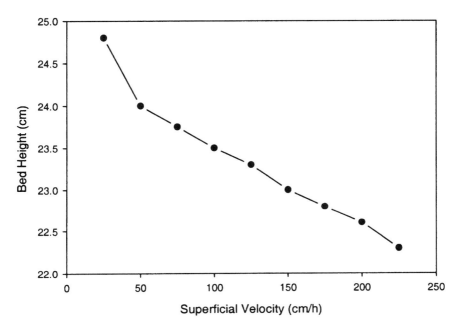

Figure 24 Bed compression in the CM Sepharose FF column with increasing superficial velocity, $d_c = 10$ cm, $L_0 = 24.8$ cm, mobile phase = H_2O.

$$\frac{dP_s}{dh} = \Delta\rho(1 - \varepsilon)g + Ku_0 - P_s k \tan \vartheta \left(\frac{4}{d_c}\right) \qquad (68)$$

where K in the Blake–Kozeny equation is estimated empirically from

$$K = K_0 \exp\left(-\frac{P}{P_{s_0}}\right) \qquad (69)$$

Equations (68) and (69) can be numerically integrated using fourth-order Runge-Kutta to generate axial pressure–flow curves for a given chromatographic matrix and column geometry. For a given media, the empirical data are generated by measuring the pressure drop and bed height of an initially gravity-settled bed at incremental superficial velocities. K_0 and P_{s_0} are determined empirically using the methods of Grace (1953) and Davies (1989). This method gave satisfactory predictions of pressure drop for size exclusion media.

An alternative method that does not require numerical integration was presented by Stickel and Fotopoulos (2001). This method involves predicting pressure drop from a series of bed compression measurements at different aspect ratios and column diameters. For a particular particle and bed aspect ratio, the maximum velocity and the critical porosity or void fraction are given by the plateau region of the ΔP–u curve. These data are linearly fit to predict critical velocities as a function of the aspect ratio,

$$u_{\text{cri}} = \frac{m\left(\dfrac{L_0}{d_c}\right) + b}{L_0} \tag{70}$$

where m and b are the linear regression coefficients.

Pressure vs. flow curves for various aspect ratios and the maximum pressure for a particular medium can be obtained from the manufacturer (Pharmacia, 2000; Janson and Hedman, 1982). The packed bed void fraction can then be obtained from the Kozeny–Carman equation or as listed in table 3–6

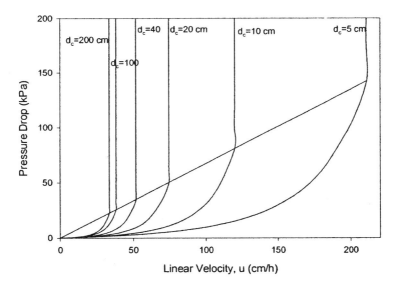

Figure 25 Predicted pressure drop for Sepharose CL-6B in columns of increasing diameter. Settled bed height = 20 cm, calculated from Kozeny–Carman equation with void fraction adjusted for bed compression.

Ion-Exchange Chromatography

of Karlson et al. (1998). Using this value as ε_c, the critical bed compression φ_c can then be calculated as

$$\varphi_c = \frac{\varepsilon_c - \varepsilon_0}{\varepsilon_c - 1} \tag{71}$$

For a given velocity, $u < u_{cri}$, the bed compression can then be calculated as

$$\varphi = \varphi_c \frac{u}{u_c} \tag{72}$$

and the void fraction as

$$\varepsilon = \frac{\varepsilon_0 - \varphi}{1 - \varphi} \tag{73}$$

The pressure drop up to u_{cri} can then be predicted using Eq. (64) as shown in Fig. 25. From this figure it is apparent that there is an initial linear region

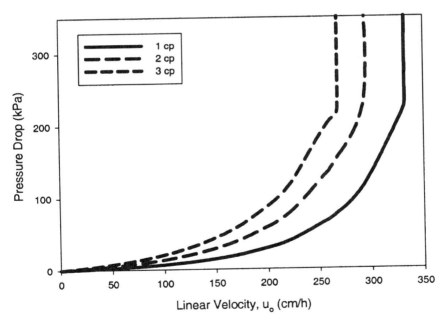

Figure 26 Predicted pressure drop for Sepharose CL-6B with increasing mobile phase viscosity, 1–3 cP. Settled bed height = 20 cm, diameter = 3 cm, calculated from Kozeny–Carman equation with void fraction adjusted for bed compression.

($u < 25$ cm/h) where wall effects are not important and hence flow is not sufficient to cause bed compression. In this region ΔP is independent of aspect ratio. Figure 25 also shows that there is little wall support at column diameters above ~20 cm.

Increasing mobile phase viscosity further reduces the maximum flow rate as shown in Fig. 26.

Thus, for large-scale fixed bed chromatography, the operating flow rate is often determined by the pressure drop. Typically, the maximum pressure drop is constrained by the equipment used at large scale. The Kozeny–Carman equation allows estimation of the maximum column length (L_{max}) for a given d_p and a backpressure (Δp_{max}) limitation (Kaltenbrunner et al., 2000).

VIII. CONCLUSIONS

This chapter presents a continuous flow plate theory model that has been shown to be very useful for further modeling, optimization, and characterization of an ion-exchange separation. In the case study presented here, reducing the ratio of protein loading and superficial velocity to bed height predictably increased resolution of the product peak from the pre-peak species. This correlation was useful for approximating alternative operating conditions to obtain a desired resolution. Using this approach, $GH-I_R$ and $K-I$ curves can be generated to reveal the elution order of the species and sensitivities to pH, ionic strength, and gradient slope. The model was then used to quickly survey operating conditions to maximize productivity with the predicted retention times within 15% of actual retention times. For sensitive separations involving few charges, the model can be used to set specification limits on buffer composition and media ionic capacity in order to maintain process consistency. Finally, it is shown that the pressure drop can be predicted for the proposed operating flow rates and column geometries using recently developed correlations, particularly those that employ simple empirical correlations.

NOMENCLATURE

a_i	Activity of species i in mobile phase (M)
$\overline{a_i}$	Activity of species i in stationary phase (M)
A	$K_e \Lambda^{|z_p|}$, parameter in Eq. (28)
A	$2\lambda d_p$, term in van Deemter equation
A_c	Cross-sectional area of the column (cm^2)

Ion-Exchange Chromatography

B	Intercept in critical velocity correlation (cm²/h)
B	Term in HETP equation
B	Counterion in mobile phase
$\overline{B}$	Counterion in bound stationary phase
B	Parameter in Eq. (28); the number of charges involved in protein adsorption
c_0	Feed concentration (g/L)
C	Term in HETP equation
C	Concentration of solute in mobile phase (M)
$\overline{C}$	Concentration of protein in stationary phase (M)
CV	Column volume (mL or L)
d_c	Column diameter (cm)
d_p	Particle diameter (cm or μm)
D_L	Longitudinal diffusion coefficient (cm²/s)
D_m	Molecular diffusion coefficient (cm²/s)
g	Acceleration due to gravity (m²/s)
g	$(I_f - I_0)/V_G$, gradient slope (M/mL)
G	gV_0, slope of the gradient normalized with respect to the column void volume (M)
GH	$g(V_t - V_0)$, slope of the gradient normalized with respect to the column stationary phase volume (M)
ΔG	Gibbs free energy change (cal/mol)
h	HETP/d_p, reduced HETP or reduced plate height
ΔH	Enthalpy change (cal/mol)
H	$(1 - \varepsilon)/\varepsilon = (V_t - V_0)/V_0$
HETP	Height equivalent to a theoretical plate (cm)
I	Dummy variable in Eq. (3) (cm · mL/mg)
I'	Dummy variable in Eq. (4) (h mL/mg)
I	Ionic strength of the buffer (mM or M)
I_E	Ionic strength of the elution buffer (M)
I_0	Ionic strength of the starting buffer (mM or M)
I_f	Ionic strength of the limit buffer (mM or M)
I_R	Ionic strength of the buffer at the peak maximum (mM or M)
k	Constant in HETP equation
k	Pressure coefficient, dimensionless
K	Blake–Kozeny constant
K_0	Permeability of uncompressed gel matrix (m²)
K	$\overline{C}/C'$, distribution coefficient
K'	Distribution coefficient for salt
K_e	Equilibrium constant

L	Bed height (cm)
L_0	Initial gravity-settled bed height (cm)
m	Slope in critical velocity correlation (cm^2/h)
m_i	molarity of species i (M)
$\overline{m_i}$	molarity of species i in the stationary phase (M)
M	Protein mass resin loading (mg/mL)
N	Total number of theoretical plates
NTU	Number of transfer units
O'	Resolution constant, Eq. (4)
P	Protein in the mobile phase
P	Productivity (g/L·h)
$\overline{P}$	Protein in bound stationary phase
Pe_z	L_u/D_L, Peclet number based on bed height
Pe_p	ud_p/D_L, Peclet number based on particle diameter
P_s	Vertical solid pressure in the bed (kPa)
P_{s_o}	Matrix rigidity (kPa)
ΔP	Pressure drop (psi or kPa)
Q	Volumetric flow rate (mL/min)
r	Column radius (cm)
R	Recovery (%)
R_s	Resolution, as defined in Eq. (47)
t	Time (s or min)
t_c	Cycle time (h)
t_F	Injection time (min or s)
t_R	Peak retention time (min or s)
$t_{R,\,LGE}$	Peak retention time in linear gradient elution (s)
u	$Q/A\varepsilon$, linear mobile phase velocity (cm/min or cm/h)
u_0	$Q/A = QL/V_t$, superficial velocity of the mobile phase (cm/min or cm/h)
u_{cri}	Critical velocity (cm/h)
v	ud_p/D_m, reduced velocity
V_e	Elution volume (mL)
V_F	Injection volume (mL)
V_g	Gel volume (mL)
V_{in}	Feed volume (L)
V_0	$A_c L\varepsilon$, void volume of the column (mL)
V_R	Retention volume (mL or L)
V_t	Total volume of the column (mL or L)
V_G	Gradient volume (mL)
W	Width of the peak at the baseline (min)

Ion-Exchange Chromatography

$W_{1/2}$	Peak width at half the peak height (min)
z	Distance from the inlet of the column (cm)
z_p	Distance of the peak position from the column inlet (cm)
Z_p	Protein valence
Z_B	Counterion valence

Greek Symbols

Λ	Effective total ion-exchange capacity
ε	V_0/V_t, void fraction of the column
ε_0	Gravity-settled void fraction of the column
γ_i	Activity coefficient of species i in the mobile phase
$\overline{\gamma_i}$	Activity coefficient of species i in the stationary phase
λ	Eddy diffusion coefficient in the HETP equation
μ_n'	n^{th} moment (min^n)
μ	viscosity (cP)
φ	Bed compression
φ_c	Critical bed compression
σ	Standard deviation of the elution curve
σ^2	Peak variance
τ_{zz}	Stress acting on the resin particles in the axial direction (Pa)
$\chi(\tau_{zz}^s)$	Expression describing contribution of porosity and specific surface area to pressure gradient
ϑ	Angle of wall friction (deg)
ξ	internal angle of friction (deg)
ϕ	Shape factor of particles (m^2/m^3)
$\Delta\rho$	Density difference between the media and the mobile phase (kPa)

REFERENCES

Arnold, FH, HW Blanch, CR Wilke. Analysis of affinity separations: II. The characterization of affinity columns by pulse techniques. Chem Eng J 30:B25–B36, 1985a.

Arnold FH, HW Blanch, CR Wilke. Liquid chromatography plate height equations. J Chromatogr 330:159–166, 1985b.

Ayazi-Shamlou, P. Handling of Bulk Solids: Theory and Practice. Butterworths, 1988.

Bidlingmeyer, BA, FV Warren. Column efficiency measurement. Anal Chem 56:1583A–1596A, 1984.

Bird, RB, WE Stewart, EN Lightfoot. Transport Phenomena, 2nd ed. New York: Wiley, 2002.

Boardman, NK, SM Partridge. Separation of neutral proteins on ion-exchange resins. Biochem J 59:543–552, 1955.

Carta, G. Linear driving force approximation for intraparticle diffusion and convection in permeable supports. Chem Eng Sci 50:887–889, 1995.

Chen, W. Separations Course. Yamaguchi Univ, Ube, Japan, 2000.

Colby, BC, BK O'Neill, APJ Middelberg. A modified version of the volume-averaged continuum theory to predict pressure drop across compressible packed beds of Sepharose Big-Beads SP. Biotechnol Prog 12:92–99, 1996.

Crank, J. The Mathematics of Diffusion. 2nd ed. Clarendon Press, 1975.

Danilov, AV, IV Vaenina, LG Mustaeva, SA Mosnikov, EY Gorbunova, VV Chrskii, MB Baru. Liquid chromatography on soft packing material under axial compression. Size-exclusion chromatography of polypeptides. J Chromatogr 773:103–114, 1997.

Davies, PA. Chem Eng Sci 44:452–455, 1989.

DeVault, D. The theory of chromatography. J Am Chem Soc 65:532–540, 1943.

Dolejs V, B Siska, P Dolecek. Modification of Kozeny-Carman concept for calculating pressure drop in flow of viscoplastic fluids through fixed beds. Chem Eng Sci 53:4155–4158, 1998.

Felinger, A, G Guiochon. Comparing the optimum performance of the different modes of preparative liquid chromatography. J Chromatogr A 796:59–74, 1998.

Forgacs, E, T Cserhati. Molecular Bases of Chromatographic Separation. Boca Raton, FL: CRC Press, 1997.

Gallant, S, A Kundu, SM Cramer. Optimization of step gradient separations—consideration of nonlinear adsorption. Biotechnol Bioeng 47:355–372, 1995.

Gallant, SR, S Vunnum, SM Cramer. Optimization of preparative ion-exchange chromatography of proteins: linear gradient separations. J Chromatogr A 725:295–314, 1996.

Giddings, JC. The role of lateral diffusion as a rate controlling mechanism in chromatography. J Chromatogr 5:46–60, 1961.

Giddings, JC. Dynamics in Chromatography. Part I. Principles and Theory, Vol. 1. New York: Marcel Dekker, 1965.

Giddings, JC. Unified Separation Science. New York: J Wiley, 1991.

Grace, HP. Chem Eng Prog 49:303–318, 1953.

Haff, LA, LG Faegerstam, AR Barry. Use of electrophoretic titration curves for predicting optimal chromatographic conditions for fast ion-exchange chromatography of proteins. J Chromatogr 266:409–425, 1983.

Horváth, C, HJ Lin. Movement and band spreading of unsorbed solutes in liquid chromatography. J Chromatogr 126:401–420, 1976.

Janson, JC, P Hedman. Large-scale chromatography of proteins. Adv Biochem Eng 25:43–99, 1982.

Jönsson, JÅ Dispersion and peak shapes in chromatography. In: JA Jönsson, ed. Chromatographic Theory and Basic Principles. New York: Marcel Dekker, 1987, pp 27–157.

Kaltenbrunner, O, A Jungbauer, S Yamamoto. Prediction of the preparative chromatography performance with a very small column. J Chromatogr A 760:41–53, 1997.

Kaltenbrunner, O, P Watler, S Yamamoto. Column qualification in process ion-exchange chromatography. Prog Biotechnol 16:201–206, 2000.

Karlson, E, L Ryden, J Brewer. Ion-Exchange Chromatography. In: J-C Janson, L Ryden, eds. Protein Purification. 2nd ed. New York: Wiley-VCH, 1998, pp 145–205.

Kennedy, GJ, JH Knox. The performance of packings in high performance liquid chromatography (HPLC) I. Porous and surface layered supports. J Chromatogr Sci 10:549–557, 1972.

Knox, JH, HP Scott. B and C terms in the Van Deemter equation for liquid chromatography. J Chromatogr 282:297–313, 1983.

Kokotov, YA. Generalized thermodynamic theory of ion-exchange isotherm. In: D Muraviev, V Gorshkov, A Warshawsky, eds. Highlights of Russian Science, Vol 1, Ion Exchange. New York: Marcel Dekker, 2000, pp 765–846.

Kopaciewicz, W, MA Rounds, J Fausnaugh, FE Regnier. Retention model for high-performance ion-exchange chromatography. J Chromatogr 266:3–21, 1983.

Kucera, E. Contribution to the theory of chromatography. Linear non-equilibrium elution chromatography. J Chromatogr 19:237–248, 1965.

Lapidus, L, NR Amundson. Mathematics of adsorption in beds. VI. The effect of longitudinal diffusion in ion exchange and chromatographic columns. J Phys Chem 56:984–988, 1952.

LeVan, MD, G Carta, CM Yon. Adsorption and ion-exchange. In: RH Perry, DW Green, JO Maloney, eds. Perry's Chemical Engineers' Handbook. 7th ed. New York: McGraw Hill, 16-40–16-43, 1997.

Levenspiel, O. Chemical Reaction Engineering. 2nd ed. New York: Wiley, 1972.

Lightfoot, EN, JL Coffman, F Lode, QS Yuan, TW Perkins, TW Root. Refining the description of protein chromatography. J Chromatogr A 760:139–149, 1997.

Lin, F-Y, C-S Chen, W-Y Chen, S Yamamoto. Microcalorimetric studies of the interaction mechanisms between proteins and Q-sepharose at pH near the isoelectric point (pI), effects of NaCl concentration, pH value, and temperature. J Chromatogr A 912:281–289, 2001.

Mann, F. Chromatography systems. Design and control. In: G Subramanian, ed. Process Scale Liquid Chromatography. Gloucestershire, UK: Stonehouse, 1995, pp 1–31.

Martin, AJP, RLM Synge. A new form of chromatogram employing two liquid phases. Biochem J 35:1358–1368, 1941.

Morton, DW, CL Young. Analysis of peak profiles using statistical moments. J Chromatogr Sci 33:514–524, 1995.

Natarajan, V, S Cramer. A methodology for the characterization of ion-exchange resins. Sep Sci Technol 35:1719–1742, 2000.

Pharmacia. Ion Exchange Chromatography. Principles and Methods. Amersham Pharmacia Biotech Uppsala, 18-1114-21 Edition AA, 1999.

Pharmacia. Ion exchange columns and media selection guide and product profile. Amersham Pharmacia Biotech UK Ltd., 18-1127-3 Edition AC, 2000.

Roth CM, KK Unger, AM Lenhoff. Mechanistic model of retention in protein ion-exchange chromatography. J Chromatogr A 726:45–56, 1996.

Schoenmakers, PJ. Optimization of Chromatographic Selectivity. New York: Elsevier, 1986.

Scopes, RK. Protein Purification. Principles and Practice. 3rd ed. New York: Springer-Verlag, 1994.

Scott, RPW. Liquid Chromatography Column Theory. Chichester, UK: Wiley, 1992.

Slaff, G. Chromatography column scale-up in an industrial environment. Presented at BioPharm '93, 1993. Cambridge, MA, June 1993.

Snyder, LR, JJ Kirkland. Introduction to Modern Liquid Chromatography, 1974.

Snyder, LR, MA Stadalius. High-performance liquid chromatography separations of large molecules: a general model. High-Perform Liq Chromatogr 4:195–312, 1986.

Sofer, G. Downstream processing in biotechnology. Center for Professional Advancement, 1995. East Brunswick, NJ. Report #9505103.

Soriano, GA, NJ Titchener-Hooker, P Ayazi-Shamlou. The effects of processing scale on the pressure drop of compressible gel supports in liquid chromatographic columns. Bioprocess Eng 17:115–119, 1997.

Stickel, JJ, A Fotopoulos. Pressure–flow relationships for packed beds of compressible chromatography media at laboratory and production scale. Biotechnol Prog 17:744–751, 2001.

Swaine, DE, AJ Daugulis. Review of liquid mixing in packed bed biological reactors. Biotechnol Prog 4:134–148, 1988.

Thömmes J, M Halfar, S Lenz, M-R Kula. Purification of monoclonal antibodies from whole hybridoma fermentation broth by fluidized bed adsorption. Biotechnol Bioeng 45:205–211, 1995.

Van Deemter, JJ, FJ Zuiderweg, A Klinkenberg. Longitudinal diffusion and resistances to mass transfer as causes of nonideality in chromatography. Chem Eng Sci 5:271–289, 1956.

Wang N-HL. Ion exchange in purification. In: JA Asenjo, ed. Separation Processes in Biotechnology, Vol 9, Bioprocess Technology. New York: Marcel Dekker, 1990, pp 359–400.

Watler, P, D Feng, S Yamamoto. Troubleshooting and productivity optimization on ion exchange chromatography. Presented at Recovery of Biologicals VIII, 1996. Tucson, AZ, October 20–26, 1996.

Watler, P, D Feng, S Yamamoto. Chromatography scale-up and optimization. Presented at 12th ASME Bioprocess Technology Seminar, 1999.

Wicke, E. Empirische und theoretische Untersuchungen der Sorptionsgeschwindigkeit-von Gasen an por sen Stoffen II. Kolloid Z 86:295–313, 1939.

Wilson, JN. A theory of chromatography. J Am Chem Soc 62:1583–1591, 1940.

Yamamoto, S. Plate height determination for gradient elution chromatography of proteins. Biotechnol Bioeng 48:444–451, 1995.

Yamamoto, S, T Ishihara. Ion-exchange chromatography of proteins near the isoelectric points. J Chromatogr A 852:31–36, 1999.

Yamamoto, S, T Ishihara. Resolution and retention of proteins near isoelectric points in ion-exchange chromatography. Molecular recognition in electrostatic interaction chromatography. Sep Sci Technol 35:1707–1717, 2000.

Yamamoto, S, Y Sano. Short-cut method for predicting the productivity of affinity chromatography. J Chromatogr 597:173–179, 1992.

Yamamoto, S, K Naknishi, R Matsuno, T Kamikubo. Analysis of dispersion mechanism in gel chromatography. Part IV. Operational conditions for gel chromatography—prediction of elution curves. Agric Biol Chem 43:2499–2506, 1979.

Yamamoto, S, M Nomura, Y Sano. Scaling up of medium-performance gel filtration chromatography of proteins. J Chem Eng Jpn 19:227–231, 1986.

Yamamoto, S, M Nomura, Y Sano. Adsorption chromatography of proteins: determination of optimum conditions. AIChE J 33:1426–1434, 1987a.

Yamamoto, S, M Nomura, Y Sano. Factors affecting the relationship between the plate height and the linear mobile phase velocity in gel filtration chromatography of proteins. J Chromatogr 394:363–367, 1987b.

Yamamoto, S, M Nomura, Y Sano. Resolution of proteins in linear gradient elution ion-exchange and hydrophobic interaction chromatography. J Chromatogr 409:101–110, 1987c.

Yamamoto, S, K Nakanishi, R Matsuno. Ion-Exchange Chromatography of Proteins, Vol. 43. New York: Marcel Dekker, 1988.

Yamamoto, S, A Okamoto, P Watler. Effects of adsorbent properties on zone spreading in expanded bed chromatography. Bioseparation 10:1–6, 2001.

Yamamoto, S, N Akazaki, O Kaltenbrunner, P Watler. Factors affecting dispersion in expanded bed chromatography. Bioseparation 8:33–41, 1999b.

Yamamoto, S, PK Watler, D Feng, O Kaltenbrunner. Characterization of unstable ion-exchange chromatographic separation of proteins. J Chromatogr A 852:37–41, 1999a.

5
A Simple Approach to Design and Control of Simulated Moving Bed Chromatographs

Firoz D. Antia
Merck & Co., Inc., Rahway, New Jersey, U.S.A.

I. INTRODUCTION

It is widely accepted that industrial separation processes, such as distillation or liquid extraction, are most effective when operated continuously, with the contacting fluids moving countercurrently in order to maximize the driving force for mass transfer between the phases. Operation of a chromatographic system in this fashion, however, is difficult because motion of the solid sorbent is likely to lead to disturbances in the packing structure, resulting in channeling and loss of efficiency. To retain the benefits of continuous countercurrent contact in a chromatographic separation, engineers in the late 1950s conceived the idea of simulated moving bed chromatography (SMB) (Broughton and Carson, 1959; Broughton and Gerhold, 1961). In SMB, a fixed bed is divided into subsections, and the flow path is switched between the sections at regular intervals to simulate countercurrent motion of solid and fluid. SMB chromatography is now used extensively in the chemical process industry, e.g., for purifying xylenes from C_8 isomers, as well as in the sugar industry for fructose/glucose separations. Worldwide installed capacity today in these industries exceeds 8 million metric tons per year (Gattuso, 1995).

Interest in SMB in the pharmaceutical and fine chemical industries has been sparked by the search for economic methods for chiral resolution. The high cost of chiral stationary phases (CSPs) has prompted exploration of SMB as a means to reduce the quantities of sorbent needed for preparative chiral liquid chromatography (Negawa and Shoji, 1992). As published examples over the past decade have shown (Nicoud et al., 1993; Küsters et al., 1995; Guest, 1997; Francotte, 2001; Schulte and Strube, 2001), SMB technology is a viable manufacturing method for chiral molecules, competing directly with classical resolution, enantioselective synthesis, and biocatalysis. Indeed, large-scale chiral resolution of pharmaceutical compounds—producing several tons per year—by SMB under good manufacturing practices (GMP) conditions has recently been reported (Cavoy and Hamende, 2000; Nicoud, 1999).

In contrast with petrochemicals or sugars, product availability is often severely limited during early development of pharmaceuticals. Most product candidates do not survive clinical trials, so for each successful drug in full-scale manufacturing there are several processes developed for compounds that fail. In this environment, time available for process optimization is minimal and the ability to switch quickly from one product to another in the same equipment is critical. Most pharmaceutical SMB separations are therefore evaluated rapidly in small benchtop or pilot units under the purview of a developmental chemist, who may be an expert in conventional chromatography but perhaps not as well versed in the operating principles of the SMB chromatograph. Because simultaneous selection of several flow rates and other parameters is required merely to achieve a separation in, let alone optimize, an SMB unit, practitioners rely on vendor-supplied software to choose and adjust conditions. These proprietary packages use detailed mathematical models to simulate dynamic behavior of the system and require estimates of the adsorption isotherm, which may be difficult to obtain in a short time.

This chapter is intended to help the reader understand how to design and operate an SMB system without detailed modeling or measurement of adsorption isotherms. First, a simple graphical scheme is presented for selection of initial separation conditions. Next, the response of an SMB to small perturbations in operating conditions is examined, and a simple set of rules is developed for diagnosing the cause of the perturbations and applying the necessary corrective action. It is shown how this set of rules can be used to design and control SMB systems. The chapter closes with a short discussion on the economics of SMB processing, with particular regard to enantiomer separations.

It is worth noting here that the set of rules, or heuristics, presented in this chapter can be codified into powerful rule-based control, design, or optimization algorithms using a structured approach such as that embodied in fuzzy logic. Although this is outside the scope of this work, it is hoped that the

Design and Control of SMB Chromatographs

approach will not only help develop a simple physical understanding of the operation and response of an SMB but also spur interest in the development of heuristic algorithms to augment currently available design and control methods.

II. TRUE AND SIMULATED MOVING BEDS

Several images have been used recently (e.g., Hashimoto et al., 1993) to convey the notion of moving bed chromatography and why it could be useful. One likens the conventional chromatographic separation to the idea of separating a hare from a tortoise by having the pair climb a staircase. Before one can introduce a second tortoise–hare couple at the bottom of the stairs for separation, one must wait for the slow-moving tortoise from the preceding couple to reach the top, or at least advance a significant way up the flight, otherwise the second hare might catch up with them and interfere with their separation. The resolution between the species is discontinuous and does not utilize the entire staircase all of the time.

Now imagine that we replace the staircase by a downward-moving escalator and introduce our competitors in the middle, rather than at the bottom, and compel them to move upward as best as they can. If the escalator speed were adjusted appropriately, the limber hare would find its way to the top, whereas the plodding tortoise would inexorably be forced to the bottom. We could then continuously drop tortoise–hare pairs at the center of the escalator, and soon nearly every rung would be filled with separating species, with the pure populations of either increasing steadily at the top and bottom. This continuous separation makes more efficient use of the staircase and is a more attractive alternative to the conventional system.

Note that for the system to work, the tortoises—which represent the more tightly bound species in chromatography—must be removed from the escalator (or sorbent) by some means so that clean rungs can be returned to the top.

A. True Moving Beds

In translating this notion into a chromatographic system, let us assume that we are attempting to separate a weakly retained species A from a more strongly retained species B and that we have at our disposal a means to "flow" the solid sorbent in a direction opposite to that of the fluid phase without encountering any of the problems that this might raise in practice. By analogy with the escalator described above, it is clear that by appropriate choice of the solid and fluid velocities, it would be possible to move either species forward or backward (with respect to the fluid flow direction), achieving separation as desired.

One way to configure such a true moving bed (TMB) device is shown in Fig. 1. This is a four-unit moving bed, a Sorbex system of the type pioneered by UOP (Ruthven and Ching, 1989). Feed is introduced continuously in the center of the system as a solution of A and B in the eluent. The two central units (or zones or sections; the terms are used interchangeably) above and below the feed point (units 3 and 2, respectively) are analogous to the top and

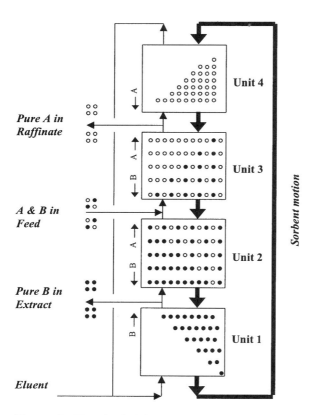

Figure 1 The classical four-unit moving bed configuration. A feed solution containing (○) A and (●) B is continuously resolved into pure A in the raffinate and pure B in the extract streams.

Design and Control of SMB Chromatographs

bottom segments of the escalator, and in these units flow rates of the moving sorbent and solvent are chosen such that the species separate; i.e., A moves forward (like the hare) in the direction of the fluid flow, and B moves backward (like the tortoise) in the direction of the solid flow. A "raffinate" stream containing pure A and an "extract" stream containing pure B are drawn off as shown. The bottom unit (unit 1) operates much as the bottom landing of the escalator that prises tortoises from the rungs of the moving stair; here, solvent flow is maintained at a rate high enough that component B moves upward despite the downward motion of the sorbent. Thus clean sorbent can be recycled as shown to the top of the unit. The top unit, on the other hand, has the solid/fluid flow ratio so adjusted that even the weakly retained component A moves downward. As a result, clean eluent free of either separating species emerges from the top of the unit and is recycled along with some makeup eluent to the bottom unit as shown. (Makeup eluent is added to compensate for that removed in the raffinate and extract streams.) The four-unit system therefore not only separates A and B continuously but also incorporates an internal recycle of both sorbent and eluent in order to enhance efficiency.

The four-unit moving-bed system—which can be compared, despite several differences, to a distillation tower with its rectifying and stripping sections fitted with a reboiler and a condenser—is widely used but is by no means the only way to set up a moving bed. For examples of other configurations and their applications, the reader is referred to the excellent review by Ruthven and Ching (1989). When modeling systems other than the four-unit one considered here, it is worthwhile to avoid the pitfalls noted by Hotier (1996).

An alternative way of describing the moving bed unit is to consider the operation as a series of adsorptions and desorptions. A species moving in the direction of the sorbent is being adsorbed by the solid from the fluid stream; a species moving in the direction of the liquid is likewise being desorbed by the fluid from the solid stream. Thus, in the four-zone system, unit 1 is a desorber for species B, units 2 and 3 are set up to adsorb B but desorb A, and unit 4 is an adsorber for species A.

Classical chemical engineering theory tells us that countercurrent operation maximizes mass transfer and thus accomplishes any adsorption or desorption operation more efficiently, i.e., with fewer theoretical plates or equilibrium stages, than any other operating configuration. Thus, for genuinely binary problems, such as the separation of enantiomers, it would appear that a properly designed four-unit true moving bed system would be a more attractive choice than a conventional batchwise process. Separation of complex mixtures can be accomplished by more intricate moving bed configurations, by coupling two binary units, or by campaigning the separation through two or more steps in the same equipment with conditions as appropriate to each step. The utility of the moving bed in such instances must, of course, be evaluated on a case-by-case basis.

B. Simulated Moving Beds

Because actual motion of solid sorbent is difficult to accomplish in a controlled and efficient manner, simulated moving beds (SMBs) have been devised. Consider the four units of the moving bed system shown in Fig. 1 redrawn as in

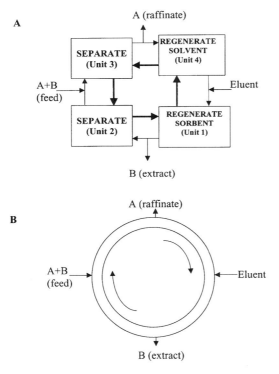

Figure 2 The four-unit moving bed of Fig. 1 progressively redrawn in panels (a) and (b) to show that it can be set up as a contiguous ring of sorbent. (b) Arrows show the direction of motion of the fluid; solid moves in the opposite direction. It is evident that motion of the solid can be simulated by moving the external ports in the direction of the fluid.

Design and Control of SMB Chromatographs

Fig. 2a, then further idealized as a circle as shown in Fig. 2b. It is apparent from this schematic that the only reference points for internal motion in the system are the two input and two output streams that connect the system to the outside world. This suggests that solid motion counterclockwise past the reference points is equivalent to motion of the reference points clockwise past the solid. It is impractical, of course, to have smooth motion of the four external flow streams around the unit, because that would require an infinite number of connection points. A more practical solution is to divide the circular bed of sorbent into a number of discrete segments, then periodically (and simulta-

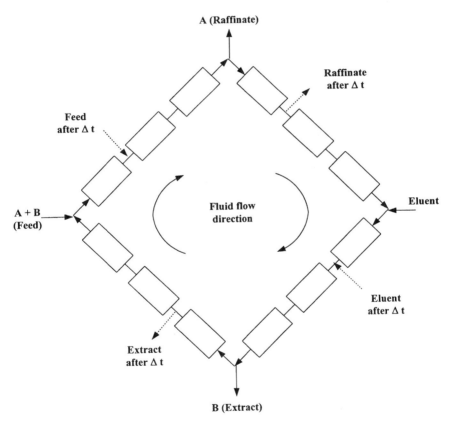

Figure 3 Schematic of an SMB unit made up of 12 columns, with three columns to each zone. The position of each of the external flow switches by one column after each switching period Δt. Flow within the column ring is driven by a pump.

neously) switch the four streams by one segment after a discrete time interval, Δt, known as the switching time. In an SMB, then, the sorbent is packed into a series of equal-sized subunits, or short columns, as shown in Fig. 3, and the feed, raffinate, eluent makeup, and extract streams are moved around the unit clockwise by one column after a time interval Δt using an appropriate arrangement of switching valves. Internal liquid flows in the unit are in a clockwise direction.

C. Equivalence Between Simulated and True Moving Beds

Clearly, the larger the number of subunits the simulated moving bed is divided into, the closer it approaches the behavior of a true moving bed (Ruthven and Ching, 1989). With fewer subunits, practical operation of the unit becomes simpler, but discontinuities are amplified; for example, concentrations in the raffinate and extract streams may oscillate widely over each switch period. Nevertheless, it has been shown that as few as six columns—two in each of the separating units (i.e., units 2 and 3) and one each in units 1 and 4, the so-called 1–2–2–1 configuration—can be sufficient for good performance (Nicoud, 1999). Eight- and twelve-column systems (2–2–2–2 and 3–3–3–3, respectively) are also common. Determining the optimal number of subunits for an SMB is best done by detailed modeling and simulation and is not covered in this chapter. In what follows, it is assumed that a sufficient number of columns are available for the SMB to approximate the behavior of a TMB.

III. SOLID AND FLUID FLOWS IN MOVING BEDS
A. Solid "Flow"

In the four-unit Sorbex SMB, all the sorbent remains within the system at all times. The volumetric rate of solid motion, or solid "flow," is thus the same in all sections of the unit. If V_c is the volume of each individual column in the unit and the total internal void fraction (i.e., the sum of the interstitial and intraparticle void volume divided by the total volume) is ε, then the total volume occupied by solid in a column is $(1 - \varepsilon)V_c$ and the equivalent solid flow rate, Q_s, in the counterclockwise direction caused by switching by one column in the clockwise direction every Δt min is

$$Q_s = (1 - \varepsilon)V_c/\Delta t \tag{1}$$

Design and Control of SMB Chromatographs

Note that the clockwise switch also carries all the liquid in the column (i.e., a volume of εV_c) in the counterclockwise direction. This is discussed in more detail later.

B. Fluid Flows

Consider a four-unit moving bed in the schematic drawn in Fig. 4 (the units are shown without division into subunits). There are eight liquid flows of concern in the system—four external flows, which include the feed, F_f, eluent makeup, F_e, raffinate, F_{raf}, and extract, F_{ext}, flows, and the internal liquid flow in each of the units, F_1–F_4. Three of the four external flows are independent and can be set to any rate desired; the fourth is determined by the fact that

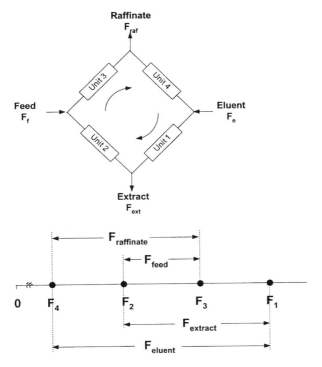

Figure 4 Schematic showing relationships between all the fluid flow rates in the moving bed system.

the sum of the two inputs must equal the sum of the two outputs. Thus if feed, eluent, and extract flows are set independently, the raffinate flow is given by

$$F_{raf} = F_f + F_e - F_{ext} \tag{2}$$

Pumps are used to regulate the external flows. Some SMB designs use three external pumps, whereas others may use a redundant fourth. For flows out of the system (i.e., for the raffinate or extract), it is important to have a pressure restrictor on the pump outlet that exceeds the internal pressure in the system, otherwise the flow is not properly regulated.

Of the four internal flows, only one is independent, and the other three can be written in terms of the independent internal flow and the external flows. If F_1 is the independently set internal flow, then

$$F_2 = F_1 - F_{ext} \tag{3}$$

$$F_3 = F_1 - F_{ext} + F_f \tag{4}$$

$$F_4 = F_1 - F_e \tag{5}$$

F_1 can be set at any flow rate subject to the maximum pressure limit of the system. An internal recycle pump is usually included within the loop of attached columns in order to define the rate and clockwise direction of the internal flow. As shown later, maximum productivity is usually attained at the maximum allowed internal flow.

C. Design Flows

As mentioned above, the column switching that is carried out to simulate the flow of solid also carries with it liquid in the counterclockwise direction. As a result, the design flow rate—or the effective clockwise liquid flow rate relative to the external reference points of the system—in any unit i, Q_i, is given by

$$Q_i = F_i - \frac{\varepsilon V_c}{\Delta t} = F_i - \frac{Q_s}{\phi} \tag{6}$$

Here, the expression for Q_s is substituted from Eq. (1) and ϕ is the so-called phase ratio, the ratio of sorbent to eluent volume in the column, i.e.,

$$\phi = (1 - \varepsilon)/\varepsilon \tag{7}$$

The design flow rate is equivalent to the liquid flow rate in an idealized true moving bed system where motion of the solid does not carry with it any en-

Design and Control of SMB Chromatographs

trained liquid. Equations (3)–(5) can be rewritten in terms of the design flow rates as

$$Q_2 = Q_1 - F_{ext} \tag{8}$$

$$Q_3 = Q_1 - F_{ext} + F_f \tag{9}$$

$$Q_4 = Q_1 - F_e \tag{10}$$

D. Graphical Representation of Fluid Flow

Figure 4 shows graphically the interrelationship between the various fluid flows in the system as determined by Eqs. (3)–(5). [Note that representation of Eqs. (8)–(10) is identical but for the offset of all points to the left by $\varepsilon V_c/\Delta t$.] The graphic provides a picture of all the fluid flows at the same time and immediately makes clear (1) what steps must be taken to change any of the flows in the system and (2) the consequences of any change in flow rate to the system. For instance, if it is desired to increase the internal flow F_2 (or the design flow Q_2) without disturbing any of the other design flows, one simply decreases the extract and feed flows by identical amounts. On the other hand, if the extract flow rate is increased without changing the feed or eluent flow rates, both F_2 and F_3 (and therefore Q_2 and Q_3) would decrease.

IV. OPERATING AN SMB UNDER DILUTE, OR LINEAR, CONDITIONS

In conventional chromatography, the migration velocity of a species depends on the velocity of the eluent as well as on the equilibrium distribution of the species between the sorbent and the eluent, i.e., the adsorption isotherm. At low concentrations of a migrating component i, the ratio between its concentration in the sorbent, q_i, and its concentration in the eluent, c_i—also known as the distribution coefficient K_i $(=q_i/c_i)$—is a constant, and the velocity of migration, u_i, is given by the expression

$$u_i = \frac{u_0}{1 + k_i'} = \frac{u_0}{1 + \phi K_i} \tag{11}$$

where A is the cross-sectional area of the column. Here k_i' is the familiar retention factor, which is equal to ϕK_i, and u_0 is the velocity of the eluent.

Now consider any section j of an ideal true moving bed (where, as mentioned above, the motion of solid entrains no liquid) where the liquid and solid flows, in opposite directions, are Q_j and Q_s, respectively. The flow velocities of liquid and solid, $u_{0,j}$ and v, respectively, are then given by

$$u_{0,j} = \frac{Q_j}{\varepsilon A}; \quad v = \frac{Q_s}{(1 - \varepsilon)A} \tag{12}$$

To an observer seated on the solid, the liquid moves by at a velocity $u_{0,j} + v$. To this observer, the motion of a species i in the section, as Eq. (11) suggests, is $(u_{0,j} + v)/(1 + \phi K_i)$, and the velocity to an observer outside the system is then

$$u_{i,j} = \frac{u_{0,j} + v}{1 + \phi K_i} - v \tag{13}$$

If the velocity in Eq. (13) is positive, species i moves in the direction of the liquid motion (the positive direction); if it is negative, the species moves in the direction of the solid flow. In other words, species i moves forward if $(u_{0,j} + v)/(1 + \phi K_i) > v$, and backward if $(u_0 + v)/(1 + \phi K_i) < v$. Simplifying, and combining this with Eqs. (7) and (12), one sees that the motion of species i is determined as follows:

If $Q_j > K_i Q_s$, species i moves forward (with liquid)
If $Q_j = K_i Q_s$, no movement (14)
If $Q_j < K_i Q_s$, species i moves backward (with solid)

As discussed earlier, in a four-unit moving bed performing a binary separation between the early-eluting species A and the late-eluting species B (so that $K_A < K_B$), each unit has a particular function; as mentioned before, zone 1 drives both species forward, zones 2 and 3 perform the separation (i.e., driving A forward and B backward), whereas zone 4 must drive both species backward. Using Eq. (14), the design criteria for successful operation of the moving bed under linear conditions are

Zone 1: $K_B Q_s < Q_1$
Zone 2: $K_A Q_s < Q_2 < K_B Q_s$
Zone 3: $K_A Q_s < Q_3 < K_B Q_s$ (15)
Zone 4: $Q_4 < K_A Q_s$

As shown in Fig. 5, the criteria can be depicted graphically on the flow relationship diagram introduced earlier. The graph contains all the information in the flow equations and design criteria and provides a simple and efficient guide to setting up the flow rates in an SMB system.

Operating an SMB under linear conditions involves the determination

Design and Control of SMB Chromatographs

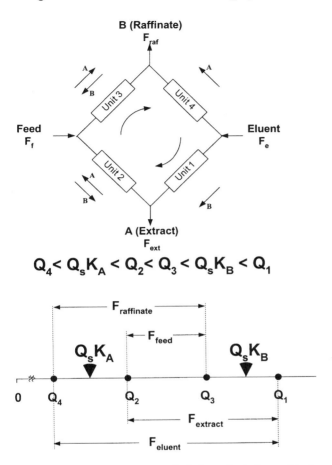

Figure 5 Depiction of the SMB design criteria in graphic form. In the upper panel, arrows depict the effective direction of motion of the components in the various units.

of the distribution coefficients K_A and K_B, selection of the column switching time Δt (and hence the solid flow rate Q_s), and, finally, the operating liquid flow rates in the system.

A. Determination of the Distribution Coefficients K_A and K_B

It is a straightforward process to determine K_A and K_B, involving analytical injections of the two components A and B into a column packed with the same sorbent and using the same mobile phase that is to be used in the SMB.

An injection of an unretained species is used to determine the void volume. If the respective retention volumes of the unretained species and of A and B are V_0, V_A, and V_B, and V_c is the empty column volume, then the void fraction is V_0/V_c, the phase ratio ϕ is $(V_c - V_0)/V_0$, and the K value for either species is given by

$$K_i = (V_i - V_0)/\phi V_0 \tag{16}$$

Choice of the mobile phase and sorbent, of course, must be determined in the same way as for any preparative separation, and the details are not discussed here. The ratio K_B/K_A is known as the selectivity α, and in general α must be maximized for maximum productivity. That being said, SMB is most useful for fairly difficult separations; if selectivities are very much higher than 2, the separation may be easy enough that there may be other means of achieving high productivity besides SMB.

A rule of thumb that many practitioners follow is to select conditions such that the k' (or ϕK) values of the species are fairly low, typically less than 5. There is further discussion of this issue later in the chapter. Another point to note here is that the system must be isocratic; gradient SMB systems are possible but lie outside the scope of this chapter.

A key requirement for success is that the column used for the determination of the parameters must be packed with the same void fraction as the SMB columns themselves. Indeed, for best results it is important that, within practical limits, all the columns used for the SMB be packed identically, with the same mass of sorbent and the same efficiency characteristics.

B. Selection of the Column Switching Time

The switching time Δt determines the solid flow rate Q_s as in Eq. (1). From Fig. 5, it is clear that the maximum design flow rate in the system is in zone 1 and that this must be chosen higher than the factor $K_B Q_s$. Thus if F_{max} is the maximum internal flow rate that one wishes to operate at in the system, Eq. (6) suggests that Q_s (and hence Δt) must be chosen such that

$$F_{max} - \frac{Q_s}{\phi} > Q_s K_B \tag{17}$$

or

$$Q_s < \frac{F_{max}}{K_B + 1/\phi} \tag{18}$$

Design and Control of SMB Chromatographs

Substituting Eq. (1) into Eq. (18), one obtains a criterion for selecting the switching time as

$$\Delta t > \varepsilon V_c \frac{\phi K_B + 1}{F_{max}} \tag{19}$$

Equation (19) indicates one reason why low k' values are often chosen for SMB operation. If high values are used, then Δt values are high and the system response time is slow, because several switches must occur before the system approaches the intended steady-state operation.

C. Selecting the Operating Flow Rates

Once Δt is selected, Q_s is calculated from Eq. (1), and all other design flow rates can be set by consulting Fig. 5, which shows the allowable values of the four internal flow rates as bounded by the design criteria. The greater the distance on the diagram by which an internal flow rate satisfies the design criteria, the more effectively that unit carries out its function. For example, in unit 1, component B will move forward more rapidly the greater the difference between Q_1 and $Q_s K_B$. In addition, the greater the distance by which the design criterion in a unit is met, the fewer the theoretical plates needed to accomplish the task of the unit.

It is worth examining two extreme circumstances. Consider first the case where there is no internal recycle pump and the flow rate in section 4, F_4, is zero, as illustrated in Fig. 6a. (Note that check valves would be needed in the internal loop to ensure that the eluent flow was directed as intended. Note also that this implies a negative design flow Q_4!). As shown in Fig. 6a, the design criteria are all satisfied. The system will work, but unit 4 is overdesigned; the criterion for the unit is met by too great a margin, and consequently a large amount of eluent is required. The example illustrates the utility of a recycle pump; eluent consumption is greatly reduced by bringing Q_4 closer to $Q_s K_A$ with the help of the pump.

A second extreme case, shown in Fig. 6b, is one in which the criteria are just satisfied. All the external flow rates—i.e., the feed, eluent, raffinate, and extract flows—are nearly equal and are given by $Q_s(K_B - K_A)$. In practice, this condition will not work; the design criteria are not satisfied by any appreciable margin, and thus an infinite number of theoretical plates would be required for a separation to occur. Nevertheless, this represents the highest theoretical feed flow and the lowest possible eluent flow—or the highest specific productivity (kilograms of product per kilogram of stationary phase per day)

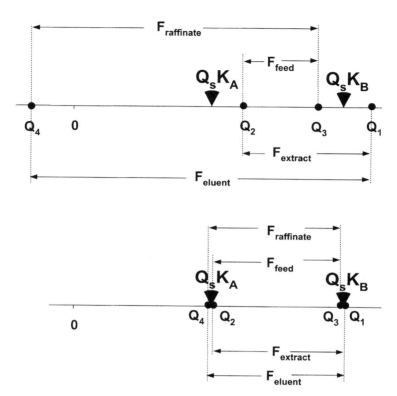

Figure 6 Two extreme cases. (a) A case with no internal recycle so that $F_4 = 0$ (and thus $Q_4 = -Q_s/\phi$); a large eluent consumption results. (b) A case where the design criteria are just satisfied; though impractical, it represents the maximum possible feed throughput with minimum solvent consumption under the prevailing conditions.

and the lowest specific solvent consumption (liters per kilogram of product purified)—under the prevailing conditions. This suggests that the maximum productivity in an SMB system is achieved at the highest possible solid flow rate, which in turn depends on the maximum liquid flow rate achievable in the system as shown in Eq. (18). Despite loss of efficiency (i.e., lower number of theoretical plates) at higher flow rate, it is generally true that maximum productivity is achieved at close to the maximum possible internal flow rate in the system.

D. Steady State

After operating for a sufficiently long period, continuous operations usually reach a steady state where there are no variations in the operating parameters with time. In the case of an SMB unit, however, because of the intermittent switching of columns, the system reaches a periodic steady state (Yun et al., 1997); parameters such as the concentrations of the products in the raffinate and extract streams change over the course of the switching period, but the same change is repeated for each switch. This can be visualized by considering the locations of the various ports in the system at the beginning and end of the switching period. As shown in Fig. 7, when the periodic steady-state condition is reached, the port returns after each switch to the same point in the internal concentration profile.

The design technique outlined above is similar to that put forward by other authors (Ruthven and Ching, 1999; Charton and Nicoud, 1995), but the graphical scheme goes further in providing a visual interpretation of the design rules and an intuitive aid to understanding the SMB system and its responses.

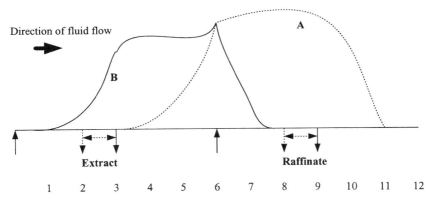

Figure 7 Internal concentration profile in an SMB unit. Over the course of one switching period, the profile moves some distance past the external ports. When the period is over, the column switches so that the ports "catch up" with the internal profile. As a consequence, the raffinate and extract concentrations vary over the course of the switching period, and the concentration history at either of these ports oscillates with a time period Δt. The system thus achieves a "periodic steady state."

V. HOW AN SMB RESPONDS TO SMALL UPSETS IN OPERATING FLOWS

Figure 5 is a starting point to begin examining the way in which an SMB system responds to small changes. In the following, small upsets in the system are considered that lead to failure of one or two units at a time. Understanding the consequences of these disturbances enables the creation of a table of responses that can serve as a guide to diagnose and correct operating problems in the system. The guide can be useful not only in system control but also in design under nonlinear conditions as well as for process optimization.

A. Single-Unit Failures

Consider a small upset in a working linear SMB system that leads to a failure in unit 3 without failures in any of the other units. As shown in Fig. 8, this can occur if for any reason Q_3 becomes greater than $Q_s K_B$. The change can be a result of, for example, a decrease in the distribution coefficient K_B arising from an increase in the feed concentration. (If the adsorption isotherm is nonlinear, K changes with concentration; in most cases when adsorption is competitive, K decreases with concentration.) The result of the failure, as can be seen in the figure, is that component B now moves in the direction of the liquid instead of the sorbent and is no longer separated from A as it was previously. Consequently, the raffinate stream is contaminated with some amount of component B. The appropriate corrective action in this case would be to reduce Q_3, without changing other internal flows, until the system returns to normal operation. Upon consulting the flow graphic, it is clear that a change to Q_3 in this manner can be accomplished by reducing the feed and raffinate flows by the same amount.

This approach can be used to consider failures in each of the other units. For example, if for any reason there were an upset leading to failure of unit 4 alone, component A in this unit would no longer be driven backward toward the raffinate stream but would move forward. Component A would thus be detected at the recycle point (i.e., the outlet of unit 4, just before it meets the fresh eluent stream) and eventually would move through zone 1 to the extract port, contaminating the extract stream. The consequences of failures in each unit, and the necessary corrective responses, are shown in Table 1.

It should be noted that failure in a unit can also occur if the internal unit flow rate in question is close to, but not actually, violating the design criterion; the condition is then not satisfied by a sufficient margin, and dispersive forces can result in failure. The corrective response would be the same

Design and Control of SMB Chromatographs

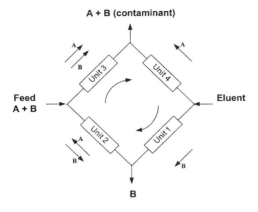

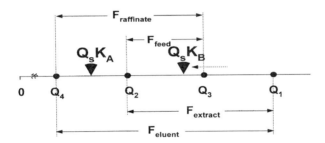

Figure 8 If a minor upset to the system causes $Q_s K_B$ to decrease or Q_3 to increase, unit 3 will fail, with consequences as shown in the upper panel.

in this case as when the design criterion is actually violated and would have the result of increasing the margin by which the design criterion is satisfied.

The analysis shows that single-unit failures can be diagnosed by monitoring the extract, raffinate, and recycle point. (The latter can be done by including a detector in the system at the appropriate location.) As Table 1 shows, each such failure has a unique symptom, making diagnosis of the observed problem quite simple.

B. Two-Unit Failures

The failure analysis can be extended to consider simultaneous failure in two units at a time. Consider, for instance, failure of units 2 and 3 together, while

Table 1 Single-Unit Failures and Appropriate Corrective Responses

Failure profile[a]				Symptoms[b]			Corrective response[c]
Unit 1	Unit 2	Unit 3	Unit 4	Raff.	Extr.	Recycle	
−	+	+	+	Bad	OK	Bad	Increase Q_1 by increasing eluent and extract flows.
+	−	+	+	OK	Bad	OK	Increase Q_2 by decreasing feed and extract flows.
+	+	−	+	Bad	OK	OK	Decrease Q_3 by decreasing feed and raffinate flows.
+	+	+	−	OK	Bad	Bad	Decrease Q_4 by increasing eluent and raffinate flows.

[a] + implies working unit; − implies failed unit.
[b] "Bad" implies contamination with a component that should not be in the stream.
[c] When responding, increase or decrease external flows by equal amounts.

Design and Control of SMB Chromatographs

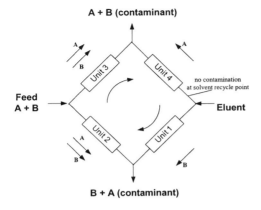

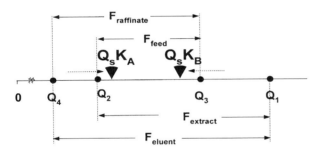

Figure 9 Schematic showing perturbations leading to failures of units 2 and 3 and their consequences.

units 1 and 4 remain operational. As shown in Fig. 9, when one is considering small perturbations only, the failure is such that in unit 2 both components move backward with the sorbent, whereas in unit 3 both species move forward with the fluid. As a result, both the extract and raffinate streams are contaminated, but there is no contaminant at the recycle point.

(One can also imagine simultaneous failures of zones 2 and 3 in a manner in which both units fail in the same direction. For example, if both Q_2 and Q_3 were greater than $Q_s K_B$, the species in both units would move forward; there would be no product at the extract stream, and all the feed would exit

Table 2 Two-Unit Failures and Appropriate Corrective Responses

	Failure profile[a]				Symptoms[b]			Corrective response
Unit 1	Unit 2	Unit 3	Unit 4		Raff.	Extr.	Recycle	
−	−	+	+		Bad	Bad	B but no A	Increase Q_1 and Q_2 by increasing eluent and decreasing feed flows. Maintain recycle flow (Q_4).
−	+	−	+					Unlikely
−	+	+	−		Bad	Bad	Both A and B	Increase Q_1 and decrease Q_4 by increasing eluent, extract, and raffinate flows.[c]
+	−	−	+		Bad	Bad	OK	Increase Q_2 and decrease Q_3 by decreasing feed, extract, and raffinate flows.[c]
+	−	+	−					Unlikely
+	+	−	−		Bad	Bad	A but no B	Decrease Q_3 and Q_4 by increasing eluent and decreasing feed flows. Maintain Q_1 by dropping recycle flow.

[a] + implies working unit; − implies failed unit.
[b] "Bad" implies contamination with a component that should not be in the stream.
[c] Increase or decrease extract and raffinate flows equally.

Design and Control of SMB Chromatographs

through the raffinate. For simplicity, however, such upsets are not considered here, because they represent a gross deviation from the norm. Indeed, as is evident from Fig. 5, simultaneous failure of units 1 and 3, or of units 2 and 4, are also gross deviations and are thus considered unlikely.)

Employing similar reasoning for failures in pairs of units 1 and 2, 1 and 4, and 3 and 4, it is clear from the corresponding symptoms shown in Table 2 that in all these cases both the raffinate and extract streams are contaminated with the incorrect species. However, each of the failures is distinguished uniquely by the condition at the recycle point. If a selective detector that could distinguish between A and B were placed at this point, immediate and accurate diagnosis of the failure would be possible.

VI. DESIGN, OPTIMIZATION AND CONTROL—A HEURISTIC APPROACH

The foregoing discussion illustrates that the consequences of small changes in operation of an SMB are easily predicted and the causes of deviations from expected results are readily diagnosed. This knowledge can be used effectively in the control or the design and optimization of an SMB unit.

A. Control

If the unit is monitored, either in real time or by offline analysis of appropriate samples taken from the raffinate, extract, and recycle streams, corrective action can be taken as necessary based on the data in Tables 1 and 2. Although terms such as "decrease flow in unit 3" are not quantitative, they give guidance to the experimentalist, who can use judgment in deciding how much to change the flows to achieve the desired result. Indeed, the system of diagnosis and response can be quite powerful when codified within a suitable control framework such as that provided by fuzzy logic or other similar heuristic control strategies (Antia, 2000).

Clearly, when designing such a strategy, gross failures of two (or more) units, as alluded to above, also have to be taken into account. Under such circumstances, the symptoms do not necessarily have a unique cause. However, a systematic approach can be taken to move the system toward a lesser failure, which then can be accurately divined. A more detailed discussion is outside the scope of this chapter.

B. Design and Optimization

In Section IV, design conditions were given for operation under so-called linear conditions, where the distribution coefficients, K_i, of the separating species are constants and independent of influence from each other. This holds true only at low concentrations, which are not, in general, suitable for high productivity separations. At the higher concentrations typical in a preparative separation, species often compete or otherwise interact with each other when adsorbing on the sorbent, so that the distribution coefficients are variables that depend on the concentration of all the species present. These are known as "nonlinear" conditions.

The above approach of diagnosis and response can be used in designing an SMB separation without detailed knowledge of the adsorption isotherm (i.e., the relationship between the K values and the species concentrations). The method is as follows.

1. Start with operation at low concentration as dictated by the linear design in Fig. 5.
2. Increase Q_1 to a large value and decrease Q_4 to a low value so that the design criteria for units 1 and 4 are satisfied by a large margin. This puts the initial focus on finding the appropriate flow rates for the critical separating units 2 and 3. (If it is known that the isotherms are favorable, i.e., concave downward in shape, it may not be necessary to increase Q_1, because the condition for unit 1 is always satisfied with the linear design criterion. This is because all values of K in the system are lower than K_B measured at low concentration.)
3. Increase the concentration of the feed in stages until system nonlinearity leads to an upset. Use corrective action from Tables 1 and 2 to adjust the system to bring it back to a condition that achieves separation. This can be repeated until the feed concentration is increased to the target level.
4. Once conditions for units 2 and 3 are set, increase Q_4 and decrease Q_1 up to, but not beyond, the point of failure. This will ensure that a minimum eluent flow rate is used and enable optimal performance.

This method has been used successfully (Küsters et al., 1995) and mirrors the approach taken by Morbidelli and coworkers in their "triangle" plots, where the focus is on units 2 and 3 (Mazzotti et al., 1997; Gentilini et al., 1998; Migliorini et al., 1999; Azevedo and Rodrigues, 1999). However, in the triangle scheme, isotherm data are required to calculate the operating region at different feed concentrations, whereas the above method can be entirely experimental or governed by an appropriately coded controller.

Design and Control of SMB Chromatographs

The method presented above is practical when the response to changes made to the operating conditions can be detected in a reasonable time frame. As a rule of thumb, five to ten complete cycles (i.e., switching over all the columns of the unit) are required before the periodic steady state is reached, although trends may be discernible before this point. As discussed earlier, operation under conditions of low retention allows for short switching times and thus more rapid system response (often on the order of a few hours). Another advantage to operating under low retention conditions is that adsorption isotherms tend to be linear or quasi-linear over a larger concentration range than at higher retention. Thus, a majority of SMB separations are operated under conditions where the retention factor, k' in conventional chromatography, would be low, usually <5.

Finding the true optimum operating conditions for an SMB unit, which involves finding the best particle size, system configuration (i.e., number of columns per section), and column length is a fairly complex procedure requiring accurate adsorption isotherm data and modeling (Biressi et al., 2000a). In the procedure outlined above, it is assumed that an SMB system is already in place, with particle size, column length, and system configuration already chosen. These parameters can also be selected by heuristic, rather than deterministic, means, but discussion of this issue is beyond the purview of this chapter. It is worth noting, however, that regardless of the choice of these parameters, optimal operation for highest productivity is almost always at the highest pressure drop allowable in the system. [Figure 5 shows that a high maximum flow rate in the system enables maximization of $Q_s(K_B - K_A)$, which is the largest possible feed rate in the linear case. This applies in the nonlinear case as well.] In this regard, recently introduced monolithic columns that allow high flow rates at reasonable pressure drops without significant losses in efficiency represent a breakthrough that could lead to extraordinarily effective SMB separation in the future (Schulte and Dingenen, 2001).

VII. ECONOMICS OF SMB OPERATION

The long history and large installed base of SMB processes in the bulk chemical and sugar industries is a testimonial to the economic benefit of using the method on the large scale. In such applications, economies of scale apply in all aspects of the process. For instance, because of the ability to perform separations with few theoretical plates, large-particle sorbents can be employed, whose costs, at least for the types of sorbents used in those industries, generally decrease with increasing particle size.

Similar operating cost advantages are not as obvious for pharmaceutical applications. The major cost for a chiral stationary phase (CSP), for example, lies in the chemistry of the particle coating and does not depend strongly on the particle size. Thus there may be no significant economic benefit from operating with large particles as is the case for simpler applications. Indeed, it appears that most pharmaceutical SMB applications operate with sorbents of less than about 40 µm. The corresponding compact bed sizes and the modest scale of pharmaceutical products usually mean that benefits that accrue from economies of scale for bulk chemicals do not apply to active pharmaceutical ingredients. Nevertheless, the higher productivity in SMB systems compared with conventional liquid chromatography has prompted considerable interest and, as mentioned earlier, at least one production scale pharmaceutical application (Cavoy and Hamende, 2000).

Because an SMB is a continuous steady-state process, yield is usually very high and is limited only by the purity of the components in the output streams. For instance, if a racemic mixture is being resolved into 99% pure (98%ee) material in both extract and raffinate streams, the yield in both streams will also be 99%. This is a significant advantage over conventional chromatography, where yields usually can be this high only when productivity is low. On the other hand, claims that sometimes appear in the literature of extraordinary productivity and low solvent usage of up to 100-fold for SMB over conventional liquid chromatography appear to be exaggerated. A well-designed conventional liquid chromatograph can be quite efficient and also can use a large proportion of the available stationary phase at all times. However, even a twofold gain in these parameters, which is usually possible with an SMB, coupled with the inherently high yield may be sufficient to justify the added capital cost.

In the pharmaceutical industry, SMB chromatography has a niche in chiral separations. When developing a strategy to move a chiral drug product through the risk-laden pharmaceutical development process, it is worth weighing the advantages of enantioselective synthesis or biocatalysis versus classical or chromatographic resolution. Whereas any one approach may ultimately prove to be the most cost-effective, there is little doubt that chromatographic separations take the least time to develop and implement. The time frame to develop an adequate conventional LC or SMB method usually ranges from a few days to a few weeks, and this rapid implementation can have significant advantages in freeing valuable personnel for other critical tasks, thus shortening the time to market.

Recently published work shows that productivity in chiral SMB chromatography can, depending on the application, range up to and even exceed 2 kg of resolved product per kilogram of sorbent per day (Nicoud, 1999). Mod-

Design and Control of SMB Chromatographs

ern chiral phases are stable and have been known to last from months to years in continuous operation. Under these productivity conditions, the operating cost of a chromatographic separation can compete favorably with classical resolution despite the high cost of the CSP.

VIII. SCALE-UP

Scale-up in chromatography is largely a matter of preserving thermodynamic parameters, such as temperature, retentivity, and selectivity, as well as the key operating parameter N, the number of theoretical plates, while increasing the volume and throughput of the system.

Simulated moving bed chromatography introduces a few additional factors into the scale-up paradigm. The ratio of liquid to solid flows in each unit, Q_l/Q_s, must be held constant, and it is advisable to keep the number of subunits, or columns, the same across scales. The simplest and most effective scaling procedure, as in other forms of chromatography, is to work on the small scale with the same type and size of particle and the same column length as is intended on the large scale. Scaling is then accomplished simply by increasing the column diameter and increasing all liquid flows in proportion to the column cross-sectional area (or the square of the diameter) and preserving the switch time Δt.

Scale-up in chromatography is predictable and straightforward as long as good flow distribution and consistent packing can be maintained on the large scale. In pharmaceutical and fine chemical applications where relatively small particle size sorbents are employed, column diameter at production scale is likely to be below 1 meter. As discussed by Colin (1993), self-packing axial compression columns that ensure good packing and flow distribution are available under this size. For larger diameter columns, distribution and packing can be challenging, as discussed in detail in Chapter 1 of this book.

IX. CONCLUDING REMARKS

The usefulness of SMB chromatography extends beyond the four-unit liquid chromatography system discussed above. For example, very effective use has been made of simulated moving bed reaction/separation systems in driving equilibrium reactions such as esterification to completion, resulting in dramatic improvements in yield and productivity (Ray and Carr, 1995; Mazzotti et al., 1996). Gas chromatography (Biressi et al., 2000b) and supercritical fluid chromatography (SFC) (DiGiovanni et al., 2001) SMB systems have also been used. In SFC, thermodynamic properties (i.e., the K values) can be manipulated favorably by changing the pressure appropriately in each unit of the SMB, providing a significant productivity boost, suggesting that SFC SMB has an attractive future despite its higher capital costs.

In addition to its widespread use in binary enantiomer and isomer separations, SMB technology has been used for more complex purifications, including those of natural products (Küsters et al., 2000; Wu et al., 1999). SMB applications in the pharmaceutical and fine chemical industries are increasing rapidly. It is hoped that this chapter has served to demystify the operating principles of the method and to spur more research and innovation in the field.

REFERENCES

Antia, FD. Towards heuristic design and control of simulated moving bed chromatographs. Presented at the Horváth Symposium, Yale University, Jan. 23–25, 2000.

Azevedo, DCS, AE Rodrigues. Design of a simulated moving bed in the presence of mass transfer resistances. AIChE J 45:956–966, 1999.

Biressi, G, O Ludemann-Hombourger, M Mazzotti, RM Nicoud, M Morbidelli. Design and optimization of a simulated moving bed unit: role of deviations from equilibrium theory. J Chromatogr A 876:3–15, 2000a.

Biressi, G, F Quattrini, M Juza, M Mazzotti, V Schurig, M Morbidelli. Gas chromatographic simulated moving bed separation of the enantiomers of the inhalation anesthetic enflurane. Chem Eng Sci 55:4537–4547, 2000b.

Broughton, DB, DB Carson. Petrol Refiner 38:130, 1959.

Broughton, DB, CG Gerhold. Continuous sorption process employing fixed bed of sorbent and moving inlets and outlets. US Patent 2,985,589, May 23, 1961.

Cavoy, E, G Hamende. Chiral chromatography: from analytical to production scale. Chim Nouvelle 18:3124–3126, 2000.

Charton, F, RM Nicoud. Complete design of a simulated moving bed. J Chromatogr A 702:97, 1995.

Colin, H. Large-scale high-performance preparative liquid chromatography. In: G Ganetsos, PE Barker, eds. Preparative and Production Scale Chromatography. New York: Marcel Dekker, 1993, pp 11–46.

DiGiovanni, O, M Mazzotti, M Morbidelli, F Denet, W Hauk, RM Nicoud. Supercritical fluid simulated moving bed chromatography. II. Langmuir isotherm. J Chromatogr A 919:1–12, 2001.

Francotte, E. Enantioselective chromatography as a powerful alternative for the preparation of drug enantiomers. J Chromatogr A 906:379–397, 2001.

Gattuso, MJ. UOP Sorbex simulated moving bed (SMB) technology. A cost effective route to optically pure products. Chim Oggi—Chem Today 13:18–22, 1995.

Gentilini, A, C Miglorini, M Mazzotti, M Morbidelli. Optimal operation of simulated moving bed units for non-linear chromatographic separations. II. Bi-linear isotherms. J Chromatogr A 805:37–44, 1998.

Guest, DW. Evaluation of simulated moving bed chromatography for pharmaceutical process development. J Chromatogr A 760:159–162, 1997.

Hashimoto, K, S Adachi, Y Shirai, M Morishita. Operation and design of simulated moving-bed absorbers. In: G Ganetsos, PE Barker, eds. Preparative and Process Scale Chromatography. New York: Marcel Dekker, 1993, pp 273–300.

Hotier, G. Physically meaningful modeling of the 3-zone and 4-zone simulated moving bed processes. AIChE J 42:154–160, 1996.

Küsters, E, G Gerber, FD Antia. Enantioseparation of a chiral epoxide by simulated moving bed chromatography using Chiralcel-OD. Chromatographia 40:387–393, 1995.

Küsters, E, C Heuer, D Wieckhusen. Purification of an ascomycin derivative with simulated moving bed chromatography. A case study. J Chromatogr A 874:155–165, 2000.

Mazzotti, M, A Kruglov, B Neri, D Gelosa, M Morbidelli. A continuous chromatographic reactor: SMBR. Chem Eng Sci 51:1827–1836, 1996.

Mazzotti, M, G Storti, M Morbidelli. Optimal operation of simulated moving bed units for nonlinear chromatographic separations. J Chromatogr A 769:3–24, 1997.

Migliorini, C, A Gentilini, M Mazzotti, M Morbidelli. Design of simulated moving bed units under nonideal conditions. Ind Eng Chem Res 38:2400–2410, 1999.

Negawa, M, F Shoji. Optical resolution by simulated moving-bed adsorption technology. J Chromatogr 590:113–117, 1992.

Nicoud, RM, G Fuchs, P Adam, M Bailly, FD Antia, R Reuille, E Schmid. Preparative scale enantioseparation of a chiral epoxide: comparison of liquid chromatography and simulated moving bed adsorption technology. Chirality 5:267–271, 1993.

Nicoud, RM. The separation of optical isomers by simulated moving bed chromatography. Pharm Technol Eur March/April (reprint 0330) 1999.

Ray, AK, RW Carr. Experimental study of a laboratory scale simulated countercurrent moving bed chromatographic reactor. Chem Eng Sci 50:2195–2202, 1995.

Ruthven, DM, CB Ching. Counter-current and simulated counter-current adsorption separation processes. Chem Eng Sci 44:1011–1038, 1989.

Schulte, M, J Dingenen. Monolithic silica sorbents for the separation of diastereomers by means of simulated moving bed chromatography. J Chromatogr A 923:17–25, 2001.

Schulte, M, J Strube. Preparative enantioseparation by simulated moving bed chromatography. J Chromatogr A 906:399–416, 2001.

Wu, DJ, Z Ma, NHL Wang. Optimization of throughput and desorbent consumption in simulated moving-bed chromatography for paclitaxel purification. J Chromatogr A 855:71–89, 1999.

Yun, T, G Zhong, G Guiochon. Simulated moving bed under linear conditions: experimental vs. calculated results. AIChE J 43:935–945, 1997.

6
Large-Scale Ion-Exchange Chromatography: A Comparison of Different Column Formats

Peter R. Levison
Whatman International Ltd., Maidstone, Kent, England

I. INTRODUCTION

Ion-exchange chromatography is an established technique used in the separation of charged molecules across a breadth of applications and industries. Chemically, ion exchange involves the exchange of solutes of like charge from a solid support bearing the opposite charge. The principles of ion-exchange chromatography and an introduction to its application in various market segments are summarized elsewhere [1]. Ion exchange is a widely used technique in bioseparations because peptides, proteins, nucleic acids, and related biopolymers are ionic and are susceptible to charge enhancement or reversion as a function of pH. On this basis, under a set of defined mobile phase conditions a biopolymer mixture may be contacted with an ion-exchange medium. Depending on the relative ionic charge of the components, some biopolymers will adsorb and others will remain in solution. Desorption of bound material can then be carried out, resulting in a degree of purification of the target biomolecule. A generic process flow for ion-exchange chromatography is presented in Fig. 1.

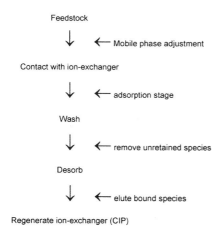

Figure 1 Ion-exchange process flow.

The key to performing ion-exchange chromatography is the contacting operation, and at large scale there are four generic approaches that might be routinely used:

1. Batch
2. Column
3. Expanded/fluidized bed
4. Suspended bed

II. CONTACTING OPERATIONS

Batch contacting is a simple and classical system whereby the ion exchanger is dispersed in a tank containing feedstock and adsorption is effected by mixing the components. The depleted feed is removed from the ion exchanger by a suitable solid–liquid separation technique such as continuous flow basket centrifugation. Washing and desorption are typically carried out in a similar manner. This technique has been described in detail and its performance relative to other modes of contacting discussed elsewhere [2,3].

Column contacting systems require that the ion exchanger be packed into a hollow device and retained by bed supports that are porous to the feedstock and other mobile phases. Column chromatography is widely used in large-scale ion-exchange processes, and a variety of column formats are available. This approach to contacting forms the basis of this chapter. It should be

Large-Scale Ion-Exchange Chromatography

noted that column contactors can be used in expanded/fluidized modes of operation or in a suspended bed mode. These areas of application are outside the scope of this chapter, but further information can be found in the literature [4–6].

III. COLUMN CHROMATOGRAPHY

Large-scale chromatography columns are available from several manufacturers, each with its own proprietary design. The systems are designed to meet different process specifications and use a variety of appropriate materials of construction. The aim of this chapter is to compare and contrast the various approaches to large-scale column-based ion-exchange processes and not to discuss the specific items of hardware or engineering capabilities of the equipment suppliers.

Ion exchange is an adsorption technique and consequently can be used in either positive or negative capture modes. These approaches are summarized in Fig. 2. At first sight, one might consider positive capture to be the preferred approach to chromatography, but that is not necessarily so. There are numerous cases of both approaches being used within the bioseparations industry, and we have reported the separation of immunoglobulins from goat serum by anion exchange [7], an example of negative capture, and the separation of monoclonal antibodies from tissue culture supernatant by cation exchange [8], an example of positive capture, to demonstrate the principles.

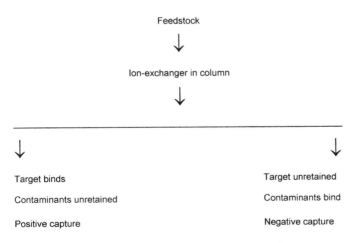

Figure 2 Approaches to ion-exchange chromatography.

Concentration of the target in the feedstock	Positive capture concentrates, negative does not. Is volume reduction necessary?
Relative amounts of target vs. impurities	If the target is the minor component, then positive capture. If impurities are the minor contaminant, then negative capture. This correlates to column size!
Mobile phase composition	If the mobile phase of the feedstock is unsuitable for adsorption of the target, then negative capture. Otherwise, mobile-phase modification is required, which has a cost!
Desorption conditions	Desorption following positive capture may not give compatible mobile phase for subsequent process/formulation steps. Is negative capture more desirable?
Validation	Negative capture offers the simplest validation.
CIP	Desorption and CIP can be one step for negative capture but two for positive capture.

Figure 3 Factors influencing the selection of positive or negative ion-exchange capture steps.

The selection of positive or negative capture steps is application-dependent, and a number of issues have to be considered, including those presented in Fig. 3. The outcome of this analysis should help facilitate the design of the process and hence its scale and economics.

In this chapter the focus is on positive capture using anion exchange, but many of the comparative issues and scale-up challenges will be similar for negative capture and also for cation-exchange chromatography systems.

IV. FEEDSTOCK

The feedstock that will be used for illustrative processes is fresh hen egg white. Egg white contains a complex mixture of proteins that have been chromatographed using a variety of ion-exchange systems [9–12]. This protein system has commercial relevance in the food industry as well as for biochemical reagents. More recently, though, the hen is being used as a host for transgenic protein systems and the egg white is the major source of the expressed transgenic protein.

Egg white is a viscous, translucent material containing 50–100 mg protein/mL that in its natural form would not be suitable for column chromatography. In our work, we dilute egg white to a 10% (v/v) suspension in 0.025 M Tris/HCl buffer, pH 7.5, and then clarify it using the fibrous weak anion-exchange cellulose CDR (Whatman International, Maidstone, UK) in batch mode. The clear feedstock containing 5–10 mg/mL total protein is suitable for column chromatography. Protocols for preparing the feedstock and assaying the components using fast protein liquid chromatography (FPLC) are described elsewhere [13].

V. ANION-EXCHANGE CHROMATOGRAPHY OF HEN EGG WHITE PROTEINS

The mobile phase composition of the hen egg white feedstock has been shown to influence its chromatography on anion-exchange celluloses [14], and the composition described above reflects a suitable system for comparative column performance testing. The chromatographic process that has been followed for column testing is outlined in Fig. 4. Various anion-exchange celluloses (Whatman International) were used according to manufacturer's instructions and recommended flow rates. Chromatography columns were obtained from the manufacturer and used according to their supplied instructions.

A typical anion-exchange separation of hen egg white proteins when carried out under the process conditions summarized in Fig. 4 is represented in Fig. 5. The chromatogram does not provide a great deal of definition, because the protein challenge is so high that the ultraviolet absorbance detector response is poor. This is a common event in process scale chromatography, where at high protein concentrations Beer's law is disobeyed and large changes in protein concentration generate small detector responses [15]. Beer's law [16] states that

$$A = \varepsilon l c$$

where

A is the absorbance of the solution at a defined wavelength
ε is the molar absorbance coefficient of the solute
l is the path length through the sample
c is the solute concentration

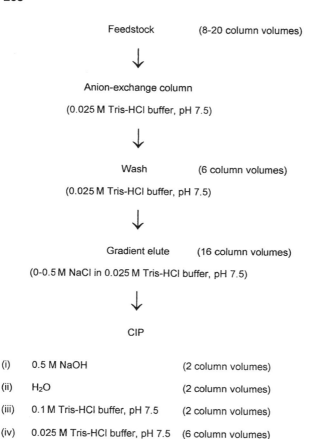

Figure 4 Process flow for the anion-exchange chromatography of hen egg white proteins.

On this basis, as the concentration of the protein solution increases, there should be a proportional increase in absorbance, typically measured at 280 nm. A plot of A_{280} versus ovalbumin concentration for a 2.5 mm path length flow cell is presented in Fig. 6. It is evident from the data that linearity is observed at low protein concentrations of up to ~10 mg/mL. This may be typical of many feedstocks, but nonlinearity is observed at higher protein concentrations that are typical of those seen during desorption and peak elution after a process loading of feedstock. This explains the differences in peak shape between Fig. 7 (see below) and Fig. 5. This will also have an adverse

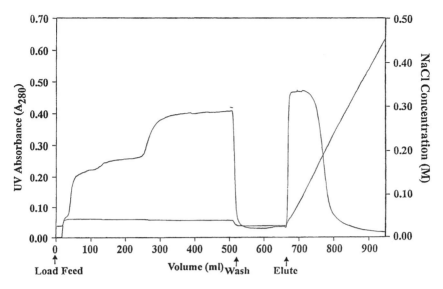

Figure 5 Chromatography of 4.2 kg hen egg white proteins on Express-Ion Q in a Moduline column (15.5 cm × 45 cm i.d.) in 0.025 M Tris-HCl buffer, pH 7.5 at a flow rate of 150 cm/h.

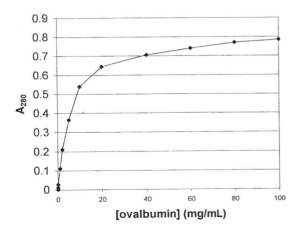

Figure 6 Absorbance of ovalbumin solutions prepared in 0.025 M Tris-HCl buffer, pH 7.5 at 280 nm using a 2.5 mm path length flow cell.

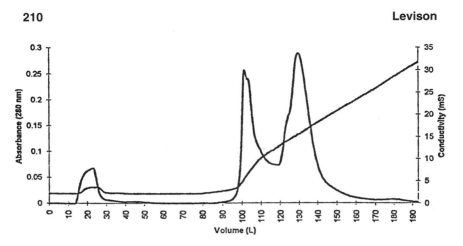

Figure 7 Chromatography of ~100 g of hen egg white proteins on Express-Ion Q in a Moduline column (15.5 cm × 45 cm i.d.) in 0.025 M Tris-HCl buffer, pH 7.5 at a flow rate of 150 cm/h.

impact on the purity information that can normally be deduced from a chromatogram.

In order to examine the composition of these major peaks we use FPLC for analysis of fractions throughout the separation. Protocols for such analyses are reported elsewhere [13].

If one applies a small quantity of egg white feedstock (0.3–0.4 column volume) referred to as an "analytical loading," then one generates a better defined separation as represented in Fig. 7. The separation reveals a peak of lysozyme eluting in the nonbound fraction, which could be a product of negative capture, and two major eluted peaks. Conalbumin elutes first followed by ovalbumin. In a process loading (Fig. 5) the large nonbound peak comprises lysozyme plus displaced conalbumin, and the salt-eluted peak is essentially pure ovalbumin [13].

VI. COLUMN DESIGN

A column may be regarded as a hollow vessel containing the chromatography medium, which is retained in place with porous bed supports. The bed supports have a porosity smaller than the size of the adsorbent particle such that the adsorbent is retained within the column unit. To ensure good chromatographic performance, a robust and reproducible column packing protocol must be de-

Large-Scale Ion-Exchange Chromatography

veloped and validated. Although this will tend to be a distillation of instructions supplied by the column manufacturer, recommendations of the chromatography medium vendor, and the experience of the process chromatographer, a number of generalities may be drawn. In order to avoid movement of the chromatography bed during use due to physicochemical interactions between ion exchanger and mobile phase, the packing pressure of the bed should be greater than the maximum operating pressure for the process. If the ion exchanger is subject to shrinkage during use, then added mechanical compression of the bed or packing in the mobile phase that gives maximum shrinkage would be suggested.

Having packed the bed, a number of simple tests may be performed to demonstrate its suitability for use. These include an estimate of the column packing density as dry grams adsorbent per unit column volume, a measure of linear flow rate at defined pressures, and determination of the height equivalent to one theoretical plate (HETP) using a tracer spike [17]. As a final qualification test, an analytical loading of feedstock run under defined process chromatography conditions (Fig. 7) gives useful information on bed integrity and chromatographic performance.

There are several column formats available to the process chromatographer. They may be broken into two groups of two. First, bed volume may be fixed or adjustable. In the former case the length of the column barrel is fixed, whereas in the latter case it may be adjusted by using a variety of design options. Second, the flow pattern through the bed may be axial or radial. In the former case the column is a tube with bed supports top and bottom, with liquid flowing axially, typically top to bottom. In the radial flow case the bed supports are two concentric tubes with the adsorbent sandwiched between. The top and bottom are sealed, and liquid flow is radial from the outer bed support to the inner one [18].

When scaling up an axial flow ion-exchange process, chromatographic bed height is maintained, as is linear flow rate, and column diameter is increased. In the case of radial flow, the bed height is effectively the thickness between the two bed support tubes, and this should be maintained. Because linear velocity cannot be readily determined in this geometric configuration, a measure of column volumes per hour is suggested. Scale-up can be achieved by increasing the length of the radial flow unit while keeping constant flow rates expressed as column volumes per hour.

In the following section we compare several different column designs and formats for the anion-exchange chromatography of hen egg white proteins. The list of columns used is not exhaustive because not all units are available to my laboratory, nor is it intended to be prescriptive as to which make and model is recommended. For a more detailed description of the experiments and their results the reader is referred to the original published reference. Column designs used in these studies are

1. Fixed volume axial flow
2. Fixed volume radial flow
3. Fixed volume Side Pack axial flow
4. Adjustable volume slurry pack axial flow
5. Adjustable volume pump pack axial flow

A. Fixed Volume Axial Flow

In this work we used a 45 cm i.d. × 16 cm unit (PREP-25; Whatman International, Maidstone, UK) that could be pump packed using a slurry of anion-exchange cellulose. The column was packed by pumping a slurry of DE52 (Whatman International) into the PREP-25 column at 15 psi and operated at ~10 psi, to give a flow rate of ~30 cm/h. An analytical loading of egg white proteins is presented in Fig. 8b, which demonstrates a 1000-fold scale-up from a laboratory scale column (1.5 cm i.d. × 15.5 cm) operated at a similar linear flow rate (Fig. 8a). Under these conditions we reported five consecutive process chromatograms each using 200 L of feedstock (10 mg/mL). During reuse we observed media fouling, and performance could be restored following overnight CIP using 0.5 M NaOH as summarized in Fig. 4. Full experimental details are reported elsewhere [15].

Although this column unit was effective for large-scale use, its major limitation was inflexibility in volume, i.e., 25 L fixed. No other production columns were readily available in this range, so it was discontinued. A key feature was the early design of pump packing, which facilitated rapid and consistent bed formation in 15–30 min. This mode of operation will be dealt with in more detail later. Column unpacking required column disassembly and manual excavation of the spent chromatography medium. Although this was easily accessible because the column barrel unit is only 16 cm deep, there are health and safety and containment issues associated with this type of operation that could be of concern in a process environment.

B. Fixed Volume Radial Flow

In this work we used a Superflo-100 and a Superflo 10 L radial flow column (Sepragen Corp, Hayward, CA) with volumes of 100 mL and 10 L and "effec-

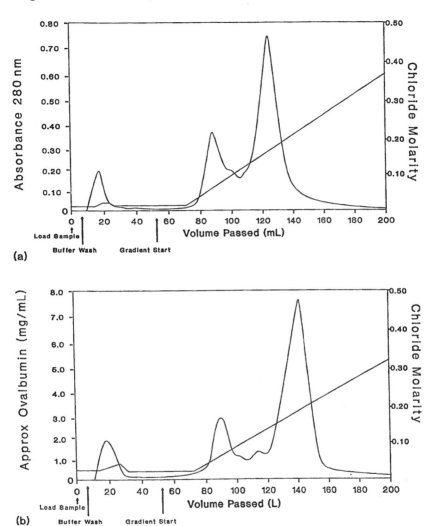

Figure 8 Chromatography of hen egg white proteins on DE52 using 0.025 M Tris-HCl, pH 7.5 at a flow rate of ~30 cm/h (a) at laboratory scale (15.5 cm × 1.5 cm i.d., 100 mg load) and (b) in a PREP-25 column (16 cm × 45 cm i.d., 100 g load).

tive bed heights" of ~3 cm and ~10 cm, respectively. In a comparative study we carried out analytical loadings of hen egg white proteins using DE52 in each of these columns and compared the performance with a 100 mL axial flow column. These data are presented in Fig. 9. Although the general peak profile is as expected for such a separation, it is apparent that the resolution between conalbumin and ovalbumin is influenced by the "effective bed height." This is anticipated, because there would be more theoretical plates with increased bed heights. In a subsequent study we demonstrated effective scale-ups between a Superflo-100, 100 mL column and a Superflo-500, 500 mL column and also between a Superflo-10L, 10 L column and a Superflo-20L, 20 L column, each pairing having similar effective bed heights [19]. A feature of radial flow columns is the ability to pump pack a slurry of ion exchanger into the column unit using axial packing ports [18,20] and also to pump-unpack the device in a similar manner. In the event, however, that fouling of the inlet bed supports was to occur due to the feedstock, then the column unit would require unpacking and disassembly. In axial flow systems it is possible to carefully replace the upper bed support with minimal disruption to the integrity of the adsorbent bed. A recent introduction for radial flow method scouting and process optimization is the WEDGE column. This can be described as a segment of the radial flow column that retains the effective bed height of the complete unit and thus lends itself to scale-down studies [20].

One major feature of radial flow columns is their ability to support higher volumetric flow rates than axial flow columns of similar volume at the same pressure. We compared the chromatographic performance of DE52 and QA52 (Whatman International) in a Superflo-100 column with a 100 mL axial flow column (4.4 cm i.d. × 6.6 cm; Millipore, Stonehouse, England). The pressure flow data generated in this study are summarized in Table 1, and it is clear that under similar operating conditions we could achieve flow rates approximately fivefold greater by using radial flow rather than axial flow. Under these operating conditions the chromatographic resolution of hen egg white proteins using gradient elution was superior with axial flow compared to radial flow. A series of comparative chromatograms using QA52 are presented in Fig. 10. This effect is anticipated because the effective bed height

---

Figure 9 Chromatography of hen egg white proteins on DE52 in 0.025 M Tris-HCl buffer, pH 7.5 using (a) a 100 mL axial flow column (6.6 cm × 4.4 cm i.d.), (b) a 100 mL Superflo-100 radial flow column at a flow rate of 15 mL/min, and (c) a 10 L Superflow-10L radial flow column at a flow rate of 1 L/min.

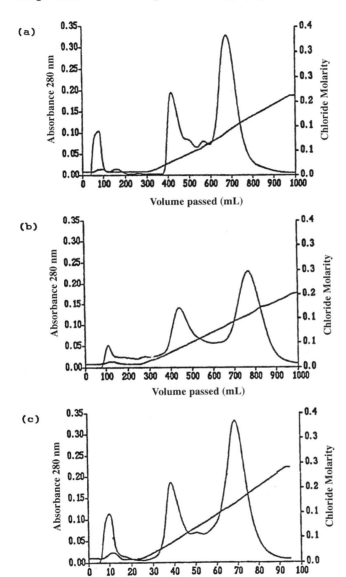

Table 1 Pressure and Flow Rate Data for Chromatography of Hen Egg White Proteins on DE52 and QA52 Under Axial or Radial Flow Conditions

Flor rate (mL/min)	Pressure (psi)			
	DE52 column		QA52 column	
	Axial	Radial	Axial	Radial
5	1	—	1	1
15	5	1	5	4
25	14	1	17	4
50	>45	1	>45	4
100	—	6	—	18
150	—	6	—	20

is less for the radial flow column than it is for the axial flow column of identical volume. It should be emphasized that we used gradient elution, and had step elution been employed, then such an effect would be minimized. Full experimental details for these studies are reported elsewhere [19,21].

C. Fixed Volume Side Pack Axial Flow

A recent innovation in column design is the Side Pack column [22]. This device is an axial flow unit that is packed by pumping a slurry through a side port such that the bed formation is bidirectional, with packing buffer emerging through both upper and lower bed supports. We evaluated a 35 cm i.d. × 16 cm Side Pack column (ProMetic BioSciences, Burtonsville, MD) using Express-Ion Q (Whatman International). The column was readily packed in ~20 min and had a packing density of 0.303 kg dry mass/L bed volume. This is higher than values of 0.218 and 0.227 kg/L that we have reported for Express-Ion Q in other process scale axial flow column designs [23,24]. This may be of benefit for the isolation of small molecules, as will be discussed later. Under these packing conditions we could operate the unit at 280 cm/h and have reported process scale separation of egg white proteins according to the

Figure 10 Chromatography of hen egg white proteins on QA52 in 0.025 M Tris-HCl buffer, pH 7.5 at flow rates of 5–25 mL/min in (a) a 100 mL axial flow column (6.6 cm × 4.4 cm i.d.) and (b) a 100 mL Superflo-100 radial flow column.

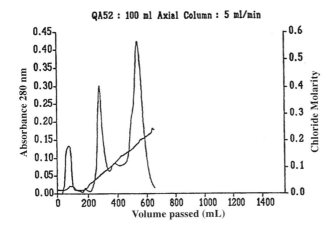
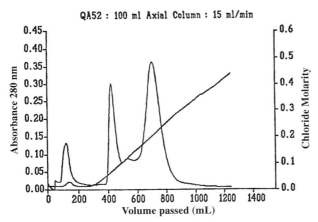
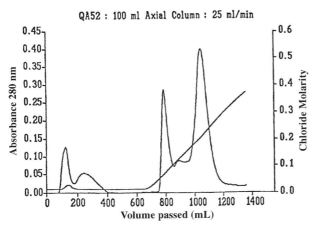

(a)

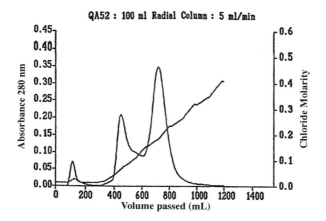
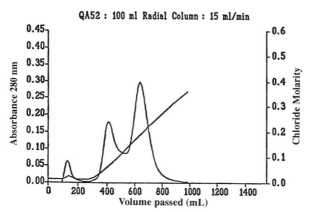
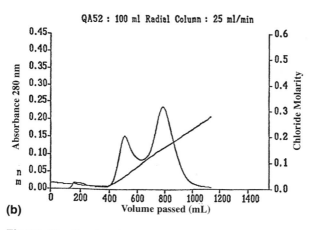

Figure 10 Continued

scheme presented earlier (Fig. 4) without a problem. These studies are described in detail elsewhere [24]. An analytical loading and process loading of hen egg white feedstock on Express-Ion Q in the Side Pack column operating at 280 cm/h are presented in Fig. 11.

Following use, the Side Pack column required disassembly to facilitate unpacking, which for reasons discussed above (Sect. V.A) may be undesirable. As with the other fixed volume designs described in the previous sections, there is inflexibility in column volume, so the process engineer must design the process flow around these fixed adsorbent volumes.

D. Adjustable Volume Slurry Pack Axial Flow

In each of the three cases discussed above the column volume was fixed, such that the feedstock volumes etc. would be tailored to meet the chromatographic performance of a fixed volume of ion exchanger. To add flexibility to the application, adjustable volume axial flow columns were designed. Although a number of designs are available from several manufacturers, the principle remains the same. Bed height can be altered by vertical adjustment of a moveable upper flow adapter that houses the top bed support. A system that we have worked with is the Moduline range of columns from Millipore (Stonehouse, UK). Our system is the G450 × 500 unit, comprising a 45 cm i.d. column barrel that is 50 cm in length. In order to pack this unit, the slurry of ion exchanger is poured into the column barrel, which may be fitted with an extension tube if the volume of slurry exceeds ~75 L. The upper flow adapter is fitted to the column assembly, and the chromatographic bed is consolidated by pumping packing buffer through the system in downflow, typically at constant pressure.

The next operation would involve depressurization, removal of headspace buffer, and positioning of the upper flow adapter on top of the bed, followed by a physical bed compression to a height similar to that established during the initial packing step. This procedure is time-consuming, typically taking at least 60 min to complete. Additionally, as column diameter increases, the upper flow adapter becomes heavy and is cumbersome to maneuver, often requiring a mechanical lowering assembly. Furthermore, its positioning can be a labor-intensive operation in order to ensure that no air is trapped under the flow adapter that could affect column performance. During initial studies using a 45 cm i.d. column of this design, we found this operation to be highly operator-dependent, and small differences of ±1 cm in bed over- or undercompression can lead to significant changes in the pressure/flow performance of the packed column [19].

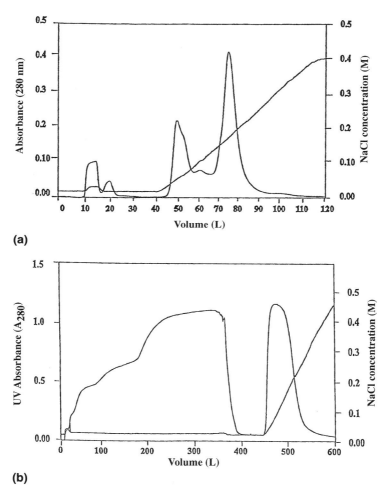

Figure 11 Chromatography of hen egg white proteins on Express-Ion Q using 0.025 M Tris-HCl buffer, pH 7.5 at a flow rate of 280 cm/h using the Side Pack column (16 cm × 35 cm i.d.) with loadings of (a) 44 g and (b) 2.2 kg total protein.

Large-Scale Ion-Exchange Chromatography

In this work we packed 20 kg of DE52 into the column at a pressure of 10 psi. This gave a relaxed bed height of 17.6 cm and packing density of 0.198 kg dry mass/L bed volume. The bed was then manually compressed using the flow adapter by up to 3 cm, and flow rate was measured at 10 psi. These data are summarized in Table 2 and presented in Fig. 12. Without adequate control this effect would have implications for process economics and likely for process validation.

Having packed columns of this design we have carried out a number of process scale studies using this system, including Express-Ion D [13] and Express-Ion Q [23], at flow rates of up to 225 cm/h using the protocol summarized in Fig. 4. The performance of Express-Ion Q in this system is represented by the analytical loadings in Fig. 7 and process loadings in Fig. 5, as described above.

In a comparative study using DE52 we ran analytical loadings of hen egg white feedstock at ~37 cm/h in both the fixed bed volume PREP-25 column (Sect. V.A) and the adjustable bed volume Moduline G450 column. In each case bed height was 16 cm. These chromatograms are presented in Fig. 13, and they are essentially identical, demonstrating the consistency of the two axial flow systems with the added flexibility of adjustable volume associated with the Moduline system.

In the slurry pack format, unpacking of the column following use is often a manual process. Typically the upper flow adapter/column top section is removed and then one manually excavates the spent adsorbent. Not only is this labor-intensive and gives rise to adsorbent loss, it is also a potential health and safety hazard because the used adsorbent is now in an open, albeit con-

Table 2 Influence of Bed Height Compression on Packing Density of a 45 cm Diameter Column Containing DE52 and Flow Rate at 10 psi in 0.025 M-Tris/HCl Buffer, pH 7.5

Bed height (cm)	Packing density (dry kg DE52/L column vol)	Flow rate at 10 psi (cm/h)
17.6	0.198	69.6
17.2	0.203	63.6
16.0	0.218	46.2
15.5	0.225	38.0
15.0	0.232	30.7
14.5	0.240	24.4

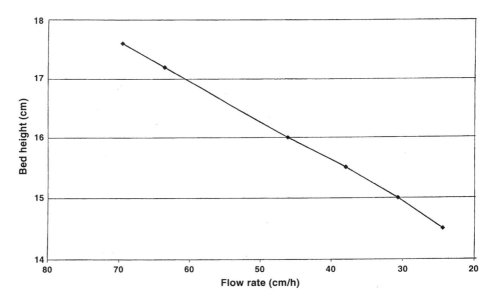

Figure 12 The influence of overcompression of a bed of DE52 using the flow adapter of the Moduline 45 cm i.d. column on the flow rate at a constant inlet pressure of 10 psi using 0.025 M Tris-HCl buffer, pH 7.5.

tained, environment. Furthermore, reaching the base of a 50 cm high column barrel can prove a physical challenge! Our experiences have shown column unpacking to take at least 60 min, an observation in keeping with the Side Pack column and PREP-25, which took two people 60 min to disassemble, unpack, clean, and reassemble.

E. Adjustable Volume Pump Pack Axial Flow

On the basis of the discussion in the previous sections on axial flow columns, a system offering the advantages of adjustable volume with the benefits of pump packing would appear attractive. Such systems have recently become available, and their design and function are described in the literature [25,26]. Not only do these columns fulfill the criteria stated above, they are also configured for in situ pump unpacking, which addresses many of the concerns expressed in the previous sections. We evaluated a 50 cm × 44 cm i.d. IsoPak column (Millipore), that had been adjusted to give a 16 cm bed. The column

Large-Scale Ion-Exchange Chromatography

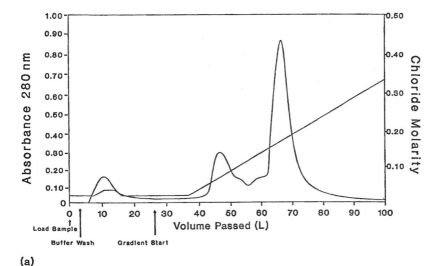

(a)

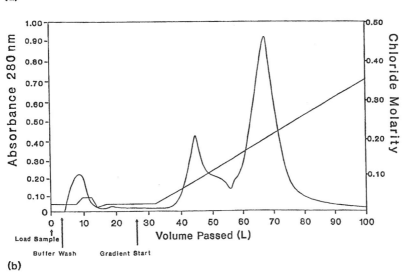

(b)

Figure 13 Chromatography of ~100 g hen egg white proteins on DE52 in 0.025 M Tris-HCl buffer, pH 7.5 at a flow rate of ~30 cm/h in (a) the PREP-25 column (16 cm × 45 cm i.d.) or (b) the Moduline column (16 cm × 45 cm i.d.).

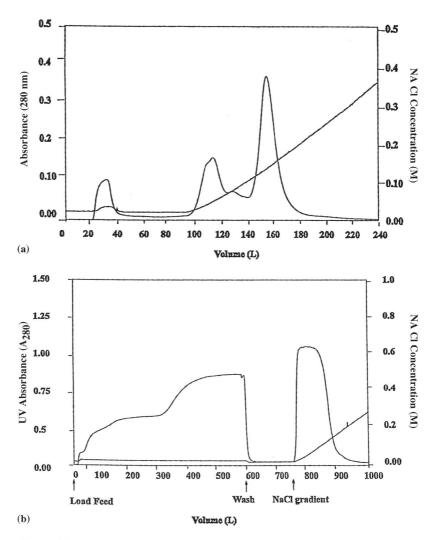

Figure 14 Chromatography of hen egg white proteins on Express-Ion Q using 0.025 M Tris-HCl buffer, pH 7.5 at a flow rate of 300 cm/h using the IsoPak column (16 cm × 44 cm i.d.) with loadings of (a) 70 g and (b) 3.4 kg total protein.

was packed with Express-Ion Q and a series of process scale separations carried out at flow rates of up to 300 cm/h according to the protocol in Fig. 4. These experiments are reported in detail elsewhere [24]. An analytical loading and process loading of hen egg white feedstock on Express-Ion Q in the IsoPak column operating at 300 cm/h are presented in Fig. 14.

The column packing process was simple to undertake, taking one person about 10 min. Furthermore, column unpacking was of similar ease, again taking one person about 10 min to complete. Using a column of this design, packing is carried out in upflow, which enables air to be displaced from the column during packing. Columns of this design are suitable for the suspended bed technique mentioned in Section I.A [5,6].

VII. COMPARATIVE PERFORMANCE OF DIFFERENT COLUMN FORMATS

The studies that have been described in the preceding sections generate several items of comparable information. This work has been ongoing since the late 1980s, and it has not been possible to compare one ion-exchange system across all column types, primarily due to the development and introduction of new ion-exchange media throughout this period. However, we are able to compare the performance of a DE52 system using the PREP-25, Superflo-10 L, and Moduline G450 and the performance of an Express-Ion Q system using the Moduline G450, Side Pack 16 L, and IsoPak 44 cm systems. From these studies a number of general observations can be drawn:

1. Pump packing is simpler than slurry packing and should be viewed as a more reproducible technique.
2. Pump unpacking is far easier and potentially less hazardous than manual unpacking.
3. If column inlet bed supports become fouled due to feedstock components and require replacement, then the replacement may be easier and less disruptive in axial flow systems.
4. Where less rigid adsorbents are used that may compress at higher pressures, radial flow columns offer superior flow performance to axial columns when used at similar pressure drops.
5. The design of the hardware system can influence the pressure/flow performance of the packed column. For example, at similar pressures the IsoPak system containing Express-Ion Q supported flow rates at least 50% greater than the Moduline system.

6. The column design and mode of packing influences the column packing density. Data generated in the various systems are summarized in Table 3. The packing densities obtained between the PREP-25 and Moduline columns are consistent. This may be anticipated even though one column is pump packed whereas the other is slurry packed, because both units are depressurized immediately following packing and prior to final use. On the other hand, the IsoPak column appears to have a slightly higher packing density, presumably due to the fact that pressure is maintained across the bed throughout all stages of packing and equilibration. The radial flow columns have a lower packing density than the axial flow columns, and the Side Pack column has a higher packing density than the other axial flow columns. We attribute these differences to the specifics of the packing processes that are unique to these column designs. Although the masses of adsorbent contained in these sets of columns differed, we did not see any significant differences in protein capacity for the comparable media. This is not unexpected; issues of pore diffusion and location of functional groups affect the dynamic binding capacities for large molecules, and it is assumed that similar accessibility to target protein, namely ovalbumin, is consistent regardless of these differences in packing density. However, this may not be the case in small-molecule purification. We have shown that when purifying a small bioactive hexapeptide by cation exchange [27], the binding capacity correlates stoichiometrically to the number of functional groups present in the column. In this case a more tightly packed

Table 3 Packing Densities of Ion-Exchange Celluloses in Various Column Formats

Adsorbent	Column	Packing density (dry kg/L)	Ref
DE52	PREP-25	0.194	15
DE52	Moduline G450	0.198	15
DE52	Superflo 10 L	0.117	—
Express-Ion D	PREP-25	0.210	13
Express-Ion Q	Moduline G450	0.218	22
Express-Ion Q	Side Pack 16 L	0.303	23
Express-Ion Q	IsoPak 44 cm	0.227	23

system such as Side Pack may offer capacity benefits. These will be offset by the packing cost because there is a need to pack more ion exchanger into the column.
7. Scale-up from small units to larger units was predictable and straightforward in all systems evaluated as described above.
8. Clean-in-place protocols were effective in all systems tested, and these are reported in more detail in the relevant publications.

VIII. CONCLUSIONS

The process chromatographer is faced with the dilemma of media selection and column hardware selection during process development. In our view, and based on studies reported here, although column designs are different, their comparative performance in large-scale ion-exchange processes is less variable. Clearly there are differences in their mode of operation and ease of use, and these may be the decision-making factors in column selection. It has been outside of the scope of this chapter to consider other protein systems or other chromatographic media, but in both our own work and that disseminated through the technical literature many of the observations and conclusions reported here are confirmed. Scaling up a process from laboratory bench through pilot scale and into production has often been viewed with some trepidation, and this was manifested by the limited availability of column formats and designs. Although the principles indicate that increasing column diameter while maintaining bed height and linear velocity, at least for axial flow (Fig. 8), is the means of scaling up, it is often regarded as a time-consuming, labor-intensive challenge. With the current innovations in column technology described above we consider that scaling up to a large column may now be regarded as a simple, rapid, and routine stage of a chromatographic separation.

REFERENCES

1. PR Levison. Cellulose-based exchangers. In: A Townshend, ed. Encyclopaedia of Analytical Science, Vol 4. London: Academic Press, 1995, pp 2278–2281.
2. PR Levison. Techniques in process-scale ion-exchange chromatography. In: G Ganetsos, P Barker, eds. Preparative and Production Scale Chromatography. New York: Marcel Dekker, 1993, pp 617–625.
3. PR Levison, SE Badger, DW Toome, ML Koscielny, L Lane, ET Butts. Eco-

nomic considerations important in the scale-up of an ovalbumin separation from hen egg-white on the anion-exchange cellulose DE92. J Chromatogr 590:49–58, 1992.
4. Expanded bed chromatography. Bioseparation 8:1–274, 1999.
5. PR Levison, AK Hopkins, P Hathi, SE Badger, F Mann, N Dickson, G Purdom. Suspended bed chromatography, a new approach in downstream processing. J Chromatogr A 890:45–51, 2000.
6. I Quinones-Garcia, I Rayner, PR Levison, N Dickson, G Purdom. Performance comparison of suspended bed and batch contactor chromatography. J Chromatogr A 908:169–178, 2001.
7. PR Levison, ML Koscielny, ET Butts. A simplified process for large-scale isolation of IgG from goat serum. Bioseparation 1:59–67, 1990.
8. G Denton, A Murray, MR Price, PR Levison. Direct isolation of monoclonal antibodies from tissue culture supernatant using the cation-exchange cellulose Express-Ion S. J Chromatogr A 908:223–234, 2001.
9. AC Awadé, S Moreau, D Mollé, G Brulé, J-L Maubois. Two-step chromatographic procedure for the purification of hen egg-white ovamucin, lysozyme, ovotransferrin and ovalbumin and characterization of purified proteins. J Chromatogr A 677:279–288, 1994.
10. AC Awadé, T Efstathiou. Comparison of three liquid chromatographic methods for egg-white protein analysis. J Chromatogr B 723:69–74, 1999.
11. LR Jacob, M Mack, G Rathgeber. Large scale purification of proteins from egg white by tentacle ion-exchangers. Int Biotech Lab, February 1994, p 14.
12. T Croguennec, F Nau, S Pezennec, G Brule. Simple rapid procedure for preparation of large quantities of ovalbumin. J Agric Food Chem 48:4883–4889, 2000.
13. PR Levison, SE Badger, DW Toome, M Streater, JA Cox. Process-scale evaluation of a fast-flowing anion-exchange cellulose. J Chromatogr 658:419–428, 1994.
14. PR Levison, DW Toome, SE Badger, BN Brook, D Carcary. Influence of mobile phase composition on the adsorption of hen egg-white proteins to anion-exchange cellulose. Chromatographia 28:170–178, 1989.
15. PR Levison, SE Badger, DW Toome, D Carcary, ET Butts. Studies on the reuse of anion-exchange cellulose at process scale. In: R de Bruyne, A Huyghebaert, eds. Downstream Processing in Biotechnology II. Antwerp: The Royal Flemish Society of Engineers (K.viv.), 1989, pp 2.11–2.16.
16. D Freifelder. Physical Biochemistry. 2nd ed. San Francisco: WH Freeman, 1982, pp 497–498.
17. O Kaltenbrunner, P Watler. Column qualification in process ion-exchange chromatography. Prog Biotechnol 16:201–206, 2000.
18. V Saxena, AE Weil, RT Kawahata, WC McGregor, M Chandler. Applications of radial flow columns for fast affinity chromatography. Am Lab, October 1987, pp 112–120.

19. PR Levison, SE Badger, DW Toome, ET Butts, ML Koscielny, L Lane. Influence of column design on chromatographic performance of ion-exchange media. In: A Huyghebaert, E Vandamme, eds. Upstream and Downstream Processing in Biotechnology. III. Antwerp: The Royal Flemish Society of Engineers (K.viv.), 1991, pp 3.21–3.28.
20. DM Wallworth. Practical aspects and applications of radial flow chromatography. In: MA Desai, ed. Downstream Processing of Proteins: Methods and Protocols. Totawa, NJ: Humana Press, 2000, pp 173–184.
21. L Lane, ML Koscielny, PR Levison, DW Toome, ET Butts. Chromatography of hen egg-white proteins on anion-exchange cellulose at high flow rates using a radial flow column. Bioseparation 1:141–147, 1990.
22. AB Alaska. Side-packed chromatographic column. US Patent 5667676, 1997.
23. PR Levison, RMH Jones, DW Toome, SE Badger, M Streater, ND Pathirana. Influence of flow rate on the chromatographic performance of agarose- and cellulose-based anion-exchange media. J Chromatogr A 734:137–143, 1996.
24. PR Levison, AK Hopkins, P Hathi. Influence of column design on process-scale ion-exchange chromatography. J Chromatogr A 865:3–12, 1999.
25. M Hofmann. A novel technology for packing and unpacking pilot and production scale columns. J Chromatogr A 796:75–80, 1998.
26. GJ Purdom. Chromatographic column and valve. PCT Patent WO9922234A1, 1999.
27. PR Levison, M Streater, JW Dennis. A scale-down study into the chromatography of a peptide using the cation-exchange cellulose Express-Ion C. J Chem Technol Biotechnol 74:204–207, 1999.

7
Development and Operation of a Cation-Exchange Chromatography Process for Large-Scale Purification of a Recombinant Antibody Fragment

Robert L. Fahrner, Stacey Y. Ma, Michael G. Mulkerrin, Nancy S. Bjork, and Gregory S. Blank
Genentech, Inc., South San Francisco, California, U.S.A.

I. INTRODUCTION

Ion-exchange chromatography is a widely used method of chromatographic purification. In cation-exchange chromatography, a positively charged protein absorbs onto a negatively charged sorbent, then an increasing salt concentration in the mobile phase selectively desorbs (elutes) the protein components as positive ions in the mobile phase buffer displace the proteins from the sorbent [1]. Cation-exchange chromatography is commonly applied to the purification of antibodies and antibody fragments [2,3].

Antibody fragments have important applications for treating human disease [4]. The chemistry and biology of antibody fragments is complicated, and we limit the discussion here to a simplified description of the characteristics that are important for development of the purification process. The type of antibody fragment we purified is an F(ab')$_2$, a molecule that consists of two light chains and the Fab portions of two heavy chains (Fig. 1). Two disulfide bonds join the heavy chains at the hinge. This F(ab')$_2$ is expressed in *E. coli*. The *E. coli* cells express the light and heavy chains separately into the

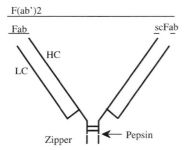

Figure 1 Schematic illustration of a F(ab')$_2$ from *E. coli*, which contains two light chains (LC) and two truncated heavy chains (HC). The leucine zipper helps form the two disulfide bonds at the hinge and is later removed by pepsin. Fab is half of an F(ab')$_2$, and scFab (single-chain Fab) is either a heavy chain or a light chain.

periplasm, and a leucine zipper added to the end of the heavy chains assists in forming the disulfide bonds between the heavy chains. Enzymatic cleavage of the heavy chains just below the hinge with the use of pepsin removes the leucine zipper during the recovery process.

The recovery process for this F(ab')$_2$ consists of several steps. After cell homogenization, the first chromatography step is expanded bed cation exchange, which allows cell debris to flow through while retaining the product. In addition to removing solids, this upstream cation-exchange step partially purifies the F(ab')$_2$. The downstream purification process for this F(ab')$_2$ consists of the use of an immobilized pepsin column to remove the leucine zipper (run in flow-through mode), cation-exchange chromatography, then hydrophobic interaction chromatography.

This chapter describes the development and operation of the downstream cation-exchange step. This step reduces the level of several product-related impurities, such as two-chain species including Fab and light-chain dimer, single-chain Fab (scFab, either light or heavy chain), and variants of the F(ab')$_2$. Although we did not fully characterize all of the product-related variants, we did have assays to measure their levels. The cation-exchange step also clears non-product-related impurities such as host cell proteins, DNA, and endotoxin, but one of the most difficult separations in process chromatography is the removal of product-related variants, and the development effort focused on the separation of the F(ab')$_2$ product from the scFab, Fab, and F(ab')$_2$ variants. To systematically develop the downstream cation-exchange step, we first defined assays and separation goals, then chose a column type

Purification of Recombinant Antibody Fragment

for development and determined its capacity, then developed wash and elution conditions, and finally scaled up the separation.

II. ASSAYS, SEPARATION GOALS, AND SCALE-UP CONSIDERATIONS

Assays play a central role in process development. They are used to characterize the load material and to compare it to the purified product, to measure mass balance, and to characterize the operation of the process [5]. It is important to have appropriate assays for measuring product quality before development begins, because without these assays it is difficult to determine the performance of a process step and to design and perform appropriate development experiments. We used four assays to judge the purification performance of the cation-exchange step during development.

The capillary electrophoresis SDS (CE-SDS) assay (Fig. 2) separates

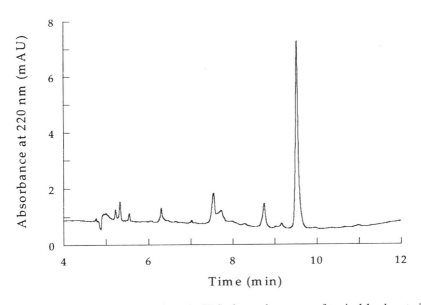

Figure 2 Capillary electrophoresis SDS electropherogram of typical load material for the downstream cation-exchange column. The assay used a fused silica capillary (50 μm × 25 cm), 15 kV field strength, reversed polarity, and injection for 8 s at 5 psi. This assay resolves scFab (6.3 min), Fab and light-chain dimer (7.5–8 min), and the $F(ab')_2$ (9.5 min).

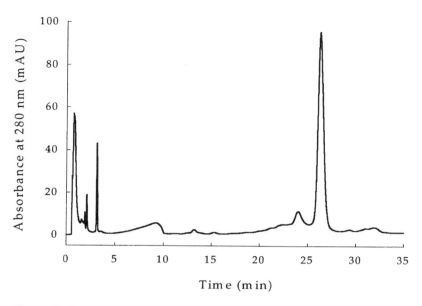

Figure 3 Ion-exchange HPLC chromatogram of typical load material for the downstream cation-exchange column. The assay uses a 4.6 × 50 mm Bakerbond CSX column at 1 mL/min, with a pH gradient from pH 6.0 to pH 6.7 over 30 min. The assay separates Fab (about 5–15 min) and F(ab')$_2$ and its charged variants (about 15–35 min). The large main peak is the F(ab')$_2$ product. Acidic F(ab')$_2$ variants elute to the left of the main F(ab')$_2$ peak, and basic F(ab')$_2$ variants elute to the right of the main F(ab')$_2$ peak.

proteins similarly to SDS-PAGE [6]. This assay has two major advantages over SDS-PAGE for process development: it is fast, and it is quantitative. The ion-exchange HPLC assay (Fig. 3) separates Fab and F(ab')$_2$ as well as the charged variants of the F(ab')$_2$. The level of host cell proteins was measured by ELISA, and the level of DNA was measured by the DNA threshold assay.

There are four main goals for the downstream cation-exchange process:

1. To significantly reduce the amount of scFab and Fab and reduce the level of F(ab')$_2$ variants and host cell proteins, while maintaining a reasonable yield (greater than 70%).
2. To remove the brown color associated with the load.
3. To purify a 9 kg batch on a reasonably sized column in a reasonable amount of time.

Purification of Recombinant Antibody Fragment

4. Because this is a process-scale separation to be used in routine manufacturing, the use of a step gradient (rather than a linear gradient) would be preferred [7], and a reliable method of peak pooling will be required.

During development and scale-up of the process, we had several considerations. With the exception of the column screen, we performed all studies with preparative loads and real feedstock. We used the same bed height, buffer components, and operating conditions during the small-scale runs as we anticipated using in the large-scale runs. One important factor that must be considered during development is the often continuous change in the separation goals, the nature of the feedstock, and the accuracy of the assays. Development must be flexible enough to handle these changes, and the process itself must be robust enough to withstand variations in feedstock, buffer makeup, and operating conditions. In addition, there may be constraints in the manufacturing plant such as limitations on column size, pump flow rates, or tank sizes. For example, our process columns had a maximum pressure rating of 3 bar, and (as we discuss later) this impacted the final operating conditions of the process.

III. COLUMN SELECTION AND CAPACITY

To conclusively compare the performance of different columns, the separation on each column needs to be fully optimized, but this approach cannot typically be used in practice because it is too time-consuming. The difficulty is, then, to design a study that with the least amount of optimization will help you choose a column that will provide the best separation. The study design will depend on the separation goals and the quality of the load material. We designed two column studies to find an appropriate column. Both studies relied on low sample loading (<1.5 g/L), which may not accurately predict the behavior under high sample loading (>20 g/L), but this approach allows a rapid and consistent evaluation of process parameters.

In an initial column screen, five different columns were loaded to about 1 g/L and a linear gradient of sodium acetate at pH 4 was run. We tested columns from Pharmacia (SP-Sepharose Fast Flow, SP-Sepharose High Performance, SP-Sepharose XL, and Source 30 S), and PerSeptive Biosystems (Poros 50 HS). The resulting chromatograms were compared qualitatively (Fig. 4), noting where the scFab, Fab, F(ab')$_2$, and the F(ab')$_2$+zipper eluted. For the SP-Sepharose, the high performance (HP) and fast flow (FF) versions

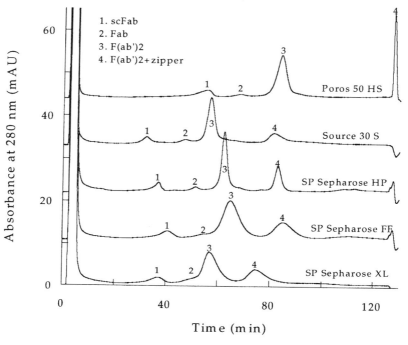

Figure 4 Comparison of different chromatographic media for separating antibody variants. The experiment used 0.46 × 10 cm columns loaded to 0.3 g/L, a flow rate of 300 cm/h, and a gradient of 80–600 mM sodium acetate (pH 4.0) over 50 column volumes (CV). The columns were regenerated with 2 M NaCl.

had peaks that eluted at essentially the same time, but the Sepharose HP had sharper peaks, as would be expected from its smaller particle size. Although the Fab was well resolved from the F(ab')$_2$ peak on the Sepharose HP, it was not well resolved on the Sepharose FF because of increased band spreading on the Sepharose FF. The Sepharose XL had peaks that were as resolved as the Sepharose FF but eluted earlier. The Source 30 S had peaks that were slightly less sharp than the Sepharose HP but were almost equally resolved. The Poros 50 HS separated all species well, especially the F(ab')$_2$+zipper, which eluted in the high-salt regeneration.

Choosing columns for further investigation relied on several considerations. Although all of the columns provided at least some separation of the scFab, Fab, and F(ab')$_2$+zipper, no separation of the F(ab')$_2$ variants was observed, except perhaps from a slight shoulder on the leading edge of the F(ab')$_2$ peak on Poros. Although the Source 30 S and SP-Sepharose HP had sharp, well-resolved peaks, we thought that since the separation would eventually be run with step instead of linear gradients this advantage would be minimal. Also, both the Source and the HP have small particle sizes, which may lead to high column pressure. On the basis of this reasoning, we chose SP-Sepharose FF and Poros 50 HS for further investigation. Both are readily available in large quantities for scale-up, and their cost is comparable and reasonable.

We then conducted a more comprehensive experiment to study the ability of SP-Sepharose FF and Poros 50 HS to separate product-related variants. After injecting a small amount (about 0.1 mg) of purified forms of the F(ab')$_2$ product and several variants individually onto Poros and Sepharose columns, we ran a linear gradient of sodium acetate at several pH values after each injection. We then determined the elution position of each peak during the gradient and constructed a graph of the elution position relative to the F(ab')$_2$ product (Fig. 5). As pH decreased, the separation of the F(ab')$_2$+zipper from the F(ab')$_2$ increased greatly, the separation of the Fab may have increased slightly, and the separation of the F(ab')$_2$ variants did not change. We tried other salts (for example, sodium citrate) and other columns, but we found no conditions under which the F(ab')$_2$ variants were well separated from the F(ab')$_2$. Based on these data, we chose Poros 50 HS for further development work because it provides better separation of the F(ab')$_2$+zipper and may provide slightly better separation of the Fab.

Several factors influenced the choice of the separation pH. At lower pH values, the separation was better, which suggests that the separation should be performed at a low pH. However, we had some concern that the product might not be stable at a pH of less than 4. In addition, at the time of development we speculated that the next step in the recovery process might be an anion-exchange flow-through chromatographic process, which requires low conductivity in the load. At lower pH values the product will elute in higher salt concentrations, increasing the conductivity of the pool and requiring dilution prior to loading on the anion-exchange column. Higher values of pH used during the cation-exchange step will have lower conductivity in the pool, but the separation may not be as good. We achieved a balance between separation and conductivity by running the separation at pH 5.0. We later used hydrophobic interaction chromatography as the next process step instead of anion exchange, but this is an example of the uncertainty that accompanies process development and the compromises that the developer is often forced to make during the development effort.

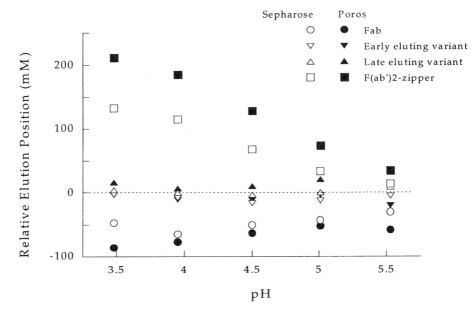

Figure 5 Effect of pH and column media on the separation of antibody variants from the F(ab')$_2$ product. For this experiment, 0.46 × 10 cm columns were loaded with 100 μg of purified forms of the F(ab')$_2$ and F(ab')$_2$ variants, and a sodium acetate gradient was run (adjusted so that the F(ab')$_2$ eluted in the middle of the gradient). Variants that elute earlier than the F(ab')$_2$ are plotted with negative values on the y axis, and variants that elute after the F(ab')$_2$ are plotted with positive values on the y axis. Open symbols, Sepharose columns; filled symbols, Poros columns. (○, ●) Fab; (▽, ▼) early-eluting variant; (△, ▲) late-eluting variant; (□, ■) F(ab')$_2$+zipper.

The batch size, column size, and throughput requirements together determine the required column capacity. High throughput requires a high flow rate, and the maximum bed height will be influenced by the flow rate requirements, the backpressure limits, and often the size of columns that are available. A 20 cm bed height and a flow rate of 200 cm/h [10 column volumes (CV) per hour] produces high throughput while maintaining a backpressure of less than 1.5 bar with the Poros media. We wanted to use a 140 cm diameter column, and with a bed height of 20 cm the column has a volume of 300 L, so to purify a 9 kg batch in a single cycle the column needs at least 30 g/L capacity.

For Poros 50 HS, the capacity was nearly constant over a wide range of conductivity at pH 4.0, whereas at pH 5.0 the capacity was significantly

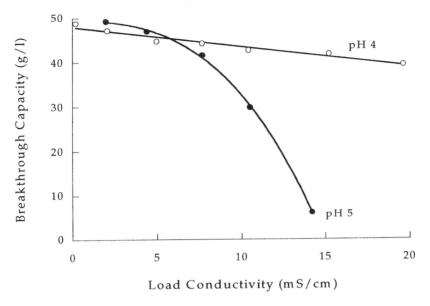

Figure 6 Effect of pH and conductivity on the breakthrough capacity of the F(ab')$_2$ on Poros 50 HS. Load was purified F(ab')$_2$ at a concentration of 1 g/L, with conductivity adjusted with 3 M NaCl and pH adjusted with acetic acid. A 0.46 × 10 cm column was loaded at 120 cm/h. Breakthrough capacity was measured at 5% breakthrough by absorbance at 280 nm.

reduced at higher values of conductivity (Fig. 6). The pool from the pepsin column is at pH 4.0 (because pepsin has high activity at pH 4), and its conductivity is typically about 12 mS/cm. If the pH of the pepsin pool were adjusted to pH 5.0 before loading onto the cation-exchange column, the load would have to be significantly diluted (to reduce the conductivity) to achieve a reasonable capacity and the cycle time would be increased. Therefore the column was loaded at pH 4.0, and the separation was performed at pH 5.0. Loading at pH 4.0 produces a capacity greater than the target of 30 g/L over a wide range of conductivity values.

Changing pH during the run can introduce several potential problems. First, it is possible that more host cell proteins will bind to the column at lower pH, thus decreasing the purity of the final pool. Second, when the pH increases from 4 to 5 immediately after loading, some protein may elute from the column if the conductivity does not decrease fast enough. Third, if a strong buffer (such as acetate) is not used, the pH changes may be slow and a large volume of buffer may be needed to effect the pH change. As we will show, none of these problems occurred in our separation, and the final, scaled-up separation loads at pH 4 and runs at pH 5.

IV. DEVELOPMENT OF WASH AND ELUTION CONDITIONS

There are many ways to elute a cation-exchange column, including linear and step gradients of increasing salt. For process applications, step gradients are preferred over linear gradients because of their easy implementation, reproducibility, and simplified pooling automation. Our concept of the separation was to use step gradients of increasing salt concentration to selectively elute portions of the protein load, fractionating the sample into Fab, $F(ab')_2$, and $F(ab')_2$+zipper. This use of multiple steps to selectively elute components is common in ion-exchange process chromatography [8]. During the step to elute $F(ab')_2$ by choosing correct elution conditions, at least some of the $F(ab')_2$ variants will be removed by pooling only a portion of the $F(ab')_2$ peak. We will name the three steps as follows: the *wash* is the step to remove the Fab, the *elution* is the step to elute the $F(ab')_2$, and the *regeneration* is the step to remove the $F(ab')_2$+zipper.

The most commonly used salt for ion-exchange chromatography of proteins is sodium chloride, but this salt is corrosive to stainless steel. For process applications using large stainless steel tanks and pipes, less corrosive salts are preferred. Our $F(ab')_2$ separation used sodium acetate, which buffers strongly at both pH 4 and pH 5, so no additional buffer is needed in the mobile phase solutions. The mobile phase solutions then contain only three components (water, sodium acetate, and acetic acid), which simplifies buffer production in manufacturing.

The wash step should reduce the amount of scFab and Fab by eluting these variants during the wash while retaining $F(ab')_2$ on the column. To determine optimal wash conditions, we washed the column with varying amounts of sodium acetate while maintaining all other variables constant (Fig. 7). As the concentration of sodium acetate in the wash increased from 140 mM to 170 mM, the amount of protein in the wash pool increased. At 140 mM sodium acetate only a small peak eluted at the very beginning of the wash phase. This small peak became larger as the salt concentration increased to 150 mM and then 160 mM. At 160 mM a second peak began eluting, and at 170 mM the second peak was larger than the first peak.

To analyze the behavior of the column under these varying wash conditions, we assayed samples of the wash pools from the study in Fig. 7 using

Purification of Recombinant Antibody Fragment

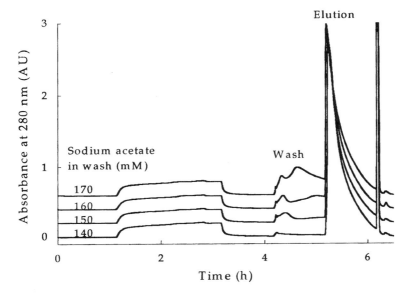

Figure 7 Stacked preparative chromatograms from the wash study. The column was a 0.66 × 20 cm Poros 50 HS run at 200 cm/h. The mobile phase buffer was sodium acetate pH 5.0. The column was equilibrated with 10 CV of 50 mM sodium acetate, loaded to 30 g/L, re-equilibrated with 10 CV of 50 mM sodium acetate, washed with 10 CV of varying sodium acetate concentrations, eluted with 10 CV of 220 mM sodium acetate, and regenerated with 2 CV of 1000 mM sodium acetate.

CE-SDS (Fig. 8). With 140 mM sodium acetate in the wash, only a small amount of Fab eluted. The pools from 150, 160, and 170 mM all had the same amount of Fab, but as the salt concentration in the wash increased, $F(ab')_2$ began to elute. These assay data are consistent with the preparative data, where the second peak in the wash phase in the preparative chromatograms is $F(ab')_2$. The CE-SDS assay is quantitative, and we found that at 150 mM or above about 85% of the loaded Fab was eluted in the wash phase. At 170 mM about 10% of the loaded $F(ab')_2$ was eluted in the wash phase. We decided to use a wash with 155 mM sodium acetate, because this will allow removal of the majority of the scFab and Fab while minimizing the loss of $F(ab')_2$ (to less than 5% loss). The use of 155 mM sodium acetate gives an operating range of about ±5 mM sodium acetate, which is within the operating ranges typically used in our manufacturing plant. Although a wash volume less than 10 CV (as used in the experiment in Fig. 7) may result in less loss of $F(ab')_2$ product

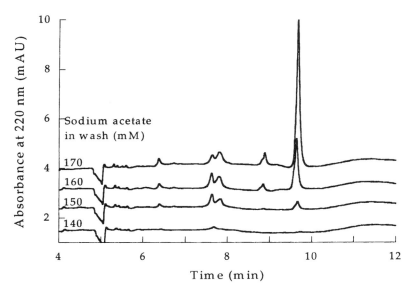

Figure 8 Electropherograms from CE-SDS of the wash pools from the preparative runs shown in Fig. 7. The Fab elutes at about 7.8 min.

(because the wash would end before the product eluted), the 10 CV wash allows for variable amounts of Fab in the load. We have purified load material with 30% Fab by CE-SDS, and the 10 CV wash was needed to completely elute the Fab during the wash.

The design of the study to determine elution conditions was assisted by further analysis of the wash study. The load material for the wash study had about 10% $F(ab')_2$+zipper. CE-SDS analysis of the load and an elution pool from the wash study showed that the $F(ab')_2$+zipper (and some heavier species) did not elute during the elution phase (Fig. 9). We found that when eluting above about 220 mM sodium acetate, $F(ab')_2$+zipper began eluting with the $F(ab')_2$. At conditions below 220 mM, the $F(ab')_2$+zipper was well separated.

We then performed a study where the concentration of sodium acetate in the elution phase was varied while all other conditions were held constant. In our study, we used concentrations of sodium acetate in the elution between 180 and 220 mM (Fig. 10). As the concentration of sodium acetate in the elution increased, the elution peak became larger and narrower. To enable a comprehensive analysis of the elution behavior, we fractionated the elution

Purification of Recombinant Antibody Fragment

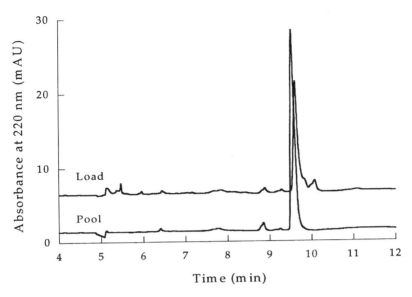

Figure 9 Electropherograms from CE-SDS analysis of the load and elution pool (from the 160 mM sodium acetate wash run in Fig. 7).

peaks into 0.2 CV fractions, then analyzed various fractions and pools using the ion-exchange HPLC assay. We first found that at all values of sodium acetate in the elution, the leading edge of the peak consisted primarily of acidic F(ab')$_2$ variants. This will allow the leading edge of the peak to be cut from the pool to remove a significant portion of the acidic F(ab')$_2$ variants.

The elution study needed to determine two process variables: the amount of sodium acetate to use for the elution and the amount of the leading edge of the elution peak to cut away. An analysis of the elution peaks by ion-exchange HPLC (Fig. 11) showed that as the concentration of sodium acetate in the elution increased, yield also increased, especially between 180 and 200 mM. These data are consistent with the preparative chromatograms, where the preparative peak is small at lower values of sodium acetate. The yield continued to increase until at 220 mM the yield of the total peak was greater than 90%. Although a sodium acetate concentration above 220 mM may result in a higher yield, it would also begin eluting F(ab')$_2$+zipper.

Purity (percent of main peak) also increased in the same trend as the yield (Fig. 11). At lower values of sodium acetate in the elution, the purity was low because the peak contained a large amount of acidic F(ab')$_2$ variants.

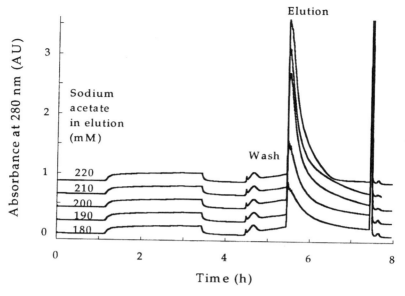

Figure 10 Stacked preparative chromatograms from the elution study. The column was a 0.66 × 20 cm Poros 50 HS run at 200 cm/h loaded to 30 g/L. The mobile phase buffer was sodium acetate at pH 5.0. The column was equilibrated with 10 CV of 50 mM sodium acetate, loaded to 30 g/L, re-equilibrated with 10 CV of 50 mM sodium acetate, washed with 10 CV of 155 mM sodium acetate, eluted with 20 CV of varying concentrations of sodium acetate, and regenerated with 2 CV of 1000 mM sodium acetate.

As the amount of the leading edge of the main peak that was removed increased, the purity increased but the yield decreased. The leading edge of the main peak contains mostly F(ab')$_2$ variants, and the percent main peak increased as it was removed. However, the leading edge also contains some F(ab')$_2$ main peak, so the yield decreased as it was removed. We achieved a balance between yield and purity by using 215 mM sodium acetate in the elution phase and cutting the first 0.5 CV of the main peak. This allows a high purity while still maintaining yield of about 80%.

The optimized process used two step gradients of sodium acetate: 155 mM for the wash, removing Fab, and 215 mM for the elution, during which the first 0.5 CV of the peak is cut to reduce the amount of F(ab') variants. The simple pooling automation waits during the elution phase for the absorbance to

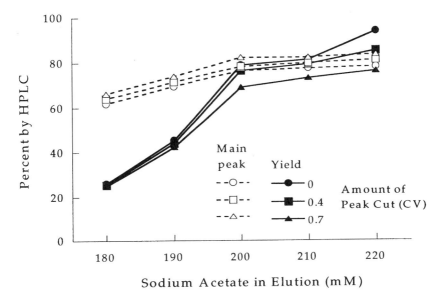

Figure 11 Ion-exchange HPLC analysis of pools from the elution study shown in Fig. 10. Pools of the fractionated main peak were made, and two values were measured. The percent main peak is the peak area of the main F(ab')$_2$ peak divided by the total peak area, and the yield is the amount of main peak in the preparative elution peak divided by the amount loaded. The amount of peak cut is the volume of the leading edge of the main preparative peak that was not included with the pool. All pools ended at 0.2 AU.

reach 0.4 AU, then begins pooling after 0.5 CV has flowed through. Pooling ends after collecting 8 CV of the peak.

The wash and elution studies used an antibody load of 30 g/L. Because the batch size varies due to titer variation and variations in extraction efficiency and upstream column yields, the cation-exchange column must be able to purify differing batch sizes, which means that the separation must be able to tolerate variations in column loading. In this type of preparative separation, which possibly relies on a self-sharpening peak (with a sharp, almost vertical leading edge) for separation of the acidic F(ab')$_2$ variants, load is of a particular concern because the self-sharpening is more pronounced at higher loads. In a study where antibody load was varied from 10 to 30 g/L (Fig. 12), the degradation of the self-sharpening effect is apparent, especially at 10 g/L,

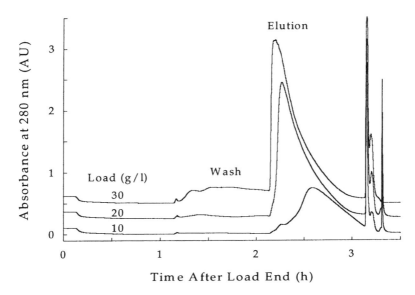

Figure 12 Stacked preparative chromatograms from the loading study. The column was a 0.66 × 20 cm Poros 50 HS run at 200 cm/h. The mobile phase buffer was sodium acetate at pH 5.0. The column was equilibrated with 10 CV of 50 mM sodium acetate, washed with 10 CV of 155 mM sodium acetate, and eluted with 10 CV of 215 mM sodium acetate.

where the peak is more Gaussian than at 30 g/L. However, even on the 10 g/L load, there is a shoulder on the front of the elution peak that is composed mostly of acidic F(ab')$_2$ variants, and when pooling begins at 0.4 AU the acidic variants are removed because the shoulder is not included with the main peak. For all these pools, purity by ion-exchange HPLC was greater than 75% and yield was greater than 75%. We anticipate that in routine operation the load will be between 20 and 30 g/L.

V. OPERATION

Using the process conditions, the separation is 7 h long, including the load, wash, elution, and regeneration (Fig. 13). Although the conductivity of the load is high, it is loaded at pH 4 to ensure protein binding. The remainder of the separation is run at pH 5.0. During the elution phase, after the absorbance

Purification of Recombinant Antibody Fragment

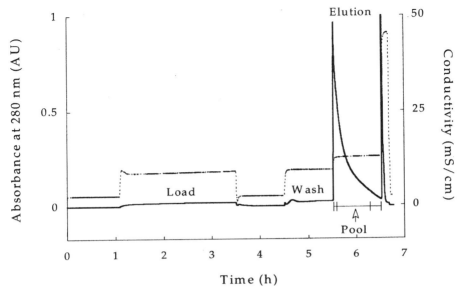

Figure 13 Chromatogram from the process separation. The column is a 20 cm long Poros 50 HS run at 200 cm/h. The mobile phase buffer was sodium acetate at pH 5.0. The column was equilibrated in 10 CV of 50 mM sodium acetate, loaded to 30 g/L, re-equilibrated with 10 CV of 50 mM sodium acetate, washed with 10 CV of 155 mM sodium acetate, eluted with 10 CV of 215 mM sodium acetate, and regenerated with 2 CV of 1000 mM sodium acetate.

reaches 0.4 AU, the first 0.5 CV of the peak is discarded, then the next 8 CV of the peak is pooled.

Analysis of the various pools by CE-SDS showed where size variants eluted (Fig. 14). During the wash, about 85% of the loaded Fab eluted, along with a small (less than 5%) amount of the F(ab')$_2$. As seen on the CE-SDS, the majority of the F(ab')$_2$ eluted in the elution phase. The collected elution pool was about 92% main peak by CE-SDS.

Analysis of various pools by ion-exchange HPLC showed where variants eluted (Fig. 15). The wash contained mostly Fab variants. The pre-pool (the first 0.5 CV of the elution peak) comprised approximately half F(ab')$_2$ and half acidic F(ab')$_2$ variants. About 40% of the loaded mass of acidic F(ab')$_2$ variants was in the pre-pool, and less than 10% of the loaded mass of F(ab')$_2$ was in the pre-pool. The collected elution pool was about 85% main peak by ion-exchange HPLC. The regeneration contained basic F(ab')$_2$ vari-

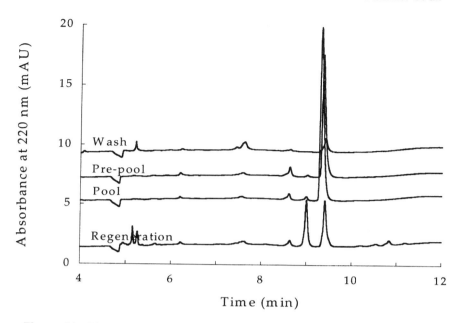

Figure 14 Electropherograms from CE-SDS analysis of pools collected during preparative chromatography in Fig. 13.

ants as well as F(ab')$_2$+zipper. About 70% of the basic F(ab')$_2$ variants was in the regeneration. The yield was about 80% by ion-exchange HPLC.

The *E. coli* homogenate is dark brown. Because not all of this color is removed during upstream processing, the load onto the downstream cation-exchange column still has some residual color that should be removed. The brown color eluted in the regeneration, and the pool was clear of color (Fig. 16). The pre-pool, which contained a large amount of acidic F(ab')$_2$ variants, was slightly purple.

The amount of host cell proteins in the collected elution pool is 300 ppm. Because the load material has about 3000 ppm host cell proteins, this represents a 10-fold clearance. The pool contains less than 100 pg/mL DNA and has less than detectable levels of endotoxin.

Thus the separation has met all of its goals: reduction in the amount of scFab, Fab, F(ab')$_2$ variants, color, and host cell proteins; high yield; good throughput; and reliable product pooling with step gradients. Well-defined assays and separation goals, systematic development, and thorough process

Purification of Recombinant Antibody Fragment

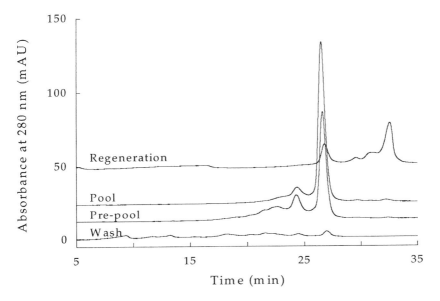

Figure 15 Ion-exchange HPLC analysis of pools from the preparative separation shown in Fig. 13.

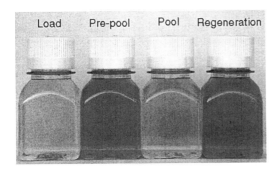

Figure 16 Photograph of the pools from the preparative separation shown in Fig. 13. The load has a slight brown color, the pre-pool has a faint purple color, the pool is clear, and the regeneration has a dark brown color.

characterization were all important elements in implementing a robust, well-understood process step.

ACKNOWLEDGMENTS

We thank Gerardo Zapata, Walter Galan, Brian Wagner, and Daljit Narindray for providing load material and technical guidance, and Christie Blackie for the photographs of the pools.

REFERENCES

1. LR Snyder, JJ Kirkland. Introduction to Modern Liquid Chromatography. New York: Wiley, 1979.
2. RK Scopes. Protein Purification: Principles and Practice. New York: Springer-Verlag, 1994.
3. E Boschetti, A Jungbauer. Separation of antibodies by liquid chromatography. In: S Ahuja, ed. Handbook of Bioseparations. San Diego: Academic Press, 2000.
4. P Carter, RF Kelley, ML Rodrigues, B Snedecor, M Covarrubias, MD Velligan, WLT Wong, AM Rowland, CE Knotts, ME Carver, M Yang, JH Bourell, HM Shepard, D Henner. High level Escherichia coli expression and production of a bivalent humanized antibody fragment. Bio/Technology 10:163–167, 1992.
5. SM Wheelwright. Protein Purification: Design and Scale Up of Downstream Processing. New York: Carl Hanser, 1991.
6. G Hunt, W Nashabeh. Capillary electrophoresis sodium dodecyl sulfate nongel sieving analysis of a therapeutic recombinant monoclonal antibody: a biotechnology perspective. Anal Chem 71:2390–2397, 1999.
7. G Sofer, L Hagel. Handbook of Process Chromatography: A Guide to Optimization, Scale-Up, and Validation. San Diego: Academic Press, 1997.
8. E Karlsson, L Ryden, J Brewer. Ion-exchange chromatography. In: J-C Janson, L Ryden, eds. Protein Purification: Principles, High-Resolution Methods, and Applications. New York: Wiley, 1998, pp 145–206.

8
Case Study: Capacity Challenges in Chromatography-Based Purification of Plasmid DNA

Sangeetha L. Sagar, Matthew P. Watson, and Ann L. Lee
Merck & Co., Inc., West Point, Pennsylvania, U.S.A.

Ying G. Chau
Massachusetts Institute of Technology, Boston, Massachusetts, U.S.A.

I. INTRODUCTION

Advances in DNA technologies offer tremendous potential for the prevention and cure of major diseases [1–11]. Gene therapy and DNA vaccines are currently under development to tackle indications of diseases such as HIV and cancer. These research efforts and the anticipated DNA-based products will require the production of large quantities of highly purified plasmid DNA [12,13]. The previously established laboratory technique for the isolation of plasmid DNA—cesium chloride/ethidium bromide density gradients—is scale-limited and cumbersome. The more recently reported plasmid DNA purification methods that are capable of producing multigram quantities for human clinical trials make use of modern chromatographic techniques, including size exclusion, anion exchange, and/or reversed-phase chromatography [12,13]. However, the cost of the chromatography step(s) at a manufacturing scale is likely to be very high owing to the use of resins, which are expensive, and to the low plasmid DNA binding capacity of these resins.

The current commercially available wide-pore resins were designed in light of the advances in recombinant protein technologies during the 1980s and 1990s. These pore sizes range from 300 to 4000 Å, sizes that have been determined to be optimal for a variety of protein separations based on mass transport considerations [14–16]. Substantial work has been done in the area of pore size and binding capacity optimization for proteins [17]. The optimal resin properties for plasmid DNA would differ greatly from those for protein because of the significant differences in molecular size and charge. Plasmid DNA is significantly larger than most proteins, with the average size of a 6 kB plasmid being 900–1300 Å. In many cases, plasmid DNA is excluded from the pores during a chromatographic separation using commercially available resins. Until larger pore size resins become available, the binding for plasmid DNA will be quite low because of the limited surface area accessible for adsorption. As a result, any means to increase the binding capacity of plasmid DNA on current resins would have a significant impact. In addition, it would be important to understand the effect of pore size on plasmid binding capacity in order to design optimal resins for plasmid chromatography. The objective of this work is to examine the impact of plasmid-binding capacity on chromatography scale-up and process economics.

To this end, this chapter begins with an examination of plasmid capacity as a function of pore size and discusses the impact of structural modulators on plasmid binding capacity. Then the effects of resin capacity changes on process times and costs are discussed.

II. BACKGROUND

Within the limits of current chromatographic resin technology and pore size, the ability to modify plasmid size may significantly increase the amount of plasmid that can bind to the resin. Of course, the modification to plasmid size would have to be reversible and independent of sequence. Fortunately, plasmid size is a function of more than the number of base pairs contained in the plasmid. Secondary and tertiary structure also affect the volume occupied by DNA. In vivo, DNA is routinely packaged into spaces smaller than its absolute length. Depending on the organism, the volume reduction can be as great as four orders of magnitude [18].

In nature DNA exists in several helical forms (secondary structures; i.e., B-DNA, A form, Z-DNA [19]). The next level of structure is imposed by the covalent circular closure of one or both strands of DNA. In the absence of circular structure, the helical double strands are referred to as linear. In general, for plasmid DNA for gene transfer and vaccine applications, it is believed that the linear form is inactive and most prone to degradation. When one of two stands is covalently closed, forming a relaxed circular structure, the plasmid is considered to be in the open circular form. This form is believed to be active but less statistically stable than the most compact supercoiled form, in which both strands are covalently closed. For DNA-based vaccinations and therapies, the supercoiled plasmid is the general form of interest, because it is the most active and is statistically stable to degradation.

In vivo, DNA compaction is achieved by an increase in supercoiling, which can occur by several mechanisms [18]. In vitro, plasmid size reduction can be achieved by volume reduction or DNA condensation. This phenomenon has been reported widely (see, e.g., Ref. 20), and the best known forms of condensed DNA are toroids. These compact ringlike structures have been detected by such methods as electron microscopy, light scattering, circular dichroism, and fluorescent microscopy. Nontoroidal shapes that have also been observed include rods, fibrous aggregates, liquid crystals, and aggregated complexes. The generation of these forms usually depends on the length of DNA or the compaction agent involved.

DNA condensation can be achieved by a number of means. Perhaps the best known is exposure to a solution containing multivalent cations. These cations reduce intersegmental electrostatic repulsion by neutralizing the negative charge density of DNA [21]. Another common condensing agent is alcohol. Whereas high concentrations of ethanol (80–90%) are commonly used to precipitate DNA, lower concentrations of ethanol, methanol, and isopropanol cause condensation. Alcohol will act in concert with some counterions to reduce the concentration and even the valency required. Ions of 2+ valence will induce condensation in alcohol–water mixtures [22]. Condensation is caused in this case by the fact that in the presence of alcohol DNA–solvent interactions are less favorable than in aqueous solutions [23]. This is also true for solutions containing neutral crowding polymers such as PEG.

By far the most common use of DNA condensation has been made in the area of gene delivery. Cationic liposomes have been used as vehicles to deliver DNA into cells. The mechanism of compaction involved in this association is likely to differ from that for multivalent cations. There is no regular morphology to these complexes. In fact, their structure has been difficult to determine, because different techniques yield different results [24].

To date, there have been limited reports on the exploitation of condensation in DNA purification. Murphy et al. [25], in the laboratory of R. Willson at the University of Houston, have used spermine and spermidine, common DNA condensing cations, to selectively precipitate salmon spermine DNA from a mixture with baker's yeast RNA. Condensation has been shown to be very selective for supercoiled plasmid DNA and therefore has a tremendous potential for use in plasmid purification.

III. EXPERIMENTAL

A. Equipment and Column Types

The analytical HPLC equipment consisted of an autosampler from Gilson Instruments and pumps, UV detector, solvent mixer, and controller from Rainin (Varian) Instruments. Semipreparative HPLC equipment consisted of the Bio-CAD system from PerSeptive Biosystems, and the preparative HPLC consisted of all components from Rainin (Varian) Instruments. Small-scale dynamic capacities were determined by using flow-packed AP minicolumns from Waters Corporation, whereas the preparative scale columns were Amicon Vantage pressure packed columns.

B. Sample Preparation

Plasmid was purified according to the process described in international patents PCT/US95/08749 (WO96/02658) and PCT/US96/07083 (WO96/36706). Following controlled heat lysis and clarification, the lysate is diafiltered to remove debris and then coarse filtered to reduce particulates. The filtered lysate is then purified by two orthogonal chromatography steps and concentrated into storage buffer. The process is presented in Fig. 1 for reference. The quality characteristics of the final purified material are

>90% covalently closed plasmid as determined by agarose gel and HPLC
<1% *E. coli* genomic DNA as determined by qPCR
<0.1% residual RNA as determined by HPLC ribose assay
<0.5 EU (endotoxin)/mg plasmid as determined by LAL
<1% protein as determined by Pierce BCA

C. Size Exclusion Chromatography

The relative size of supercoiled plasmid was determined by size exclusion on a 7.8 mm ID × 30 cm L GDNApw column from Tosohaas connected to a Rainin analytical HPLC system. Fifty microliters of purified supercoiled plasmid at a concentration of 0.016 mg/mL was injected, and the mobile phase was run at 0.2 mL/min. A size reference curve was constructed by running purified supercoiled plasmid of size ranging from 2.7 to 7.7 kb (Gibco BRL) in 0.1 M NaCl, 50 mM Tris, pH 8.0. The salt concentration was recommended by the manufacturer to counteract any electrostatic interaction with the column matrix. To determine the effect of salt on the size of process-purified 6.6 kb

Chromatography-Based Purification of Plasmid DNA

	Step Yield (%)	Genomic DNA (%)	RNA (mg/mg)	Protein (mg/mg)	Endotoxin (EU/mg)
E-coli fermentation					
Controlled heat lysis	100	10-15	82	6.7	10^6
RNAse treatment	99	3	0.71	0.78	10^6
Diafiltration	85	3	0.28	0.046	842
AEX	58	<1	0.04	<0.001	<0.1
RPC	58	<1	0.04	<0.001	<0.1
UF/DF	58	<1	0.04	<0.001	<0.1

Figure 1 The schematic representation of the chromatography-based purification process for DNA (top to bottom, left-hand column) along with the corresponding contaminant clearances and step yields. AEX = anion exchange; RPC = reversed-phase chromatography; UF/DF = ultra filtration/diafiltration.

plasmid (described above), NaCl concentration in samples containing the 6.6 kb plasmid was adjusted from 0.1 M to 2.0 M. Elution was carried out with 50 mM Tris pH 8.0 buffer containing the same NaCl concentration as the corresponding sample. To determine the effect of isopropanol on the size of the pure 6.6 kb plasmid, percent IPA in samples containing the plasmid was adjusted from 1.2% to 20% v/v. Elution was carried out with 0.1 M NaCl, 50 mM Tris pH 8.0 buffer containing the same percent IPA as the corresponding sample.

D. Light Scattering

The radius of gyration for all samples was monitored by using a high resolution multiangle light scattering (MALLS) photometer from Wyatt. All measurements were taken at 25°C using a 5 mW helium-neon laser at a wavelength of 632.8 nm. All samples, buffers, and reagents were filtered using a 0.22 μm filter prior to analysis, and the samples were prepared under exactly the same conditions as in the SEC experiments. The data were interpreted on the basis of a random coil model for the structure of plasmid DNA. However, the light scattering data were used only to determine relative size changes in the plasmid; the same trend in size would be observed irrespective of the model used.

E. Static Capacity Determination

The appropriate amount of resin for a 1 mL packed column was resuspended in 50 mM Tris at pH 8.0. The slurry was centrifuged for 5 mins at 1500 rpm, and the supernatant was then removed. This procedure was repeated twice to rid the slurry of shipping solvent. For each resin being investigated, four aliquots were prepared as above. In aliquot A, 10 mL of 20× diluted column load material was added and mixed with resin. In aliquot B, 10 mL of 2× diluted column load material was added and mixed with resin. In aliquot C, 10 mL of column load material was added and mixed with resin. In aliquot D, 40 mL of column material was added and mixed with resin. Column load material contained a total plasmid concentration of 470 mg/mL as determined by anion-exchange HPLC assay. Incubation was carried out at room temperature for 24 h with continuous mixing. After incubation, each aliquot was centrifuged for 5 min at 1500 rpm to collect the resin. Supernatant from each aliquot was filtered through a 0.45 µm syringe filter and then analyzed by anion-exchange HPLC assay for plasmid content.

F. Dynamic Capacity Determination

AP minicolumns were flow packed at two times the operating pressures and equilibrated with the appropriate buffer (50 mM Tris HCl at pH 8.0 for anion exchange and 100 mM ammonium bicarbonate with 1.2% v/v IPA at pH 8.5 for reversed-phase). For anion exchange, material was loaded onto the columns at a flow rate of 5 mL/min.

In the reversed-phase case, the plasmid was adjusted to 0.1 M NaCl, 1.2% v/v IPA at pH 8.5 and 0.7 mg of supercoiled plasmid per milliliter of resin was loaded. In both cases, the load effluents were tested for plasmid content by the HPLC assay.

G. HPLC Anion-Exchange Assay

The HPLC assay for quantitation of supercoiled plasmid DNA in RNase-treated samples employs a Rainin HPLC system with Gilson autosampler. A Poros Q/M (PerSeptive BioSystems, Boston, MA) column is equilibrated with 0.605 M NaCl in 50 mM bis-tris propane (Sigma Chemical Company, St. Louis, MO) buffer at pH 7.5, 100 µL of sample is injected onto the column, and after a wash a 50 CV gradient to 1.0 M NaCl is performed to elute the supercoiled plasmid. The plasmid generally elutes between 0.7 and 0.8 M NaCl and is separated from the linear and open-circle plasmid forms. Plasmid

Table 1 Pore Sizes Determined by Mercury Intrusion Porosimetry

Resins used in evaluation	Particle size (μm)	Pore size (Å)	
		Reported	Measured
Polymer Labs PLSax	15–25	1000	1689
Polymer Labs PLSax	15–25	4000	3874
Polymer Labs PLRPS, 100 Å	50–70	100	100
Polymer Labs PLRPS, 300 Å	50–70	300	300
Polymer Labs PLRPS, 1000 Å	50–70	1000	1099
Polymer Labs PLRPS, 4000 Å	50–70	4000	5896

content of the sample is determined by comparison of the eluting peak area to a standard curve.

H. Resin Properties

Reported pore size values were provided by Polymer Laboratories. Mercury intrusion porosimetry was used to measure resin pore size (see Table 1). The analysis was performed by Porous Materials Inc. of Ithaca, NY. In this method, mercury is applied to the resin bead and forced into its pores by increasing pressure incrementally. The change in mercury volume that intrudes into the pore at each pressure increment is equal to the volume of pores whose diameters fall within a specific range. Each range corresponds to a particular pressure interval.

IV. RESULTS AND DISCUSSION

A. Laboratory Scale Studies

In an effort to examine the effect of pore size on plasmid binding capacity of the chromatography resins, dynamic binding capacity measurements were performed. The Polymer Labs resins were selected for this study because they are available in different pore sizes with a fixed chemistry and particle size. Examining the reversed-phase resin, supercoiled plasmid DNA (6.6 kb, pUC derived) was loaded onto various PLRPS (Polymer Labs) columns with reported pore size ranging from 100 Å to 4000 Å. The results (each data point represents the average of three experiments) are summarized in Table 2, with the corresponding vendor-reported protein binding capacities. For a fixed col-

Table 2 Dynamic Binding Capacity on PLRPS Columns with Different Pore Sizes

Resin	Dynamic binding capacity for protein (mg/mL)	Dynamic binding capacity for plasmid (mg/mL)
PLRPS 50–70µ, 100 Å	3.5	0.06
PLRPS 50–70µ, 300 Å	8.5	0.04
PLRPS 50–70µ, 1000 Å	11.0	0.11
PLRPS 50–70µ, 4000 Å	6.0	0.39

umn size, the surface area decreases with increasing nominal pore size. The protein binding capacity of the resin [data obtained from the vendor for bovine serum albumin (BSA)] initially increases as a function of pore size. This initial increase in capacity for BSA with increasing pore size can be explained by the greater accessible surface area of the larger pore resin to the protein. With the smaller pore sizes, although the total surface area is large, the BSA cannot fully access the pores and therefore, the accessible surface area is limited. At ~1000 Å, an optimal pore size is reached where the BSA can access most of the pores, and further increases in the pore size merely reduce the overall surface area of the resin. The plasmid binding capacity is significantly lower than that of BSA owing to the larger size of the plasmid, which is inaccessible to the pores. For plasmid binding, it appears that only at 1000 Å is there any penetration into the pores for binding. Because the plasmid is such a large macromolecule relative to proteins, we have not yet reached the optimal pore size where the plasmid can fully access the pores.

Similar experiments were performed with Polymer Labs anion-exchange resin PLSAX (100 Å and 300 Å pore size resins of the PLSAX chemistry were unavailable in the 50–70 µm resin bead and therefore were not included in this study). The plasmid DNA was loaded onto the columns in 50 mM Tris-HCl buffer at pH 8. The resulting capacities (each data point represents the average of three experiments) are presented in Table 3.

As was observed with the reversed-phase resins, a similar trend is seen for the plasmid binding capacity of anion-exchange resins. A general conclusion that can be drawn from these trends is that the optimal pore size for plasmids under the conditions examined is greater than 4000 Å. Extensive studies on the impact of mass transfer limitation and intraparticle diffusion in protein and small-molecule chromatography can be found, for example, in the classic works of Horvath and coworkers [26], Regnier and coworkers [17],

Table 3 Dynamic Binding Capacity on PLSAX Columns with Different Pore Sizes

Resin	Dynamic binding capacity for plasmid (mg/mL)
PLSAX 15–25µ, 1000 Å	0.7
PLSAX 15–25µ, 4000 Å	3.0

and Van Deemter [27], but the mass transport and resin parameters involved in optimizing plasmid DNA chromatography are not as well investigated. Extrapolating from the data at hand for proteins, there appears to be an optimal pore size for BSA (see Table 2) that maximizes binding capacity at ~1000 Å for the 50 µm particle size. Assuming that BSA is a spherical protein that would bind in perfectly cylindrical pore, the ratio of the diameter of BSA (~80 Å) to the optimal 1000 Å pore is 12.5. By analogy, for a "spherical" plasmid, the optimal pore size that would maximize binding capacity (based on an order-of-magnitude type of analysis) would occur with a 12,500 Å or 1.2 µm pore size. This approximation is for the extreme case where the plasmid is viewed as a sphere; however, it can serve to provide a gross approximation of the "optimal" pore size for plasmid DNA. According to the work of Horvath and coworkers, the gigaporous particles that have the largest pores are in general available in a pore diameter to particle diameter ratio $(d_{pore}/d_{particle}) > 10^{-2}$. The approximation (based on an order-of-magnitude type of analysis) for plasmid DNA suggests a $d_{pore}/d_{particle}$ of 2.5×10^{-2}. However, resins with such a large porous area may not be sufficiently stable to the high pressure and high flow rates needed at large scale. Consequently, there is a need for a novel absorbent technology capable of providing controlled and sufficiently large pores to render more of the internal surface area accessible to the plasmid without sacrificing resin stability for large-scale use. In the absence of these kinds of adsorbents, however, an effective chromatographic operations strategy would be to optimize plasmid binding capacities for currently available resin technology. When considering the pore accessibility of a plasmid, the key property that can be adjusted is its physical size. By changing the solution properties of the plasmid, we were able to modulate the apparent size relative to the pore size. In order to measure the relative size changes of the plasmid under various solution conditions, we used an HPSEC method to determine the respective plasmid sizes.

Size exclusion chromatography was selected as a tool to probe the size of the plasmid in a porous resin. The GDNA pw column from TosoHaas provides excellent resolution of plasmids in the 1–10 kb range. For these studies, a 7.8 mm × 30 cm column was operated at a flow rate of 0.2 mL/min. By injecting various samples of supercoiled plasmids of sizes ranging from 2.7 to 7.7 kb onto the column and determining the volume for elution, a quantitative standard curve for plasmid size versus retention time was generated. This curve is shown in Fig. 2.

High performance size exclusion chromatography was then used to examine the effects of additives on the plasmid apparent size. The first additive selected was NaCl. Salt was selected as a potential modulator of plasmid size, because DNA is a highly charged polymer. Inherently, counterions such as salt would reduce the charge repulsion between DNA segments. For this study, a broad range of NaCl concentrations from 0.1 to 2 M were used. These salt concentrations were used for sample makeup as well as for the elution buffer. As the salt concentration increases, the elution volume from the SEC column

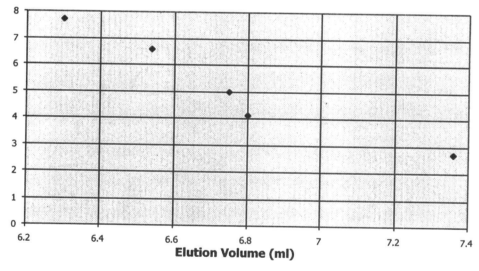

Figure 2 Size exclusion of supercoiled plasmids of varying sizes. Each data point shown in the chart represents the mean value of duplicates. Fifty microliters of supercoiled plasmid at a concentration of ~0.016 mg/mL was injected onto a 7.8 mm ID × 30 cm long GDNA pw column. The plasmid was eluted isocratically at 0.2 mL/min with 50 mM Tris, 0.1 M NaCl, pH ~8.

increases (as can be seen in Fig. 2). An increase in retention time in SEC corresponds to a decrease in the apparent size of the supercoiled plasmid. Comparing the elution volume of plasmid in the presence of NaCl with the standard curve shows that 1 M salt can modify the apparent size of a 6.6 kb plasmid to below 4 kb (see Fig. 3).

The next potential size modulator examined was alcohol. Alcohols such as isopropyl alcohol (IPA) are known to modify plasmid structure (see, for example, Ma and Bloomfield [22]). To determine the effect of IPA on elution volume, a broad range (1.2–20% v/v) of IPA concentrations were selected. The plasmid remained soluble in 20% v/v IPA. The samples were prepared with the various IPA concentrations, and the SEC column was run using a buffer with the same percent IPA. As the IPA content of the sample increased, the volume in which the plasmid eluted also increased. This increase in elution volume reflects a decrease in the apparent size of the supercoiled plasmid. A comparison of these results to the standard curve shows that IPA can modify the apparent size of a 6.6 kb plasmid to between 4 and 5 kb. Figure 4 illustrates the decrease of apparent plasmid size with the increase in IPA content.

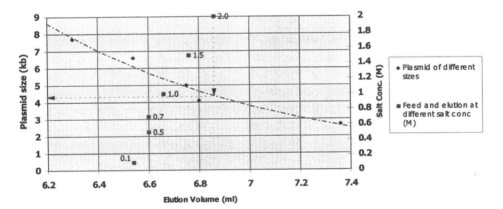

Figure 3 Size exclusion of a 6.6 kb supercoiled plasmid. Each data point shown on the chart represents the mean value of duplicates. Fifty microliters of supercoiled plasmid at a concentration of ~0.016 mg/mL was injected onto a 7.8 mmID × 30 cm long GDNA pw column. The plasmid was eluted at 0.2 mL/min with 50 mM Tris, [NaCl] as noted, pH = 8. The results are overlaid with the reference curve. In the presence of 2 M NaCl, the apparent size of the plasmid decreases to less than 4 kb. (♦) Plasmids of various sizes; (■) feed and elution at various salt concentrations.

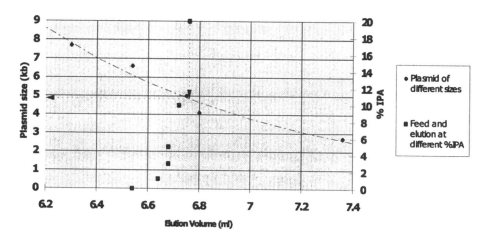

Figure 4 Size exclusion of a 6.6 kb supercoiled plasmid at varying percentages of IPA. Each data point shown in the chart represents the mean value of duplicates. Fifty microliters of supercoiled plasmid at a concentration of ~0.016 mg/mL was injected onto a 7.8 mmID × 30 cm long GDNA pw column. The plasmid was eluted at 0.2 mL/min with 50 mM Tris, IPA at percentages noted, pH ~8. The results are overlaid with the reference curve. In the presence of 20% IPA, the apparent size of the plasmid decreases to between 4 and 5 kb. (♦) Plasmids of various sizes; (■) feed and elution at various percentages of IPA.

The physical effect of salt and IPA on apparent plasmid size was also measured by using dynamic laser light scattering. High resolution multiangle light scattering experiments were performed in collaboration with Professor Denis Wirtz at Johns Hopkins University using a Wyatt DAWN DSP MASLS photometer to monitor the radius of gyration. The data generated for the radius of gyration for plasmid as a function of NaCl concentration are shown in Fig. 5.

Light scattering data show the trend in the apparent size of the plasmid, which first increases and then decreases with increasing NaCl concentration. (Note: The light scattering data were interpreted using an assumption that the plasmid is spherical; this may not be the case for all salt concentrations.) Similar size changes in DNA as a function of higher salt concentration were initially identified in earlier data by Hagerman [28], who predicted a decrease in chain flexibility at high salt concentrations, resulting in an inflexible rodlike structure for DNA. Aside from neutralizing the phosphate charges, increases in ionic strength can overwhelm electrostatic repulsion and bring DNA chains

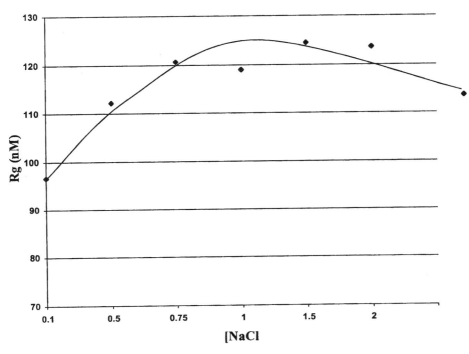

Figure 5 Laser light scattering of a 6.6 kb supercoiled plasmid at varying salt concentrations.

closer together to achieve a more condensed DNA form. This type of structural modification is further supported by atomic force microscopy (AFM) studies [29], which revealed that as ionic strength was increased (through increase in NaCl), the structure of the supercoiled plasmid became less relaxed and rounded and more superhelical and rodlike, hence assuming a more collapsed form. In the case of IPA, the light scattering data showed a slight but significant change in apparent size with increasing amounts of IPA (data not shown). AFM data in the presence of propyl alcohol [30] show that DNA in alcohol also undergoes a conformational change. However, unlike that which was observed with salt, the change is due to dehydration of the molecule from a hydrated, wider B-DNA to a dehydrated, narrow A-DNA form. Therefore, both salt and alcohol can serve as structural modifiers of DNA, through decreasing chain flexibility and potentially collapsing or condensing the molecule.

Table 4 Dynamic Binding Capacity on PLRPS 1000 Å Columns

NaCl concentration (M)	Saturation binding capacity (mg plasmid/mL PLRPS)
0.1	0.11
1.0	>0.19

From the results of SEC, light scattering, and AFM data, it was concluded that the addition of apparent size-modulating agents such as NaCl and alcohol to plasmid could improve plasmid accessibility to pores in porous resins. To demonstrate this effect, the plasmid DNA binding capacities of the Polymer Labs resins were again measured. This time, the reversed-phase (PLRPS) mobile phase contained 1 M NaCl, and the anion-exchange (PLSAX) mobile phase contained 5% IPA. The results presented in Tables 4 and 5 indicate that for a 1000 Å and 4000 Å PLRPS resin the plasmid binding capacity was doubled through the addition of 1 M NaCl. Similarly, for both the 1000 Å and 4000 Å PLSAX resins, the binding capacity increases slightly with the addition of IPA. Tables 6 and 7 show the plasmid binding capacity on PLSAX resins at 0% vs. 5% IPA.

Table 5 Binding Capacity of PLRPS 4000 Å Columns

Nacl concentration (M)	Saturation binding capacity (mg plasmid/mL PLRPS)
0.1	0.39
1.0	0.86

Table 6 Binding Capacity of PLSAX 1000 Å Columns

Percent IPA	Saturation binding capacity (mg plasmid/mL PLSAX)
0	0.74
5	1.40

Table 7
Binding Capacity of PLSAX 4000 Å Columns

Percent IPA	Saturation binding capacity (mg plasmid/mL PLSAX)
0	3.0
5	3.4

The extent of change is less pronounced for the PLSAX with 4000 Å pore size than for that with 1000 Å pores. This is likely because the plasmid is already capable of accessing more of the pores in the resin with the larger pores. Consequently, the change in plasmid size achieved in an anion-exchange resin with larger pores is insufficient to markedly access more pores. In the resin with 1000 Å pores, the plasmid binding capacity doubles in the presence of IPA, because the change in plasmid size makes it more accessible to the internal surface area.

Because the structural changes induced by NaCl are different from those of alcohol, the latter being a dehydration and the former an electrostatic neutralization, the effects on the biophysical properties of the molecules are also likely to be different. The size exclusion data suggest that the 6.6 kb plasmid in 5% IPA has an apparent size of a ~5 kb plasmid. This apparent change is sufficient to double the plasmid capacity of 1000 Å pore size resin. However, for an anion-exchange resin with 4000 Å pores, enough surface area is already accessible to the plasmid that a reduction in apparent size from 6.6 to 5 kb is insufficient to impact capacity. In the case of reversed-phase resin, the addition of 1 M NaCl only changes the apparent size of a 6.6 kb plasmid to 6 kb, and there is apparently a twofold increase in capacity at both 1000 and 4000 Å. These data may seem incongruous at first, but a look at the pore size distribution of the resins is revealing. From mercury porosimetric determinations of the average pore size and the distribution, it is clear that the pores in anion-exchange resin have a tighter size distribution than those of the reversed-phase resin. This is particularly true for the 4000 Å pore size resins. The median pore diameter (the diameter when the measured pore volume reaches 50% of the total) is more comparable to the pore diameter at the maximum peak in the pore distribution curve for the anion-exchange resins, as shown in Table 8. The reversed-phase resins with a 4000 Å reported pore size have significantly larger pores than the counterpart anion-exchange resin. In fact, in the reversed-phase PLRPS resin at least 50% of the pores are greater than 4700 Å in size and the majority are closer to 6000 Å. Taking this into consider-

Table 8 Measured Pore Sizes

Resin	Median pore size[a] (Å)	Pore size (Å)
PLSAX, 1000 Å	1637	1689
PLSAX, 4000 Å	3524	3874
PLRPS, 1000 Å	1102	1099
PLRPS, 4000 Å	4735	5896

[a] Median pore size is defined as the diameter when the measured pore volume reaches 50% of the total.

ation, the change in the apparent size from a 6.6 kb plasmid to that of a 6 kb plasmid generated by the addition of 1 M NaCl may be sufficient to access these larger pores in the PLRPS resin and hence contribute to a twofold rise in capacity. Similarly, with the anion-exchange resin, the median pore size is very close to 4000 Å and the majority of the pores are also around 4000 Å. Consequently, even the dramatic change observed in the apparent size of the plasmid from 6.6 kb to ~5 kb observed with the addition of IPA is insufficient to allow the plasmid access to the 4000 Å pores.

B. Implementation and Scale-Up

Results from small-scale experimental studies were used to scale up the chromatography steps for the preparative purification of plasmid DNA. The approach of maintaining the ratio of feed volume to column volume (as described in Chap. 1) was adopted. This ratio was relatively easy to maintain upon scale-up because of both consistent resin capacities and consistent fermentation starting materials. Two issues arising from the column housings chosen for this work made it difficult to use the conventional approach for scale-up by maintaining a constant bed height. First there are limited options in commercially available column hardware that can withstand pressure drop. In addition, the pressure limitations of the larger scale column housings are much lower than those of bench-scale housings. This generated the need to use shorter, fatter columns than were used in the small-scale experiments. Linear flow rate was maintained within a range determined in small-scale experiments as the scale-up parameter. Figures 6 and 7 show the A_{260} profiles for typical lab- and pilot scale anion-exchange and reversed-phase chromatography steps. The step behaviors as well as the shapes of the profiles are very similar at the two scales. Overall, scale-up to the preparative scale was successful in terms of

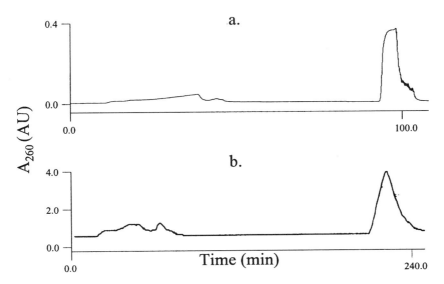

Figure 6 A comparison between (a) laboratory-scale and (b) pilot scale anion-exchange plasmid purification. Lab equipment consisted of a Biocad 700E workstation and a 0.5 cm ID Q HyperD F column. Pilot scale equipment consisted of a Rainin SD 1 system and a 25.2 cm ID column.

both product yield and quality. Table 9 summarizes the capacities observed from lab to pilot scale.

The purification process as shown in Fig. 1 with the exception of adding 1 M NaCl size modulator to the reversed-phase chromatography has been routinely used to manufacture plasmid DNA at the tens of grams scale for cGMP clinical supplies. The result of adding salt is a two-fold increase in capacity. The breakdown of overall cost of raw materials for the purification process is shown in Fig. 8. Over 56% of the raw materials cost is due to the operation of the chromatography step. The most expensive step is the reversed-phase chromatography step because of the low plasmid binding capacity of the resin, contributing to more than 80% of the chromatography costs. Nevertheless, the reversed phase serves as a final polishing step and plays a critical role in the process to ensure overall product purity. The addition of 1 M NaCl to the reversed-phase step results in a twofold increase in binding capacity, which results in a 28% overall reduction in raw materials cost. In terms of cycle time, the longest step in the process is the reversed-phase chromatography step. When operating at a linear flow rate of 50–75 cm/h (maximum flow

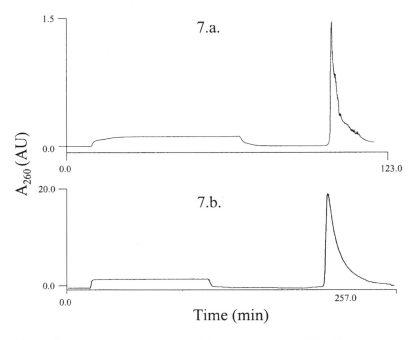

Figure 7 A comparison between (a) laboratory-scale and (b) pilot scale reversed-phase plasmid purification. Lab equipment consisted of a Biocad 700E workstation and a 0.5 cm ID Poros R 1 50 column. Pilot scale equipment consisted of a Rainin SD 1 system and a 25.2 cm ID column.

Table 9 Change in Capacity of Reversed-Phase Resin upon Scale-Up

	Column plasmid binding capacity (mg/mL)	
Scale	No addition of NaCl	[NaCl] = 1 M
Bench (~1 L column volume)	1.5–2.5	4.5–5.0
Pilot (10–14 L column volume)	N/A[a]	4.5–5.0

[a] NaCl concentration has always been adjusted to 1 M for pilot scale processing.

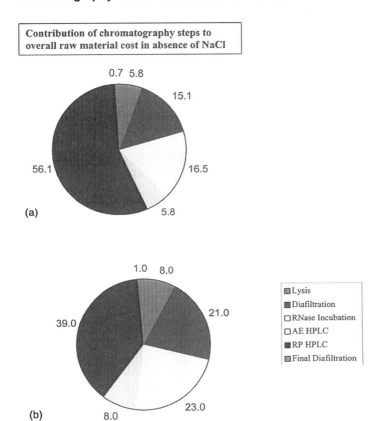

Figure 8 The breakdown of total process raw material costs, the corresponding contributions due to the reversed-phase chromatography step, and the impact of the addition of 1 M NaCl. (a) Without NaCl; (b) with 1 M NaCl. Individual contributions of each process step are indicated by the numbers next to the pie sections.

rate because of column pressure limitation of 100 psi), the reversed-phase chromatography without employing the NaCl modulator would require 32 h from cleaning through elution. By doubling the plasmid binding capacity of the resin through the addition of 1 M NaCl, the column size was reduced by 50% thereby reducing the step cycle time by 50%. Table 10 summarizes the

Table 10 Impact on Process Economics

Process attribute	Reversed-phase step operated without NaCl	Reversed-phase step operated with 1 M NaCl
Fold change in resin plasmid binding capacity	1×	2×
Step cycle time	32 h	16 h
% Raw materials cost due to reversed-phase chromatography step	56	39

contribution of the addition of the salt as a size modulator on the process economics in terms of both raw material costs and cycle time.

V. CONCLUSIONS

Commercially available chromatography resins have been designed for protein separations and are therefore not optimal for the binding and separation of plasmids. Working with resins of different pore sizes, it was identified that the plasmid binding capacity of resins is lower than that for proteins owing to limited accessible surface area. The larger plasmid cannot access the same total surface area of the resin as a smaller protein. Novel technology needs to be developed to generate stable absorbents with sufficient pore sizes to maximize plasmid binding capacities. In the meantime, we have identified structural modifiers that can be used with the existing resin technologies to increase plasmid binding capacities. Through the addition of IPA and NaCl in SEC studies, we have shown that the apparent size of plasmid DNA can be reduced. Through biophysical characterization, the effect of structural modifiers was shown to collapse and condense DNA size. This change in apparent size resulted in greater plasmid binding capacities of porous resins owing to surface area accessibility. An increase in plasmid binding capacity can further translate to reduced process costs and cycle times for purification of plasmid DNA.

REFERENCES

1. LA Babiuk, PJ Lewis, G Cox, S van Duren Littel-can den Hurk, M Baca-Estrada, SK Tikoo. DNA immunization with bovine herpesvirus-1 genes. Ann NY Acad Sci 772:47–63, 1995.

2. J Cohen. Naked DNA points way to vaccines. Science 259:1691–1692, 1993.
3. JJ Donnelly, F Arthur, D Martinez, DL Montgomery, JW Shiver, SL Motzel, JB Ulmer, M Liu. Preclinical efficacy of a prototype DNA vaccine: enhanced protection against antigenic drift in influenza virus. Nature Med 1:583–587, 1995.
4. ML Lagging, K Meyer, D Hoft, M Houghton, RB Belshe, R Ray. Immune responses to plasmid DNA encoding the hepatitis C virus core protein. J Virol 69: 5859–5863, 1995.
5. JW Shiver, HC Perry, ME Davies, MA Liu. Cytotoxic T lymphocyte and helper T cell responses following HIV polynucleotide vaccination. Ann NY Acad Sci 772:198–208, 1995.
6. E Manickan, R Rouse, Z Yu, WS Wire, BT Rouse. Genetic immunization against herpes simplex virus. Immunol 155:259–265, 1995.
7. RE Spier. Nucleic acid vaccines. Vaccine 13:131–132, 1995.
8. B Wang, J Boyer, V Srikantan, K Ugen, L Gilbert, C Phan, K Dang, M Merva, MG Agadjanyan, M Newman, R Carrano, D McCallus, L Coney, WV Williams, DB Wiener. Induction of humoral and cellular immune responses to the human immunodeficiency type 1 virus in nonhuman primates by in vivo DNA inoculation. Virology 211:102–112, 1995.
9. NJ Caplen, X Gao, R Hayes, R Elaswarapu, G Fisher, E Kinrade, A Chakera, J Schorr, B Hughes, JR Dorin, DJ Porteous, EWFW Alton, DM Geddes, C Coutelle, LH Williamson, C Gilchrist. Gene-therapy for cystic-fibrosis in humans by liposome-mediated DNA transfer: the production of resources and the regulatory process. Gene Therapy 1:139–147, 1994.
10. GJ Nabel, A Chang, EG Nabel, G Plautz, BA Fox, L Huang, S Shiu. Immunotherapy of malignancy by in vivo gene transfer into tumors. Human Gene Therapy 3:399–410, 1992.
11. EG Nabel, G Plautz, GJ Nabel. Transduction of a foreign histocompatibility gene into the arterial wall induces vasculitis. Proc Natl Acad Sci USA 89:5157–5161, 1992.
12. J Schorr, P Moritz, T Seddon, M Schleef. Plasmid DNA for human gene therapy and DNA vaccines. Ann NY Acad Sci 772:271–273, 1995.
13. NA Horn, JA Meek, G Budahazi, M Marquet. Cancer gene therapy using plasmid DNA: purification of DNA for human clinical trials. Human Gene Therapy 6: 565–573, 1995.
14. G Leaver, JR Conder, JM Howell. Adsorption isotherms of albumin on a crosslinked cellulose chromatographic ion exchanger. I. Chem Eng Symp Ser 118:1–9, 1990.
15. LL Lloyd, FP Warner. Preparative high-performance liquid chromatography on a unique high-speed macroporous resin. J Chromatogr 512:365–376, 1990.
16. KA Tweeten, TN Tweeten. Reversed-phase chromatography of proteins on resin-based wide-pore packings. J Chromatogr 359:111–119, 1986.
17. NB Afeyan, SP Fulton, FE Regnier. Perfusion chromatography packing materials for proteins and peptides. J Chromatogr 544:267–279, 1991.

18. VF Holmes, NR Cozzerelli. Closing the ring: links between SMC proteins and chromosome partitioning, condensation and supercoiling. Proc Nat Acad Sci USA 97:1322–1324, 2000.
19. H Berman. An Introduction to Nucleic Acids. *http://ndbserver.rutgers.edu/ NDB/archives/NAintro/index.html*, 1996.
20. VA Bloomfield. DNA condensation. Curr Opin Struct Biol 6:334–341, 1996.
21. J Bednar, P Furrer, A Stasiak, J Dubochet. The twist, writhe and overall shape of supercoiled DNA change during counterion-induced transition from a loosely to a tightly interwound superhelix. J Mol Biol 235:825–847, 1994.
22. CL Ma, VA Bloomfield. Condensation of supercoiled DNA induced by $MnCl_2$. Biophys J 67:1678–1681, 1994.
23. VA Bloomfield. DNA condensation by multivalent cations. Biopolymers 44: 269–282, 1997.
24. VA Bloomfield. DNA condensation. Curr Opin Struct Biol 6:334–341, 1996.
25. JC Murphy, JA Wibbenmeyer, GE Fox, RC Willson. Purification of plasmid DNA using selective precipitation by compaction agents. Nature Biotechnol 17: 822–823, 1999.
26. DD Frey, E Schweinheim, C Horvath. Effect of intraparticle convection on the chromatography of biomacromolecules. Biotechnol Prog 4:273–284, 1993.
27. JJ van Deemter, F Zuiderweg, A Klinkengberg. Chem Eng Sci 5:271, 1956.
28. PJ Hagerman. Flexibility of DNA. Ann Rev Biophys Biophys Chem 17:265–286, 1988.
29. YL Lyubchenko, LS Shlyakhtenko, T Aki, S Adhya. Atomic force microscopic demonstration of DNA looping by GalR and HU. Nucl Acids Res 25:873–876, 1997.
30. HG Hasma, M Bezanilla, F Zenhausern, M Adrain, RL Sinsheimer. Atomic microscopy of DNA in aqueous solutions. Nucl Acids Res 21:505–512, 1993.

9
Case Study: Purification of an IgG$_1$ Monoclonal Antibody

Paul K. Ng
Bayer Corporation, Berkeley, California, U.S.A.

I. PROCESS SUMMARY

Monoclonal antibodies have enormous potential as therapeutic and diagnostic agents because of their remarkable specificity and affinity for their targets. Tumor necrosis factor (TNF) has been shown to be involved in the pathogenesis of septic shock. Because antibodies to TNF offer measurable and significant protection against the effects of sepsis in several experimental systems [1–3], a process was developed to produce a monoclonal antibody against TNF. The cell line chosen was a murine hybridoma, A10G10, which expressed an IgG$_1$ with anti-TNF property. The purification process consisted of three different ion-exchange chromatography steps followed by one size exclusion chromatography step. Also included in the process was one DNA reduction step and one independent viral inactivation step.

II. DESCRIPTION OF UPSTREAM PROCESS

The production cell line, A10G10, was obtained from a frozen ampule of the manufacturer's working cell bank. Cells were thawed, placed in medium, incubated, and expanded. The cells were then cultivated in a deep-tank stirred suspension fermentor under optimal production conditions with respect to growth profile, cell viability, growth consumption, and product productivity. A dilute stream containing the product was removed continuously by a proprietary cell retention system, while fresh medium was simultaneously replenished at the same rate to the fermentor. Carryover cells and cell debris were separated by dead-end filtration. The filtrate was concentrated by an ultrafiltration system with a nominal molecular weight cutoff value of 10,000 or 30,000.

III. DESCRIPTION OF DOWNSTREAM PROCESS

The antibody was processed through a multistep purification scheme. A list of the major components in the fermentor harvest fluid is given in Table 1. Physical properties of the components are comparable to those studied by others in a cell culture containing T-cell-binding monoclonal antibody [4]. The components in the medium were selected and optimized so that the process designer did not have the burden of having to remove bovine-derived proteins present in the medium [4,5]. During process development, differences in charge, size, and affinity were exploited to purify the antibody from potential contaminants including cell substrate nucleic acid, host cell proteins, media components, and the possible presence of endogenous and/or adventitious virus associated with the cell substrate and its fermentation. The purification strategy is designed to result in an intravenous concentrate that meets in-house standards derived from U.S. Food and Drug Administration guidelines. Figure 1 is a schematic diagram of the purification process consisting essentially of cation-exchange chromatography, anion-exchange chromatography, hydroxyapatite batch adsorption, size exclusion chromatography, viral inactivation and formulation. The logic behind selecting an ion-exchange approach was that murine IgG_1 is a poor binder to protein A. Utilizing high salt to make it bind might create problems such as a local concentration gradient or a solubility

Table 1 Components in Fermentor Harvest Fluid Containing Murine IgG_1

Component	Molecular weight	Isoelectric point
Murine IgG_1	150,000	6.3–7.0
Human serum albumin	65,000	4.9
Human transferrin	90,000	5.9
Insulin	5,700	5.3
Fatty acids, salts, sugars, and amino acids	<2,000	—
Host cell proteins	Vary	Mostly negative
Nucleic acids	1,000–1,000,000	Highly negative
Endotoxins	>100,000	Highly negative

Purification of an IgG₁ Monoclonal Antibody

Ultrafiltered clarified tissue culture fluid
↓
Cation Exchange Chromatography (S Sepharose®) — Removes bulk of cellular proteins and other contaminants
↓
Anion Exchange Chromatography (Q Sepharose®) — Removes additional contaminants
↓
Hydroxyapaptite (HAPT) batch adsorption — Removes DNA
↓
Size Exclusion Chromatography (S200 HR Sephacryl®) — Removes aggregates
↓
pH 4 hold (2-8C) — Inactivates virus
↓
Ultrafiltration/diafiltration — Buffer exchange and formulation
↓
Sterile filtration
↓
Fill/Freeze Drying

Figure 1 TNF MAb purification flow chart.

limit for the sodium chloride. More importantly, antigenicity and immunogenicity due to residual protein A in the product was not fully understood in the initial phase of process development.

The following describes details of each individual step in the process.

1. Early Purification—Polyethylene Glycol Precipitation. In the

Table 2 Recovery Across PEG Precipitation of UF-cTCF

Lot number	Grams of IgG_1 in thawed UF-cTCF	Grams of IgG_1 in PEG pellets	Recovery (%)
3147–1	76.9	55.9	73
3147-2a	54.2	39.0	72
3147-2b	37.2	44.5	120
3147–3	95.3	69.7	73

initial design, a classical method using precipitation followed by ion-exchange chromatography was chosen. The antibody was first precipitated by polyethylene glycol (PEG 3350) after ultrafiltration of clarified (cell-free) harvests. Operating conditions such as temperature, pH, PEG concentration, and protein concentration were optimized. As a gross purification step, precipitation [6] enabled an early concentration of the product while it simultaneously served to remove contaminants. pH of the protein solution at an A_{280} of 20–25 was initially adjusted to 6.1–6.4. PEG 3350 was added to 17.5% at 2–8°C. The precipitate was recovered by centrifugation, and the pellets were dissolved in Tris/saline buffer at pH 8.5. This IgG_1 concentrate was stored frozen at −35°C until further processing. As shown in Table 2, step recoveries in four representative runs were >70%.

Although PEG precipitation has been implemented in large-scale processing that resulted in clinical materials, it presented shortcomings such as product insolubility, difficulty in eliminating the polymers from the product, and aerosol generation during continuous centrifugation in a GMP area. We therefore developed a chromatography step, as discussed below.

2. Cation-Exchange Chromatography. An alternative process step to PEG precipitation, cation-exchange chromatography, was developed. In cation exchange, contaminants with low isoelectric point below the pH of the buffer do not bind to the column. Therefore, the step allowed a significant purification of IgG_1 from contaminating proteins. The S Sepharose cation exchanger was equilibrated in acetate buffer (0.01 M sodium acetate, pH 4.6). The column was then washed with 0.01 M sodium acetate, 0.05 M NaCl, pH 5.5. IgG_1 was eluted with 0.01 M Tris(hydroxymethyl)aminomethane (Tris), 0.05 M NaCl, pH 7.2. Under these conditions IgG_1 could be quantitatively recovered [7].

3. Anion-Exchange Chromatography. Following cation-exchange chromatography, an anion-exchange chromatography step was implemented for further removal of contaminants. During the early phase of process development, carryover albumin was used as a marker for contaminant proteins. A significant amount of IgG_1 was separated from albumin by anion exchange, because the latter was more tightly bound by virtue of its more acidic isoelectric point (see Table 2). The Q Sepharose anion exchanger was equilibrated in Tris buffer (0.05 M Tris, 0.05 M NaCl), pH 8.5. The IgG_1-rich anion exchanger was washed in equilibration buffer. The IgG_1 was eluted with a gradient from 0.05 M to 0.20 M NaCl and 0.05 M Tris, from pH 8.5 to pH 7.0.

To further decrease the contaminant levels, the anion-exchange chromatography step was repeated (Q Sepharose) using step elution.

4. Hydroxyapatite Contact. Batch contact of the IgG_1 solution with hydroxyapatite (DNA grade Bio-Gel HTP gel) was performed to provide additional assurance of residual host cell DNA clearance. It was deemed necessary to include this step because neither cation-exchange chromatography nor anion-exchange chromatography consistently removed DNA to an acceptable level. Details of the mechanism for DNA separation are available in the manufacturer's package insert (Bio-Rad Laboratories, Hercules, CA). Binding of DNA and proteins was a function of phosphate concentration and hydroxyapatite concentration. By adding radiolabeled cell substrate DNA to the protein solution, optimum phosphate (0.1 M) and hydroxyapatite (0.8%) concentrations were established at which IgG_1 would not bind. To test the utility of this approach, DNA contents of samples before and after hydroxyapatite treatment were monitored during purification development. By dot blot hybridization, DNA was reduced from a mean of 0.53 pg/mg IgG ($n = 6$) to a mean of 0.18 pg/mg IgG ($n = 6$). Although the reduction was not as dramatic as the spike experiment demonstrated, the step did give a lower amount of DNA prior to size exclusion. After hydroxyapatite batch contact was fully integrated into the purification train, DNA content in the product was routinely measured at <100 pg per 500 mg IgG_1.

5. Size Exclusion Chromatography. To remove large molecular weight aggregates and other contaminants with molecular weights less than that of IgG_1, a size exclusion chromatography (SEC) step was employed. The choice of Sephacryl S200 as the matrix was based on high selectivity (fractionation range 5000–250,000 Da) and high yield. Although SEC is more vulnerable to axial compression, thereby giving rise to slower linear velocity, it is a generally accepted technique for the final polishing step [8,9].

6. Low pH Hold Step. A hold step at pH 4 was implemented into the process for reduction of potential viral contaminants.

7. Formulation. Tumor necrosis factor MAb was stabilized with glycine and maltose, and the final adjustment of concentrations was achieved by membrane ultrafiltration or diafiltration.

8. Sterile Filtration, Filling, and Freeze-Drying. The product was sterile-filtered through an absolute 0.2 μm filter, aseptically filled into glass vials, and freeze-dried.

IV. PERFORMANCE OF CHROMATOGRAPHY STEPS

A. Purity Improvement Across Chromatography Steps

Immunoglobulin G_1 was purified in five steps from ultrafiltered clarified tissue culture fluid (UF-cTCF). The first step employed cation-exchange chromatography by taking advantage of the fact that the pI of the protein is 6.3–7.0 and therefore it is positively charged at pH 4.6. By a progressive pH gradient wash from 4.6 to 5.5 followed by elution at pH 7.2, a greater than twofold increase in purity could be achieved (Table 3). The protein was further purified by two anion-exchange chromatography steps. Here, we again exploited the binding strength of the protein to the column by selecting an operating pH of 8.5, which is above the pI of IgG_1. The purity after the two columns was 90%. A batch adsorption step for DNA removal with hydroxyapatite gave a slight improvement in purity. A final polishing step using size exclusion chromatography essentially brought the product to a purity of 100%.

B. Clearance of Cellular Proteins Across Chromatography Steps

To compare the efficiency of each process step for the removal of host cell proteins, we studied the clearance factor, which is the ratio of the host cell protein content before the step to the host cell protein content after the step. As shown in Fig. 2, much of the clearance was actually completed in the first two columns, i.e., cation exchange and the first anion exchange. The three log reduction over the two chromatography steps is akin to that of a protein A step

Table 3 Step Improvement Across the Purification Process

Purification step	% Purity IgG_1/total protein ($n = 6$)	
	Mean	S.D.
UF-cTCF	18	1
Post cation exchange	39	8
Post anion exchange 1	61	4
Post anion exchange 2	91	3
Post HAPT	95	3
Post size exclusion	104	3

Purification of an IgG$_1$ Monoclonal Antibody

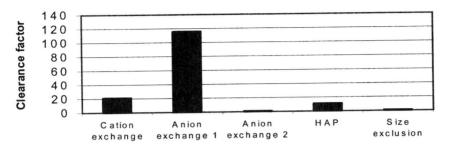

Figure 2 Clearance of contaminant across entire process.

that was evaluated for scale-up operations. Details of an alternative process implementing the protein A step are beyond the scope of this report.

C. Comparisons of Chromatographs

Chromatographs of small-scale runs and full-scale runs were assessed to compare the operational differences. The size exclusion step presented a noneventful profile, because a single peak was routinely obtained. In the case of the ion-exchange steps, the starting material is of lower purity, and consequently multiple peaks were obtained in the profiles. An example of a comparison is shown in Fig. 3. A slight disparity was seen in the S Sepharose chromatography profiles that was due in part to the potential for greater resolution on the small-scale column. Furthermore, variation in the starting material and differences in the optical path length of the UV detectors also contributed to the observed disparity.

IV. SCALE-UP CRITERIA USED FOR SCALING UP THE CHROMATOGRAPHY STEPS

All of the chromatography steps were scaled up linearly by maintaining the column bed heights and the linear flow rates constant and increasing the column surface areas appropriately. The process cycle time therefore remained essentially constant. In the case of size exclusion with Sephacryl S200 resins, it was determined that we need to divide the column bed height into three sections of equal height. This was in consideration of gel deformation by hydrodynamic forces if the bed height were to remain as a single long column.

To ensure full automation and production reliability, a fully automated

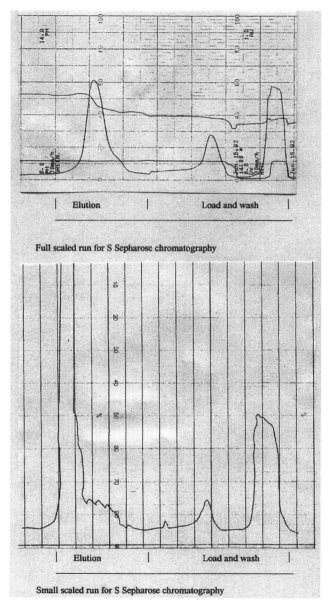

Figure 3 Comparison of S Sepharose chromatography profiles. (Top) Full-scale run; (Bottom) Small-scale run.

process was developed on the Pharmacia BioProcess System. Specifically, much effort was instituted to control the first anion-exchange step where gradient elution was employed. In a preset program, the gradient was formed by adding an upper limit buffer (0.05 M Tris, 0.2 M NaCl, pH 7.0) with mixing into 15 column volumes (CV) of equilibration buffer (0.05 M Tris, 0.05 M NaCl, pH 8.5). The rate of addition was half the column flow rate. The IgG_1 peak was collected when two criteria were met: A_{280} was above 10% of the baseline, and conductivity was greater than 8.5 mS/cm. To control the end of elution, the eluate was programmed to waste when A_{280} was within 10% of the baseline. Incidentally, the pH of the upper limit buffer was chosen to be 7.0, a pH much closer to the pI than that of the lower limit buffer, thereby reducing the amount of salt to effect the elution of IgG_1 from the column. The IgG_1 could then be applied to a second anion-exchange chromatography step without the burden of high salt concentration in the starting pool.

V. COMPARISON OF CHROMATOGRAPHY STEPS AT SMALL SCALE AND AT PILOT SCALE

On the basis of the small-scale experimental results, a method was designed for the purification of large amounts of IgG_1 in the pilot plant. Retrospectively, small-scale runs were performed to show comparability with the scale-up runs. Such an analysis was necessary because the data were needed prior to validation studies such as column lifetime and viral clearance. The general guidelines we followed are summarized below. A number of such guidelines have been published [10,11], and they are illustrative of the kind of scale-up/scale-down modus operandi one normally takes.

1. Chemicals of identical grade and identical gels were used.
2. For the ion-exchange columns, the protein loading per liter of gel was kept constant. In our case, we projected an increase in throughput for subsequent scale-up; we therefore maintained the loading at 100–150% of a predetermined loading value. The value was initially optimized or fine-tuned by range-finding experiments at the bench level. It was then further verified at the pilot plant level, leading to batches used in phase I clinical studies.
3. For the size exclusion column, the ratio of sample volume to column volume was kept constant. Similarly to the ion-exchange columns, the sample volume was maintained at 100–150% of a predetermined loading value.

4. An identical linear flow velocity was maintained for each chromatography column.
5. The bed diameter was varied, and the bed height was kept constant.
6. For gradient elution, the same ratio of total gradient volume to column volume was maintained.

Chromatography columns selected for the small-scale runs are compared to the pilot scale runs from two campaigns in Table 4. Apart from the longer small-scale size exclusion column, bed heights of the ion-exchange columns are comparable at both scales. Using the subscripts s and p to describe parameters at small and pilot scale, respectively, a scale-down factor or the inverse of a scale-up factor can be calculated by the equation derived earlier [8]:

$$\text{Scale-down factor} = \frac{V_{\text{column},s}}{V_{\text{column},p}} = \frac{L_b D_{c,s}^2}{L_l D_{c,p}^2}$$

where V_{column}, L, and D_c are the column volume, length, and diameter, respectively. For the columns studied, scale-down factors fall in the range of 1/357 to 1/1641.

The performance of the small-scale and pilot scale runs was assessed

Table 4 Comparison of Small-Scale Columns and Pilot Plant Scale Columns

		Pilot plant scale	
Process	Small scale	Campaign 1	Campaign 2
Cation exchange			
Gel volume (L)	0.039	62	64
Diameter (cm)	1.1	44	44
Bed height (cm)	41	41	42
Anion exchange 1			
Gel volume (L)	0.219	88	88
Diameter (cm)	2.2	44	44
Bed height (cm)	57.6	58	58
Anion exchange 2			
Gel volume (L)	0.098	35	40
Diameter (cm)	2.2	44	45
Bed height (cm)	25.7	23	25
Size exclusion			
Gel volume (L)	0.2	238	241
Diameter (cm)	1.6	63	63
Bed height (cm)	99.5	77	78

Purification of an IgG$_1$ Monoclonal Antibody

Table 5 Performance of Columns at Pilot Scale and at Small Scale

Run number[a]	Cation-exchange chromatography		Anion-exchange chromatography 1		Anion-exchange chromatography 2		Size exclusion chromatography	
	Cellular protein (μg/mg)	Step yield (%)	Cellular protein (μg/mg)	Step yield (%)	Cellular protein (μg/mg)	Step yield (%)	Cellular protein (μg/mg)	Step yield (%)
PS1	57	81	1.95	92	0.91	91	<0.02	100
PS2	155	67	2.51	89	3.54	100	<0.02	100
PS3	80	73	3.89	79	3.00	100	<0.02	74
PS4	78	80	3.39	92	3.39	100	<0.02	80
PS5	54	82	2.58	89	3.09	100	<0.02	100
PS6	89	90	2.62	85	2.31	94	<0.02	93
PS7	18	103	0.34	80	0.74	89	<0.02	98
PS8	27	90	0.24	87	0.15	89	<0.02	99
PS9	23	87	0.10	83	0.08	98	<0.02	95
PS10	20	91	0.12	83	0.13	92	<0.02	99
PS11	22	94	0.09	90	0.12	83	<0.02	98
PS12	22	85	0.26	90	0.12	92	<0.02	97
Mean	54	85	1.50	86	1.50	94	<0.02	94
Acceptance criteria (Mean ± 3 S.D.)	0–177	58–100	0–5.85	72–100	0–5.90	77–100	<0.02	68–100
SS1	47.5	80	0.55	95	0.12	100	<0.02	96
SS2	43	78	0.45	100	0.16	100	<0.02	94
SS3	39	91	2.64	92	0.19	100	<0.02	82
SS4	42	88	0.97	82	0.25	78	<0.02	93
SS5	46	90	0.49	93	0.35	86	<0.02	83
SS6	36	82	0.58	89	0.35	84	<0.02	79

[a] PS = pilot scale; SS = small scale.

by two parameters. These are the step yield, which measured the total IgG_1 before and after each step, and host cell protein content, which normalized the host cell protein concentration to the IgG_1 concentration. As shown in Table 5, 12 large-scale runs and six small-scale runs were compared. In the analysis of the pilot scale runs, we determined a priori that the data of the small-scale runs should fall within three standard deviations (S.D.) of the mean. The choice was based on our control chart analyses that 3 S.D. would cover our operating experience. This is consistent with statistical analysis of a normal distribution curve, which has 99.7% of its values within ± 3 S.D. of the statistical measure of interest [12]. Under the acceptance criteria established, the general observation that the small-scale runs did not change from the variation limit is very encouraging. It permits further studies of any follow-up validation work.

VI. SCALE-UP ISSUES

A. Process Modification

Modification in the purification process used for clinical trials requires biological validation to ensure comparability with previous methods. In this case, cation-exchange chromatography was introduced to replace PEG precipitation. To study the impact of this change, a series of biological tests were performed on materials prepared by both methods. These tests included pharmacokinetics, acute toxicity, and safety pharmacology. Details of the studies are beyond the scope of this chapter. Results of the studies indicated that the process modification did not significantly affect the antibody's biological activity. This demonstration gave reassurance for subsequent implementation of the cation-exchange chromatography step into the purification process.

B. Less Than Monomer (LTM) in the Product

The most vital biochemical issue in operating the process was the discovery of a protein we designated as LTM. By SEC HPLC, it was seen as a secondary eluting peak with a molecular weight of 110 kDa. LTM was also characterized as an $F(ab)_2$-like fragment. During the course of purification, it was converted from the IgG_1 molecule under pH conditions of 4–5. These conditions were present in the cation-exchange chromatography, gel filtration, and acidic pH hold steps. A cathepsin-like acid protease, activated by acidification, had been identified as the proteolytic agent responsible for degrading IgG_1 into LTM.

High performance liquid chromatographic data from the large-scale preparative analyses over a span of two years are shown in Table 6. In the trials,

Purification of an IgG$_1$ Monoclonal Antibody

various starting materials were used. Monomer contents were above 90%. LTM ranged from 1% to as much as 8%. This variation appeared to be weakly linked to the quantity of protease in the final container material (data not shown). However, a low level of protease, when given sufficient time, would elicit a disproportionate formation of LTM if the temperature was also artificially elevated to 37°C. This suggested that processing time needs to be well controlled or minimized when the product is most susceptible to protease degradation conditions, which included cation-exchange chromatography, gel filtration, and pH hold [13].

There was very little information available regarding LTM formation before we initiated scale-up. However, the general observation that monomer levels in 20 clinical runs (Table 6) exceeded the specification of >90%, with only one batch at the borderline, was encouraging. It was imperative that by following the purification scheme with minimal deviation in terms of pH and

Table 6 Monomer and Less Than Monomer (LTM) in Large-Scale Lots

Lot no.	% Monomer[a]	% LTM[b]
PR3147	97	2
PR3152	94	4
PR3163	97	4
PR3166	96	2
PR3176	95	2
PR3177	97	1
PR3190	94	4
PR3196	94	2
PR3198	92	6
PR3200	92	5
PR3209	92	6
PR3210	91	5
PR3212	93	5
PR3214	90	8
PR3215	92	6
PR3219	94	3
PR3220	93	5
PR3222	97	1
PR3223	97	1
PR3224	96	2

[a] Specification: minimum 90%.
[b] Specification: maximum 10%.

temperature, a reproducible process with >90% monomer was attainable. In fact, this would be required to make the process commercially viable.

C. Pyrogen Removal

Large-scale processing of TNF MAb requires multiple use of the aforementioned chromatography columns. Matrices used in the preparation of these columns are subject to potential microbial contamination. Sterilization by steam autoclaving is not feasible for working columns. Gas sterilization with formaldehyde or preservation with thimerosal is problematic owing to residual toxic effects. Our experience indicates that the use of acid–base washes in the ion-exchange columns and a 0.01 M NaOH wash in the size exclusion columns were effective against bacterial growth. The column buffer eluates were regularly checked by Limulus amebocyte lysate (LAL) testing to ascertain that they were endotoxin-free (<0.03 EU/mL, or below detection level). These procedures resulted in final products that routinely passed the USP rabbit pyrogen test.

D. Virus Removal

The risk of acquiring hepatitis or HIV from plasma products was a serious concern in the past. As demonstrated by the recent safety record of the products, stringent donor screening and ensuing viral inactivation during manufacturing have dramatically minimized such risk. The presence of human-derived viruses from bloodborne diseases is essentially eliminated in products manufactured by recombinant DNA technology. Nevertheless, concern for viral safety in the production process still exists because of potential contamination from the raw materials and potential endogenous viruses in cell culture. Many preventive measures are in place, of which the following are highlighted:

1. Every operating fermentor was tested at regular intervals for suspect virus or adventitious virus, for microbial sterility, and for mycoplasma.
2. In fermentation, production equipment was dedicated for each unit operation. Between production runs, steam sterilization or caustic decontamination was used where applicable.
3. Production operators were trained to use hygienic practices that prevent viral contamination, such as aseptic techniques where applicable, protective clothing, shoe covers, and surface decontamination of materials transferred into the production facilities.
4. Bulk product prior to filling was tested for adventitious agents.

In addition to the aforementioned measures, the purification process was designed and validated to clear a panel of viruses that include retroviruses and three other types of viruses of different taxonomic groups and sizes. For the viral inactivation/clearance study, viral spikes were introduced at logical points representing independent mechanisms. The overall effect of the purification process on the virus would thus be the sum of the deductions obtained at the groups of steps encompassed by each viral spike. The reduction factor for retrovirus was ≥ 16.9 log, which is comparable in magnitude to those obtained in other reports [14,15]. This reduction exceeded the theoretical risk calculated from the number of retrovirus-like particles in tissue culture fluid and the volume of tissue culture fluid needed to generate a therapeutic dose.

VIII. CONCLUSION

Monoclonal antibody to TNF was produced by a hybridoma cell line, A10G10. A purification scheme was devised that used chromatography on cation-exchange, anion-exchange, and size exclusion gels. The process was scaled up to purify gram quantities of monoclonal antibody with a $>50\%$ overall recovery. The product is $>95\%$ purity by protein mass and $>90\%$ monomer by HPLC. Chromatographic comparability with regard to step yield and host cell protein content was also demonstrated in both small-scale and pilot scale batches.

ACKNOWLEDGMENTS

I thank the following for their contributions:

>R. Carrillo, M. Shearer, and M. Connell for their reviews of the chapter
>The staff of Purification Development, GMP Clinical Development and Validation [J. Andersen, B. Bautista, R. Casuga-Padilla (currently at Baxter International Corp., Hayward, CA), N. Cheung, M. Coan, V. Hamrick, M. Kitchen, V. Lee, P. O'Rourke, G. Roldan, M. Shearer, and M. Wong] for their work in many phases of the project
>The Quality Assurance staff for various assays
>R. Irwin for a discussion regarding the effect of process modification on biological assays
>M. Wong for a discussion regarding the generation of less than monomer

And a special thanks to Anurag Rathore (Pharmacia Corp., Chesterfield, MO) for comments during the final phase of writing.

REFERENCES

1. B Beutler, IW Milsark, AC Cerami. Passive immunization against cachectin/tumor necrosis factor protects mice from lethal effect of endotoxin. Science 229: 869–871, 1985.
2. LB Hinshaw, P TeKamp-Olson, AC Chang, PA Lee, FB Taylor, CK Murray, GT Peer, TE Emerson Jr, RB Passey, GC Kuo. Survival of primates in LD_{100} septic shock following therapy with antibody to tumor necrosis factor-alpha (TNF alpha), Circ Shock 30:279–292, 1990.
3. CJ Walsh, HJ Sugerman, PG Mullen, PD Crey, SK Leeper-Woodford, GJ Jesmonk, EF Ellis, AA Fowler. Monoclonal antibody to tumor necrosis factor α attenuates cardiopulmonary dysfunction in porcine gram negative sepsis. Arch Surg 127:138–144, 1992.
4. LS Hanna, P Pine, G Reuzinsky, S Nigam, DR Omstead. Removing specific cell culture contaminants in a MAb purification process. BioPharm 433–437, 1991.
5. W Jiskoot, JJCC Van Hertrooij, JWTM Klein Gebbinck, T Van der Velden-de Groot, DJA Crommelin, EC Beuvery. Two-step purification of a murine monoclonal antibody intended for therapeutic application in man. J Immunol Methods 124:143–156, 1989.
6. A Polson, C Ruiz-Bravo. Fractionation of plasma with polyethylene glycol. Vox Sang 23:107–118, 1972.
7. M Coan, VW Lee. Antibody Purification Method. US Patent 5110913, 1992.
8. R Anurag, A Velayudhan. An overview of scale-up in preparative chromatography. In: Scale-Up in Preparative Chromatography: Principles and Industrial Case Studies. New York: Marcel Dekker, 2001.
9. P Ganon. Purification Tools for Monoclonal Antibodies. Tucson, AZ: Validated Biosystems, 1996.
10. GK Sofer, LE Nystrom. Process Chromatography: A Practical Guide. San Diego, CA: Academic Press, 1989.
11. J Bonnerjea, P Terras. Chromatography systems. In: BK Lydersen, NA D'Elia, KL Nelson, eds. Bioprocess Engineering. New York: Wiley, 1994, pp 160–186.
12. B Kimball. Control charts for bioprocess control. BioPharm 6:46–48, 1998.
13. T Cui, M Tomic, M Coan. Development of a Cathepsin D-like Protease Activity Assay and the Implications to Purification of IgG Molecules. Bayer Rep No. CB-9745, 1997.
14. EM White, JB Grun, CS Sun, AF Sito. Process validation for virus removal and inactivation. BioPharm 5:34–39, 1991.
15. G Mitra, M Wong, J Bettencourt, G Tsay, P Ng. Protein purification in preparative scale of mammalian cell culture derived products. Ann NY Acad Sci 782: 422–431, 1996.

10
Case Study: Normal Phase Purification of Kilogram Quantities of a Synthetic Pharmaceutical Intermediate

Larry Miller and James Murphy
Pharmacia Corporation, Skokie, Illinois, U.S.A.

I. INTRODUCTION

Preparative liquid chromatography (PLC) has been used in the pharmaceutical industry for years to purify small molecular weight molecules [1–7]. Early in the development of a potential pharmaceutical compound, speed is a critical issue. It is important to minimize the time required from discovery to administration of the compound to humans in Phase I clinical trials. Often the most time-consuming process is the development of a synthetic method that produces material of suitable purity for clinical trails. One approach is to use a nonoptimized synthetic method and purify the final product using preparative chromatography. This approach can reduce development time by eliminating the need to optimize each chemical reaction and develop a final recrystallization or other purification method.

In the early 1990s Searle had an angiotensin II receptor antagonist program in development. The synthesis of the final drug substance used compound A (Fig. 1) as a synthetic intermediate. During the synthesis of compound A, an impurity (compound B, Fig. 2) was generated at the 10% level.

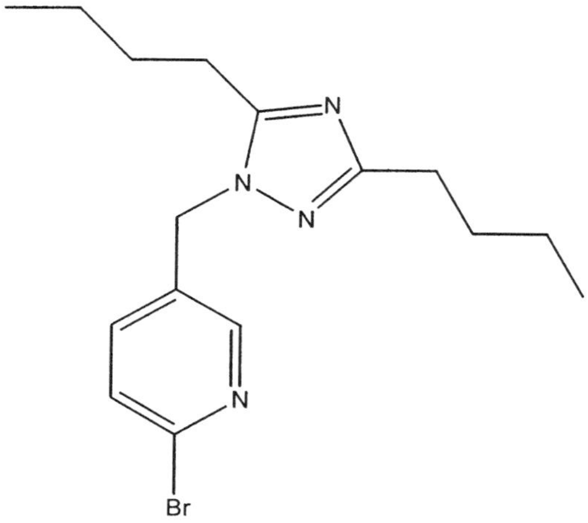

Figure 1 Structure of compound A.

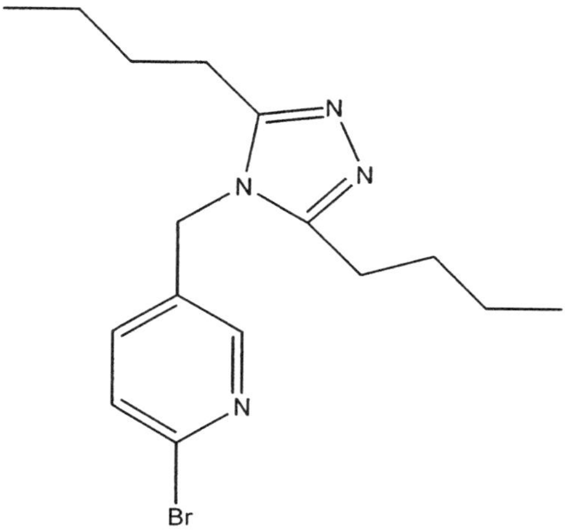

Figure 2 Structure of compound B.

Purification of a Synthetic Pharmaceutical Intermediate

This impurity could not be removed by recrystallization; therefore, PLC was needed to reduce compound B to acceptable levels.

Two different approaches can be used for the development of preparative chromatographic methods. The first involves the scale-up of an analytical high performance liquid chromatography (HPLC) method [8–10]. This technique is referred to as preparative HPLC. This technique has the advantage of being able to use identical stationary phases for method development and purification. In addition, small-particle stationary phases can be used, which often provide the best efficiencies for separation. The main disadvantage is the need to use analytical HPLC equipment for method development, which can increase cost and the time required for screening stationary phases and mobile phases. Also, all components in the sample mixture may not elute from the column or may not be detected with the detectors being used. The second approach is to develop a separation on thin layer chromatography (TLC) plates and scale it up to a preparative chromatographic method [11–14]. This technique is often referred to as column chromatography. This approach uses less expensive equipment for method development, and numerous stationary phases and mobile phase systems can be screened at the same time. Also, all components of the mixture are "visualized" on the TLC plate. The disadvantages are that only a few phases are available as TLC plates, limiting the number of phases that can be screened. This technique is also limited to large particles (40 μm and larger).

This chapter reports on the development of a preparative chromatographic method using scale-up from a TLC method. The use of TLC to screen mobile phases and the use of pilot columns to develop the final purification conditions are presented.

II. EXPERIMENTAL

A. Equipment

The large-scale preparative liquid chromatograph was either a SepTech 800A preparative chromatograph (Wakefield, RI) or a Milton Roy (Ivyland, PA) 3 L/min dual piston pump. For the pilot columns, mobile phase was pumped with a Waters 590 pump (Milford, MA). The column effluent was fractionated using a Gilson Model FC202 or FC80 fraction collector (Middleton, WI). The preparative columns varied in size from 60 mm × 10 mm i.d. to 183 cm × 23.7 cm i.d. and were obtained from YMC (now Waters Corp., Milford, MA).

B. Materials

The bulk packings were Merck silica gel 60, 40–63 µm, 60 Å irregular silica gel from EM Science (Cherry Hill, NJ). All chemicals for purification were synthesized in the laboratories of Searle (Skokie, IL). The solvents and reagents were reagent grade or better and obtained from a variety of sources.

Thin layer chromatographic separations were achieved on Merck silica gel 60 F_{254} TLC plates (Darmstadt, Germany). Spots were detected by placing the plates in a chamber saturated with t-butyl hypochlorite vapors followed by spraying with a starch–potassium iodide solution. Spots were detected under visible light.

III. RESULTS AND DISCUSSION

A. Twenty-Six Kilogram Campaign

The initial Active Pharmaceutical Ingredient (API) demands for Phase I supplies required the purification of 26 kg of compound A. Small quantities (<100 g) of compound A had been previously purified. Prior to scale-up to kilogram quantities, process optimization was required to develop a robust, efficient purification process.

1. TLC Method Development

The first step when developing a preparative method is to use TLC to screen various stationary and mobile phases. When developing a TLC method for scale-up to preparative loadings it is desirable to have an R_f of 0.2–0.5 for the desired component. This corresponds to capacity factors (k') of 1–4 when scaled up to a preparative column. Capacity factors can be calculated from TLC plates by using the equation

$$k' = (1 - R_f)/R_f \tag{1}$$

From the respective k', selectivity (alphas) can also be calculated for the components of the mixture to be separated.

During the initial TLC method development for compound A a total of 11 solvent systems were investigated. The results are listed in Table 1. The results were evaluated using the following criteria: (1) k' for compound A between 2 and 5; (2) maximum separation between compounds A and B; (3) maximum solubility of the reaction mixture in the mobile phase. Based on these criteria, the best solvent system was ethyl acetate–heptane. This solvent

Table 1 Thin Layer Method Development for Separation of Compounds A and B[a]

Solvent system	R_f		k'		Alpha
	Compound A	Compound B	Compound A	Compound B	
70/30 ethyl acetate/heptane	0.24	0.01	3.16	99	31.5
50/50 ethyl acetate/heptane	0.15	0	5.67	N/A	N/A
30/70 ethyl acetate/heptane	0.05	0	19	N/A	N/A
99/1 ethyl acetate/NH$_4$OH	0.50	0.10	1.0	9	9
80/20/1 ethyl acetate/heptane/NH$_4$OH	0.45	0.05	1.22	19	15.6
70/30/1 ethyl acetate/heptane/NH$_4$OH	0.30	0.02	2.33	49	21
60/40/1 ethyl acetate/heptane/NH$_4$OH	0.30	0.02	2.33	49	21
60/40/0.5 ethyl acetate/heptane/NH$_4$OH	0.25	0.02	3.0	49	16.3
30/70 acetone/heptane	0.26	0.03	2.85	32.3	11.3
50/50/0.5 methylene chloride/heptane/NH$_4$OH	0	0	N/A	N/A	N/A
40/60/0.5 methylene chloride/heptane/NH$_4$OH	0	0	N/A	N/A	N/A

N/A = not applicable.
[a] TLC plates: E. Merck silica gel

system was used as a starting point for development of a large-scale chromatographic method.

2. Development of Chromatographic Process

When the solvent system had been finalized, pilot columns were performed to optimize the preparative method. For these pilot columns a 10 mm i.d. × 60 mm column containing approximately 2.5 g of silica gel was used. This column mimics the larger columns in our laboratories and is used to study the effects of silica type, sample loading, flow rate, solvent composition, and equilibration volume. In our laboratories, two types of 40 μm silica gel are available (E. Merck and ICN). Using identical chromatographic conditions, these two silicas were compared. The results are summarized in Table 2. The limit for compound B in purified compound A was <0.5%. For all experiments, fractions were combined such that the yield of compound A was >95%. From this combined fraction, the level of compound B was determined by HPLC analysis. E. Merck silica gel produced significantly higher purity material and was used for all further experiments. The effect of sample load on the separation was then investigated. The results from these experiments are summarized in Table 3. As the sample load increased, the purity decreased. Based on these results, a sample load of 909 mg sample per gram silica gel was selected.

The next parameter investigated was flow rate. During scale-up the linear velocity must be kept constant. To scale from columns of differing internal diameters, the following equation is used:

$$F_2 = F_1(R_2/R_1)^2 \quad (2)$$

where F_2 is the flow rate on the larger column, F_1 is the flow rate for the smaller column, and R_2 and R_1 are the radii of the larger and smaller columns, respectively. Flow rates of 3 and 6 mL/min on the 10 mm i.d. column (which correspond to 700 and 1400 mL/min on a 15 cm i.d. column) were investi-

Table 2 Effect of Silica Gel on Purification of Compound A

Silica type	% Compound B in purified fraction
E. Merck	<0.1
ICN	5.2

Preparative conditions: Flow rate; 3 mL/min; load, 909 mg crude compound A per gram silica gel; mobile phase; 30/70 ethyl acetate/heptane.

Purification of a Synthetic Pharmaceutical Intermediate

Table 3 Effect of Sample Load on Purification of Compound A

Sample load (mg sample/g silica gel)	% Compound B in purified fraction
909	<0.1
1612	3.4
1887	3.8

Preparative conditions: Flow rate; 3 mL/min; E. Merck silica gel; mobile phase; 30/70 ethyl acetate/heptane.

gated. The results are summarized in Table 4. Product of suitable purity was generated at both flow rates. At the higher flow rate the level of compound B was only slightly less than the acceptable limit. To maintain a more robust process, the lower flow rate was selected.

The next parameter investigated was mobile phase composition. Table 5 summarizes these results. Either 30 or 50 vol% of ethyl acetate in heptane gave acceptable purity. The decision was based on economics. Because 50 vol% ethyl acetate required lower solvent requirements and resulted in a higher throughput method, it was selected. The final parameter investigated was equilibration volume. For this process the silica gel is dry packed into a column. The column is then wetted with solvent prior to introduction of sample. To minimize solvent consumption, it is preferable to equilibrate the column with only the required volume of solvent. Equilibration with 1, 2, 3, and 4 column volumes (CV) was investigated. The results are summarized in Table 6. Equilibration volume has only a slight effect on product purity. Two column volumes was selected for equilibration.

Table 4 Effect of Flow Rate on Purification of Compound A

Flowrate (mL/min)	% Compound B in purified fraction
3.0	<0.1
6.0	0.45

Preparative conditions: E. Merck silica gel; sample load, 909 mg sample/g silica gel; mobile phase; 30/70 ethyl acetate/heptane.

Table 5 Effect of Mobile Phase Composition on Purification of Compound A

Mobile phase (v/v)	% Compound B in purified fraction
30/70 ethyl acetate/heptane	<0.1
50/50 ethyl acetate/heptane	<0.1
Ethyl acetate	3.4

Preparative conditions: E. Merck silica gel; sample load; 909 mg sample/g silica gel; flow rate; 3 mL/min.

Based on all small-scale experiments, the final purification conditions were

Column: 10 mm i.d. × 60 mm
Silica gel: E. Merck
Sample load: 909 mg sample/g silica gel
Flow rate: 3 mL/min
Elution: 50/50 ethyl acetate/heptane

3. Scale-Up to 15 cm i.d. Column

The separation was scaled to a 15.2 cm i.d. × 100 cm column. The flow rate for the separation was set at 700 mL/min using Eq. (2). The sample load for the larger column was calculated using the equation

Table 6 Effect of Equilibration Volume on Purification of Compound A

Equilibration Volume (CV[a])	% Compound B in purified fraction
1	0.19
2	0.09
3	0.13
4	<0.1

Preparative conditions: E. Merck silica gel; sample load; 909 mg sample/g silica gel; flow rate; 3 mL/min; mobile phase: 50/50 (v/v) ethyl acetate/heptane.
[a] Column volume, defined as equivalent to silica gel contained in the column.

Purification of a Synthetic Pharmaceutical Intermediate

$$\text{Load}_2 = \text{Load}_1 \left[\left(\frac{R_2}{R_1}\right)^2 \left(\frac{L_2}{L_1}\right) \right] \tag{3}$$

where R is the radius of the column and L is the length of the column. For columns of identical length the load is directly related to the square of the column radius. Using this equation, a load of 2.27 g on a 10 × 60 mm column corresponds to a load of 8.8 kg on a 15.2 × 100 cm column. The material for purification was prepared in the Searle pilot plant. The amount of crude compound A prepared in each lot was 9.7 kg. This quantity was too large to be purified in one injection. To maintain lot integrity, the 9.7 kg was purified using two injections. The final purification conditions were

    Column: 152.6 mm i.d. × 1000 mm stainless steel column
    Silica gel: E. Merck, 9.5 kg, 40 μm
    Flow rate: 700 mL/min
    Equilibration: 18 L of 50/50 (v/v) ethyl acetate/heptane
    Sample size: 6.5 kg dissolved in 3.5 L of 50/50 (v/v) ethyl acetate/heptane
    Sample load: 684 mg sample per gram silica gel
    Elution: 72 L of 50/50 ethyl acetate/heptane

In total, 26 kg of compound A was purified using four injections. During the purification the level of compound B was reduced from 9% to <0.1%.

B. One Hundred Sixty Kilogram Campaign

As the project moved through development, additional quantities of API were needed. To meet chemical demand, 160 kg of compound A had to be purified. At this point additional process development work was performed with a goal of developing a cost-effective process for future chemical needs and ultimately for manufacturing. One of the goals of this work was to develop a single-solvent mobile phase for purification. A single solvent would be more cost-effective at manufacturing scale, would be easier to operate and control, and would eliminate the need for gradient formation, leading to a more robust process.

1. TLC Method Development

Additional TLC method development using single-solvent mobile phases was performed. The results are summarized in Table 7. Evaluation of these results showed three solvents (ethyl acetate, methyl *t*-butyl ether, and acetonitrile)

Table 7 Thin Layer Method Development for Separation of Compounds A and B

Solvent system	R_f		k'		Alpha
	Compound A	Compound B	Compound A	Compound B	
Ethyl acetate	0.30	0.02	2.33	49	21.0
n-Butyl acetate	0.20	0.03	4	32.3	8.08
Acetone	0.65	0.25	0.54	3	5.55
Methyl ethyl ketone	0.55	0.15	0.82	5.67	6.91
Methyl n-propyl ketone	0.45	0.10	1.22	9	7.38
Methyl isobutyl ketone	0.35	0.06	1.86	15.6	8.39
Methyl isoamyl ketone	0.35	0.05	1.86	19	10.22
Isobutyl alcohol	0.47	0.33	1.13	2.03	1.80
t-Amyl alcohol	0.42	0.18	1.38	4.56	3.30
Methyl t-butyl ether	0.22	0.01	3.55	99	27.89
Tetrahydrofuran	0.55	0.25	0.82	3	3.66
Dioxane	0.60	0.25	0.67	3	3.48
Methylene chloride	0.00	0.00	N/A	N/A	0
Acetonitrile	0.50	0.06	1	15.67	15.67

TLC plates: E. Merck silica gel.

Purification of a Synthetic Pharmaceutical Intermediate

Table 8 Effect of Mobile Phase on Purification of Compound A

Mobile phase	% Compound B in purified fraction
Acetonitrile	0.9
Methyl *t*-butyl ether	0.1

Preparative conditions: E. Merck silica gel; sample load, 1000 mg sample/g silica gel; flow rate, 3 mL/min.

that gave good separation of compounds A and B, had acceptable retention, and exhibited good solubility of the reaction mixture. Additional factors to be considered at this scale were (1) availability of solvent, (2) cost of solvent, and (3) toxicity of solvent. Ethyl acetate was eliminated based on work done for the 26 kg campaign (Table 5).

2. Development of the Chromatographic Process

Pilot experiments were performed using methyl *t*-butyl ether and acetonitrile. These results are summarized in Table 8. These experiments showed that methyl *t*-butyl ether produced higher purity material than an acetonitrile mobile phase. The final parameter to be investigated prior to processing was the effect of flow rate using methyl *t*-butyl ether as the mobile phase. The chromatographic system to be used for large-scale processing had a maximum flow rate of 3 L/min. Flow rates equivalent to 1.7 and 2.5 L/min on a 23 cm i.d. column were investigated. These results are in Table 9. Product of acceptable purity was obtained at both flow rates. At the higher flow rate, product

Table 9 Effect of Flow Rate on Purification of Compound A

Flow rate (mL/min)	% Compound B in purified fraction
3.0	0.1
4.5	0.4

Preparative conditions: E. Merck silica gel; sample load, 1000 mg sample/g silica gel; mobile phase, methyl *t*-butyl ether; column, 10 mm i.d. × 60 mm.

purity was close to unacceptable levels. To maintain a more robust process, the lower flow rate was selected.

3. Scale-Up to 23 cm i.d. Column

The separation was scaled to a 23.6 cm i.d. × 180 cm column. The flow rate for the separation was set at 1700 mL/min using Eq. (2). Using Eq. (3), a load of 2.8 g on a 10 × 60 mm column corresponds to a load of 47.0 kg. The amount of chemical to be purified was 160 kg. To maintain lot integrity the 160 kg was purified using four injections. The final purification conditions were

> Column: 236.5 mm i.d. × 1800 mm stainless steel column
> Silica gel: E. Merck, 40 kg, 40 µm
> Flow rate: 1700 mL/min
> Equilibration: 40 L of methyl t-butyl ether
> Sample size: 40 kg dissolved in 20 L of methyl t-butyl ether
> Sample load: 1000 mg sample per gram silica gel
> Elution: 320 L of methyl t-butyl ether

In total, 160 kg of compound A was purified using four injections. During the purification the level of compound B was reduced from 10% to <0.1%.

IV. CONCLUSIONS

Preparative chromatography is a cost-effective purification technique. The use of TLC to screen stationary and mobile phases is fast and cost-effective. TLC methods can be efficiently scaled up to preparative chromatographic methods; good results are obtained during these scale-ups. The use of TLC coupled with pilot columns gives the chromatographer a powerful tool to quickly screen and optimize methods for large-scale purifications.

REFERENCES

1. AJ Mical, MA Wuonola. Prog Pharm Biomed Anal 1:26, 1994.
2. RM Ladd, A Taylor. LC-GC 7(7):584, 1989.
3. JG Turcotte, PE Pivarnik, SS Shirali, HK Singh, RK Sehgal, D Macbride, NI Jang, PR Brown. J Chromatogr 499:55, 1990.
4. L Miller, D Honda, R Fronek, K Howe. J Chromatogr 658(2):429, 1994.

5. R Dixon, R Evans, T Crews. J Liq Chromatogr 7(1):177, 1984.
6. L Weidolf, KJ Hoffman, S Carlsson, KO Borg. Swed Acta Pharm Suec 21(4): 209, 1984.
7. J Adamovics, S Unger. J Liq Chromatogr 9(1):141, 1986.
8. RG Bell. J Chromatogr 590:163, 1992.
9. P Painuly, CM Grill. J Chromatogr 590:139, 1992.
10. JV Amari, PR Brown, PE Pivarnik, RK Sehgal, JG Turcotte. J Chromatogr 590: 153, 1990.
11. BA Bidlingmeyer, ed. Preparative Liquid Chromatography. Amsterdam: Elsevier, 1987, p 46.
12. M Verzele, C Dewale. Preparative High Performance Liquid Chromatography— A Practical Guideline. Ghent: TEC, 1986, p 68.
13. PC Rahn, M Woodman, W Beverung, A Heckendorf. Preparative Liquid Chromatography and Its Relationship to Thin Layer Chromatography. Technical Literature. Waters Corp., Milford, MA, 1979.
14. L Miller, R Bergeron. J Chromatogr A 658:489, 1994.

11
Case Study: Development of Chromatographic Separation to Remove Hydrophobic Impurities in Alatrofloxacin

Sam Guhan and Mark Guinn
Pfizer Global Research and Development, Groton, Connecticut, U.S.A.

I. BACKGROUND

Alatrofloxacin mesylate is the L-alanyl-L-alanyl prodrug of trovafloxacin mesylate (the structure is shown in Fig. 1). It is intended for administration by intravenous infusion. Following intravenous administration, the alanine substituents in alatrofloxacin are rapidly hydrolyzed in vivo to yield trovafloxacin.

Trovafloxacin is a fluoronaphthyridone related to the fluoroquinolones with in vitro activity against a wide range of gram-negative and gram-positive aerobic and anaerobic microorganisms. The bactericidal action of trovafloxacin results from inhibition of DNA gyrase and topoisomerase IV. DNA gyrase is an essential enzyme that is involved in the replication, transcription, and repair of bacterial DNA. Topoisomerase IV is an enzyme known to play a key role in the partitioning of the chromosomal DNA during bacterial cell division. The mechanism of action of fluoroquinolones including trovafloxacin is different from that of penicillins, cephalosporins, aminoglycosides, macrolides, and tetracyclines. Therefore, fluoroquinolones may be active against pathogens that are resistant to these antibiotics.

Parenteral compositions are more affected by trace contaminants than other pharmaceutical compositions. This is so because parenteral composi-

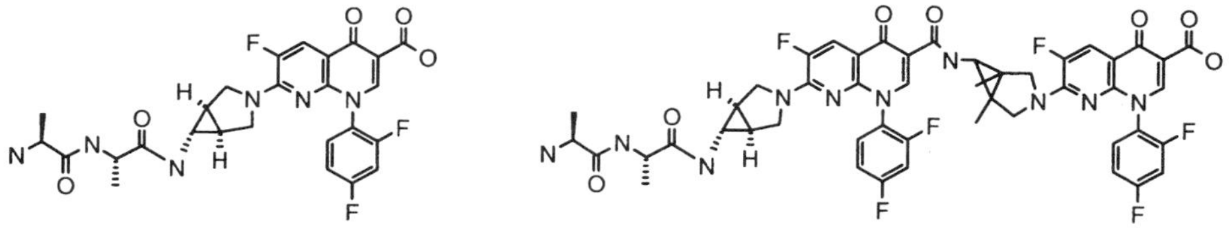

Figure 1 Structure of (left) alatrofloxacin and (right) dimer impurity.

tions, whether diluted or reconstituted, are liquid dosage forms whose quality standards are evident upon visual inspection. Visual inspection of parenteral compositions can detect less polar impurities as low as 10 ppm. The presence of visible impurities affects the public's confidence in the safety, stability, and use of a product. It is therefore important to remove as many less polar impurities as possible from parenteral compositions, because less polar impurities are generally less water-soluble and more likely to precipitate from the formulation. Furthermore, most industrial nations require that parenteral compositions under consideration for drug approval meet minimum standards of purity and absence of particulate contamination. The minimum standard in the United States is that the parenteral composition "be essentially free from particles that can be observed on visual inspection."[i] During the development of the intravenous formulations for alatrofloxacin, a problem was seen with development of turbidity, especially when the formulations were stored for several weeks. Simple crystallizations or solvent extractions could not solve the turbidity problem. It was previously reported that the quinolone antibiotic ciprofloxacin also possesses impurities that render the product unsuitable for parenteral formulation. European patent publication 287,926 describes the purification of ciprofloxacin by treatment with diatomaceous earth to yield a product that is suitable for parenteral formulation. This treatment did not yield a product of alatrofloxacin that was amenable to IV formulation; consequently, we were asked to develop a robust separation technique that would significantly reduce the hydrophobic impurities in the alatrofloxacin so that the bulk drug substance could be formulated into parenteral formulations.

II. METHOD DEVELOPMENT

The intravenous formulation of alatrofloxacin was seen on occasion to form a slight haze. A small amount of the haze was isolated, and analysis indicated the presence of a pseudodimer impurity (structure shown in Fig. 1 along with

Removal of Hydrophobic Impurities

that of alatrofloxacin), along with drug product, silicone oil, etc. Because it was suspected that this dimer might be involved in the haze formation and was not easily purged by crystallization, it was decided to determine the feasibility of using preparative chromatography for removal of the dimer. Typical chromatograms of crude alatrofloxacin bulk made by two different synthetic routes are shown in Fig. 2. As can be seen in the figure, there are a large number of impurities that elute after the main band (alatrofloxacin elutes at 1–2 min, and dimer elutes at about 10 min). The majority of this work was done with feed similar in quality to that shown in Fig. 2a, because that was considered to be the worst-case scenario for dimer levels.

Given the separation of the drug and the dimer on a reversed-phase HPLC system, it was decided to examine various low pressure hydrophobic resins for their ability to purge the dimer.

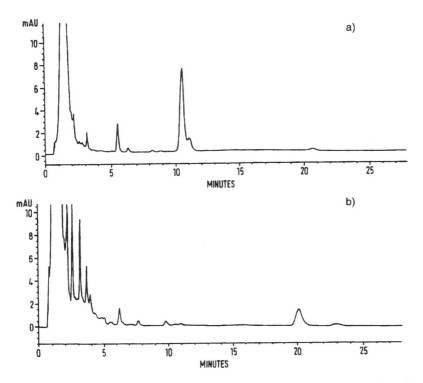

Figure 2 Crude alatrofloxacin from two different synthetic procedures. Alatrofloxacin elutes in the 1–2 min range, and the dimer impurity elutes at about 10 min.

Table 1 Summary of Resin Physical Properties

Resin	Material	Particle size (μm)	Pore size (Å)	Surface area (m²/g)
HP-20	Styrenic	250–600	300–600	500
CG-161md	Styrenic	50–100	110–175	900
XAD-16	Styrenic	300–1200	200–250	800
HP-21	Styrenic	250–600	120–200	600
CG-71md	Methacrylic	50–100	200–300	500
CG-300md	Styrenic	50–100	~300	700

A. Resin Screening

For the initial screening of resins, batch experiments were performed. The list of resins tested in the initial screen along with their physical properties is shown in Table 1. A solution of 70 mg/mL of alatrofloxacin was made in either water or 0.05% aqueous methanesulfonic acid (MSA). The resins were added to the solutions (the ratio of resin to alatrofloxacin is shown in Table 2) and stirred using a magnetic stir bar. At 2, 4, 6, and 24 h a sample was removed and filtered and assayed. The results from the batch experiments are summarized in Table 2. As can be seen in the table, the greatest reduction in the dimer content was seen with the HP-20, CG-300, and CG-161 resins. It was decided to pursue the CG-161 and HP-20 resins because the reduction in the dimer was greatest for these two resins and they offered a reasonable range in physical properties. CG-300 had about the same reduction as HP-20, but because its particle size and manufacture were the same as those of CG-161 and the performance was poorer, it was decided to pursue HP-20. The kinetics of the process are shown in Fig. 3. As can be seen in the figure, the adsorption was significantly faster for the CG-161 resin than for the HP-20 resin.

B. Development of Preparative Separation

One of the main purposes of the initial laboratory study was to produce representative material so that it could be tested in the formulations to confirm that the haze did not develop. Other goals were to identify the better resin for this separation between HP-20 and CG-161 and to optimize both a batch process and a column process so that they could be compared. Both resins are hydrophobic, polystyrene divinylbenzene–based low pressure resins suitable for production scale operations. The advantages of the HP-20 are its larger particle size and thus easier batch handling along with lower cost (about 7–10 times

Removal of Hydrophobic Impurities

Table 2 Summary of Batch Results

Resin	Solution solvent	Ratio of drug to resin	ppm of dimer[a] before treatment	ppm of dimer[a] after resin treatment	Reduction in dimer in solution[b]	Percent of alatrofloxacin remaining in solution
HP-20	Water	1:1	700	21.2	39-fold	85
HP-20	0.05% MSA	1:1	700	15.0	55-fold	85
HP-20	Water	1:1	700	27.7	30-fold	85
HP-20	0.05% MSA	1:1	700	11.4	70-fold	85
HP-20	Water	2:1	700	ND	30-fold	ND
HP-20	0.05% MSA	4:1	700	ND	13-fold	ND
HP-20	0.05% MSA	1:1	7000	239	34-fold	85
CG-161	Water	1:1	700	6.21	132-fold	85
CG-161	0.05% MSA	1:1	700	6.18	133-fold	85
CG-161	0.05% MSA	2:1	700	ND	31-fold	ND
CG-161	0.05% MSA	4:1	700	ND	18-fold	ND
CG-161	0.05% MSA	1:1	7000	11.1	77-fold	82
CG-161	0.05% MSA	1:1	700	15.3	54-fold	85
CG-300	Water	2:1	700	ND	26-fold	ND
CG-300	0.05% MSA	4:1	700	ND	11-fold	ND
CG-300	0.05% MSA	1:1	700	ND	10-fold	ND
XAD-16	Water	1:1	700	ND	14-fold	ND
XAD-16	0.05% MSA	1:1	700	ND	14-fold	ND
HP-21	Water	1:1	700	ND	14-fold	ND
HP-21	0.05% MSA	1:1	700	ND	12-fold	ND
CG-71	0.05% MSA	1:1	700	ND	13-fold	ND

[a] ppm of dimer is defined relative to alatrofloxacin (μg dimer/g alatrofloxacin).
[b] Reduction of dimer in solution is defined as concentration of dimer in the feed divided by the concentration of the dimer in solution at the end of the experiment (it does not take into account the yield of alatrofloxacin).

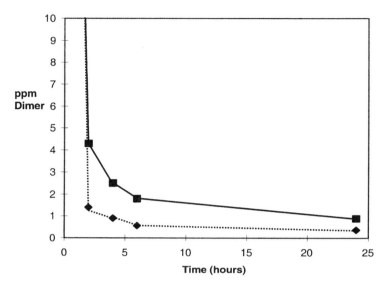

Figure 3 Batch resin kinetics. Alatrofloxacin concentration is 70 mg/mL; dimer is present at 700 ppm with respect to alatrofloxacin. Data taken at room temperature. ♦ Data for CG-161; ■ data for HP-20.

lower than CG-161). The advantages of CG-161 are better purge of dimer impurity and faster kinetics (hence it will probably work better in column mode).

1. Scale-Up Using HP-20 Resin in Batch Mode

Based on the resin screening results, a 1:1 resin-to-drug ratio was chosen along with 0.05 methanesulfonic acid solution. Fifty grams of alatrofloxacin bulk containing about 0.07% dimer (compared to drug) was dissolved in 714 mL of a 0.05% methanesulfonic acid solution and stirred for 24 h with 50 g of HP-20 adsorbent resin. The resulting effluent was assayed to contain 18.7 ppm dimer (compared to drug). The yield across this step was >80%. This material was then isolated in the zwitterionic form, then crystallized as the methanesulfonic acid salt, at which point the dimer levels dropped to <1 ppm in the finished goods.

2. Scale-Up Using HP-20 Resin in Column Mode

In order to determine the performance in the column mode, a 15 mL column was packed with 10 g of HP-20 adsorbent. Twenty grams of alatrofloxacin

Removal of Hydrophobic Impurities

containing 0.07% dimer was dissolved in ~280 mL of 0.05% aqueous MSA. The column was then loaded at 5 bed volumes (BV) per hour, and line samples were withdrawn periodically to determine the breakthrough. Breakthrough was observed after <100 mL. To compensate for the slow kinetics, another column run was made in which the flow rate was decreased to as low as 0.5 BV/h. However, breakthrough of the dimer (~20 ppm) was still seen after only 150 mL of feed had passed over the column. Thus, the HP-20 resin in the column mode was judged not suitable for dimer reduction because of poor dynamic loading performance.

3. Scale-Up Using CG-161 Resin in Column Mode

Because CG-161 had a much smaller particle size, it was decided to test it only in the column mode. As was done for the HP-20, a 15 mL column was packed with 10 g of CG-161 resin. About 50 g of feed was dissolved in 714 mL of water and pumped over this column at a 5–6 BV/h flow rate. Line samples of the column effluent were taken periodically and monitored for dimer. No breakthrough of the dimer was seen after all the feed and a 60 mL water wash had been pumped over the column. The effluent and wash were then isolated in the same fashion as the HP-20-treated solution. The final material was assayed to contain <1 ppm dimer. The yield across the resin step was >95%. The dynamic loading performance of CG-161 was much better than that of the HP-20 for dimer reduction and was chosen for further scale-up work.

4. Summary of Batch and Column Resin Experiments

Owing to its better purge of dimer, greater product yield, faster kinetics, and possible reuse of resin, the CG-161 in the column mode was chosen as the optimum resin for dimer removal from alatrofloxacin bulk. Chromatograms of the purified alatrofloxacin are shown in Fig. 4. The chromatograms in Figs. 4a and 4b correspond to the feed samples in Figs. 1a and 1b, respectively. The approximate concentrations of various feed impurities are given in Table 3. Bulk lots produced from both the batch resin treatment and the column treatment did not show the haze in the intravenous formulations.

C. Development of Cleaning Protocol for Resin Column

To optimize process economics, the resin column needed to be regenerated and used for more than one batch. Accordingly, different solutions were investigated for their ability to regenerate the column after one pass. Initial testing with pure methanol and acetonitrile produced very broad clean-up peaks, pos-

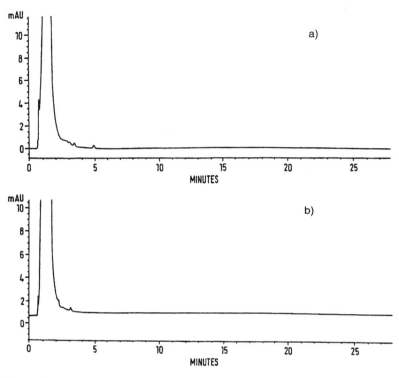

Figure 4 Chromatograms of purified alatrofloxacin. Chromatograms (a) and (b) correspond to chromatograms (a) and (b), respectively, of Fig. 1.

sibly due to the relatively poor solubility of the mesylate salts of the alatrofloxacin derivatives in the pure solvents. Therefore, mixtures of small amounts of water (with acid and base modifiers) in methanol and acetonitrile were investigated. The optimum regeneration solution was found to be 90/10 ACN/water with 0.05% MSA. After 10–15 BV of this solution, the amount of dimer in the effluent was <1 ppm. Lab experiments indicated that the resin column could be regenerated and reused for at least three to five cycles without breakthrough of dimer in the product.

D. Use of Alternative Feeds

1. Dimer Purge from Crude, Precrystallized Feed

All the previous work was done on samples of crystalline bulk. If the resin purification had adequate robustness, ideally the crystallization step could be

Removal of Hydrophobic Impurities

Table 3 Concentrations of Impurities Pre- and Post-Chromatography

Retention time	Feed lot A		Feed lot B	
	Concentration before resin	Concentration after resin	Concentration before resin	Concentration after resin
2.1	28	ND	214	ND
2.5	8	ND	250	1
2.8	5	ND	ND	—
3.1	38	5	197	5
3.6	ND	ND	94	ND
3.9	ND	ND	34	ND
4.8	ND	ND	16	ND
5.0	5	ND	17	ND
5.5	124	13	17	ND
6.2	22	ND	83	ND
7.6	ND	ND	27	ND
8.4	8	ND	ND	ND
9.8	ND	ND	30	ND
10.5	700	ND	8	ND
10.9	97	ND	10	ND
20.1	22	ND	211	ND
23.0	ND	ND	29	ND

ND = not detected.

skipped in the synthesis and the cruder reaction mixture applied directly to the column. However, samples of precrystallized feed contained 1.5–2 times higher levels of dimer. To determine the feasibility of direct treatment of these cruder feed samples, several runs were made on the precrystallized feed material.

These samples were assayed to contain between 0.11% and 0.13% dimer compared to 0.07% or less in the crystallized samples. In addition to the dimer, the crystallization purged other impurities, as evidenced by the fact that the overall purity of these crude samples was found to be 90–93% of the crystallized feed.

Fifty grams of the 0.11% impurity feed was dissolved in 715 mL of water and passed over a 15 mL CG-161 resin column at a flow rate of ~5 BV/h. The dimer level in the effluent product solution was <1 ppm (relative to drug), showing that the column was compatible for use with the higher dimer lots.

2. 7000 ppm Dimer Lots

To determine the bounds of the feed quality, the use of the resin to treat a lot containing 10 times as much impurity was evaluated. A bulk lot containing 7000 ppm (0.7%) dimer (10 times the normal amount) was dissolved in water at a concentration of 14.3 g water per gram of solids and treated using the CG-161 resin. However, product breakthrough was seen almost immediately. In order to compensate for the effects of kinetics, the flow rate was reduced to 0.5 BV/h. Even at these reduced flow rates, breakthrough of ~0.05% dimer in the product effluent was observed.

Pilot Scale Purification of High Dimer Lot. This process was scaled to the pilot plant to reduce dimer level in the high dimer lot. About 20 kg of bulk was treated using 2.8 kg of CG-161 at a flow rate of 0.5 BV/h to reduce the dimer level from 0.7% to 0.07% after the first pass. The column was then packed with fresh resin and equilibrated with water. The product solution after the first pass (containing 0.07% dimer) was passed over this new column at 3–5 BV/h. The dimer level in the effluent solution was only reduced from 0.07% to 0.045%.

It is unknown at this point why these extremely high levels of dimer changed the chromatographic behavior of the alatrofloxacin bulk. Even after reducing the dimer level to that of the crystallized feed (~700 ppm), the dimer could not be purged when treated on a fresh CG-161 column.

Removal of Other Impurities. Because the resin column was able to successfully remove dimer from lots containing ~700 ppm (0.07%) dimer, we investigated the use of this technique to remove other impurities that might be present in the feedstreams. Two other structurally related compounds were chosen such that the first had a hydrophobicity between that of alatrofloxacin and dimer and the second was more hydrophobic than the dimer.

In order to determine the purge of these impurities in the resin process, pure aqueous alatro-floxacin solution was spiked with 1000 ppm of one of the compounds and the feed sample was treated with CG-161 polystyrene resin under the standard column conditions developed for dimer purge.

For each column run, 50 mg of the compound was first dissolved in DMSO at a concentration of 2.5 mg/mL and added to 50 g alatrofloxacin in 715 mL water. Based on HPLC assays of column effluents, both compounds were retained on the resin and purged from the column effluent. There was a very minor breakthrough (<20 ppm with respect to drug) of the less hydrophobic impurity in the effluent. This was not surprising, because this impurity is more polar than the dimer and would be expected to elute earlier on a hydrophobic resin. On the other hand, the more hydrophobic compound was completely purged under the column conditions used, as expected.

Removal of Hydrophobic Impurities

Table 4 Alternative Resin Evaluation

Resin	Particle diameter (μm)	Pore size (Å)	Surface area (m²/g)	Performance
CG-161md	50–100	110–175	900	Benchmark resin. No dimer breakthrough at a loading of 12–15 g feed/g dry resin. Used in all scale-up studies to date.
CG-161cd	80–160	110–175	900	Resin coarser than the benchmark resin. Settles more easily and costs marginally less than CG-161md. No breakthrough of the dimer was seen under the loadings studied.
CG-161sd	20–50	110–175	900	Resin significantly (three times) more expensive than the benchmark resin. It also settles more slowly and will have higher pressure drops. No breakthrough of the dimer seen across this column.
HP-20df	50–100	300–600	500	Breakthrough of the dimer (15–20 ppm with respect to drug) seen across this resin column.
HP-20ss	75–150	300–600	500	No noticeable amount of dimer eluted in the effluent.
HP-20s	100–200	100–180	500	Small breakthrough of dimer (5–10 ppm with respect to drug) seen across this resin column. However, this resin is the least of all the resins in this table.

E. Evaluation of Alternative Resins for Dimer Reduction

Because all of our experimental column work to date had been with the CG-161 (medium grade) resin, we evaluated alternative resins with properties closer to those of CG-161 than those of HP-20 for their ability to purge the dimer from alatrofloxacin bulk streams. The main reasons for the further investigation of resin types were to ensure that the optimum process was chosen and also to have an alternative source of the stationary phase in case the supply of CG-161 was interrupted. The feed used for evaluation was the "crude" precrystallized solids (containing 0.13% dimer). A column of 15 mL was packed with each resin. Forty grams of the feed was added to 572 mL of water, and the feed solution was pumped over the column at ~5 BV/h. Line samples were taken every hour and examined by HPLC for dimer breakthrough. Table 4 details the results from these resin experiments. As can be seen in the table, all six of the resins provided good resolution of the dimer from the alatrofloxacin product. Based on results from the batch experiments (Sect. II, B.1), where ~19 ppm in the effluent purged to below 1 ppm in final product, all six of the resins could be used for processing the bulk.

III. cGMP ASPECTS

Because this resin process was to be used for the production of a pharmaceutical product and hence for manufacture under cGMP conditions, part of the process development effort focused on resin suitability from a regulatory point of view. Discussions with the manufacturer of CG-161 revealed that TosoHaas had done extensive work in evaluating cleaning methodologies for the CG-161 resin and had developed a protocol that made the resin suitable for use in cGMP manufacturing. This protocol had been filed with the FDA as part of a drug master file. This washing protocol was used prior to all cGMP manufacturing.

IV. DEMONSTRATION AT PILOT SCALE

In order to use this process for the manufacture of alatrofloxacin, the process had to be run at pilot scale and the resulting purified product placed on an ICH stability program. To provide adequate scale and material, the CG-161 resin process was scaled in the pilot plant to produce 15–35 kg lots of dimer-free drug. The isolation procedure is detailed in the flow sheet shown in Fig. 5. The equipment used for this process is shown in Fig. 6.

Removal of Hydrophobic Impurities

2 kg of dry CG-161 resin
Pre-treated by swelling in 12% IPO solution
Defined by settling in water and decanting the fines
↓
Resin packed in Amicon VA250x500 column (~7.6 liter column volume)
Column Washed to Remove Extractables
Column conditioned by 10BV water wash
↓
Feed makeup using 14.3 kg water per kg of feed
Feed solution pumped over column at 5 BV/hour flow rate
Effluent collected and line samples assayed by HPLC
5 BV water wash after feed loading
↓
Effluent + Wash assayed for purity and column yield
Column treated material was then precipitated in zwitterionic form
and then the methane sulfonate salt was formed and cystallized.

Figure 5 Flowsheet of alatrofloxacin purification.

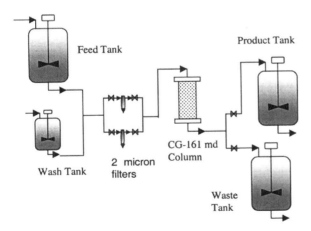

Figure 6 Schematic of column setup.

Table 5 Pilot Scale Lots Produced

Feed amount (kg)	Amount of resin used (kg)	Dimer level in feed (ppm)[a]	Dimer level in product after resin step (ppm)[a]	Yield (%)
15.7	1 (~4 L CV)	600	<1.0	~87
24.55	2 (7.6 L CV)	600 ppm	<1.0	>95
26.1	2 (7.6 L CV)	600 ppm	~1.0	94
31.6	2.5 (9.5 L CV)	600 ppm	<1.0	>95
26.1	2 (7.6 L CV)	600 ppm	~1.0	>94

[a] With respect to drug.
Source: United States Pharmacopeia, R-246, 1994.

A total of five pilot lots of purified drug substance were produced to enable the stability studies required for the NDA submission. A summary of these lots is shown in Table 5.

V. TRANSFER TO PRODUCTION SCALE

The resin process was transferred to Pfizer Global Manufacturing and has been used for the commercial production of alatrofloxacin.

12
Case Study: Process Development of Chromatography Steps for Purification of a Recombinant *E. coli* Expressed Protein

Anurag S. Rathore
Pharmacia Corporation, Chesterfield, Missouri, U.S.A.

I. INTRODUCTION

Discovery and production of recombinant proteins has been on an increasing trend. These proteins are generally obtained from fermentation of microorganisms and the protein of interest is released from the cells. In most cases, the resulting solution contains species such as other bacterial host cell proteins (HCPs), nucleic acids (DNA and RNA), endotoxin, and other host cell impurities [1,2]. These, combined with the other product-related impurities that have very similar physicochemical properties that are present in the process stream, make the task of purification of the target molecule a challenge. However, purification of the protein from all these impurities is necessary not only to reduce the negative impact that their presence in biopharmaceutical therapeutics may cause to the safety of the product, but also to demonstrate good manufacturing procedures (GMPs) to the regulatory agencies [3]. Unless these impurities have been cleared during processing and reduced to acceptable levels (typically ng/mg for HCPs and pg/mg for DNA), the product is unlikely to be acceptable for use in clinical or commercial purposes [1,4].

In view of the above-mentioned complexities, designing a purification process is often a multiphase procedure that involves planning and a careful consideration of the numerous factors that may have an impact on the quality and quantity of the final product [5–11]. This chapter briefly describes the approach that was followed during the development of the chromatographic purification steps for purification of a recombinant *E. coli* expressed protein. Further, it describes in detail the procedures and results of the process development studies that were performed.

The protein of interest is a dual agonist of both the granulocyte colony stimulating factor (G-CSF) and fetal liver tyrosine kinase 3 (flt-3) receptors. It is under development as a component of cancer vaccine protocols to stimulate dendritic cell proliferation. The objective of this work was to develop a robust and efficient purification process that would generate material of adequate purity and quantity.

II. PHASES OF PROCESS DESIGN

This section describes the steps that were undertaken to design a robust and characterized manufacturing process. The approach taken was to start with the general purification scheme based on our prior experiences with the molecule and experience acquired while manufacturing early clinical supplies. The general scheme that was used as the starting point is illustrated in Fig. 1.

The recombinant *E. coli* were grown on media containing salts, trace metals, glucose, and yeast and then induced with nalidixic acid to express the protein subunit. The protein subunit was produced as insoluble inclusion bodies within the cells that were then isolated by repeated homogenization and centrifugation. The protein subunit was solubilized with urea and refolded. The refold solution was filtered and directly loaded onto a cation-exchange (CE) column for removal of endotoxin, recombinant organism cell protein (ROCP), misfolded monomer, and aggregate species of the product. This was followed by an ultrafiltration step for removing salts, concentrating protein, and buffer exchange. Then separation was performed on an anion-exchange (AE) column to remove endotoxin, ROCP, aggregate species, and product-related impurities. This was again followed by a UF step to remove salts and concentrate the protein solution and finally by a 0.2 µm filtration. The resulting bulk protein solution was frozen and stored at or below $-70°C$.

Once the order of unit operations that would be required for processing this product had been decided, a thorough resin screening process was per-

Case Study

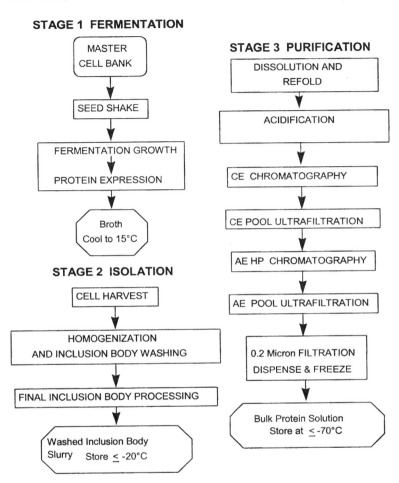

Figure 1 Overall process flow chart. Process development of chromatography steps for purification of a recombinant *E. coli* expressed protein.

formed for the two chromatography steps that are present in this process. This was followed by optimization of operating conditions through a series of process development studies, resulting in an optimized purification process. This chapter describes the procedures utilized and the results obtained during the process development studies that were carried out for this project.

A. Resin Screening

Over the years, quality and safety guidelines for biotechnology products have grown increasingly stringent and have made chromatography one of the most important and most commonly used operations in downstream processing. Although any number of chromatography resins may be suitable for a particular separation, physicochemical differences between the resins can significantly affect their performance for a given problem. Selection of chromatographic resins is therefore a key component of process optimization. An approach to screening resins for chromatographic optimization in the purification process of a protein drug substance was published recently [13]. This approach, illustrated in Fig. 2, utilizes three steps to facilitate a rapid and efficient comparison of different resins. In the first step, studies were performed to examine the

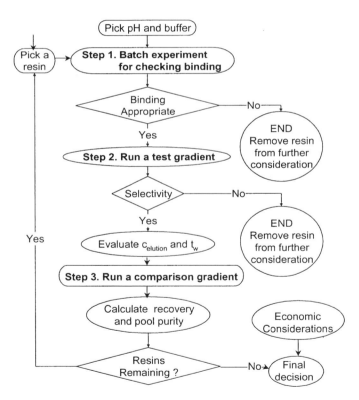

Figure 2 Resin screening protocol. (Reprinted courtesy LCGC North America, Advanstar Communications Inc.)

Case Study

Table 1 Resin Screening for SP Column

Resins[a]	Binding[b]	Selectivity[b]	Recovery[c] (mAU/mL)	Pool purity (%)	Final decision[b]
Resin 1	✓	X			
Resin 2	✓	✓	4.2	87.0	X
Resin 3[d]	✓	✓	N/A	N/A	X
Resin 4	✓	✓	6.9	79.0	✓

[a] Resins 1–4 (not in order) include BioRad Macro S, Whatman S, Pharmacia CM FF, and Pharmacia SP FF.
[b] ✓ = Satisfactory column performance; X = unacceptable column performance.
[c] Based on analysis by cation exchange HPLC (CE-HPLC).
[d] None of the fractions met the pooling criteria.
Source: Reprinted courtesy LCGC North America, Advanstar Communications Inc.

binding between the resin and the target molecule. Next, an identical shallow gradient was performed over all the resins under consideration to generate information on the selectivity and interactions between the product and the resin. The third and final step used the data from preceding steps to provide a fair comparison of the performance of the different resins. This procedure

Table 2 Resin Screening for Q Column

Resins[a]	Binding[b]	Selectivity[b]	Recovery[b,c] (mAU/mL)	Pool purity[c] (%)	Final decision[b]
Resin 5	X				
Resin 6	✓	X			
Resin 7	✓	X			
Resin 8	✓	X			
Resin 9	✓	X			
Resin 10	✓	✓	X		
Resin 11	✓	✓	75.4	96.5	X
Resin 12	✓	✓	25.3	100	X
Resin 13	✓	✓	81.5	97.7	✓

[a] Resins 5–13 (not in order) include BioRad DEAE, BioRad High Q, TosoHaas Q650M, Whatman Q, Whatman QA52, Whatman DE53, Pharmacia DEAE FF, Pharmacia Q FF, and Pharmacia Q HP.
[b] ✓ = Satisfactory column performance; X = unacceptable column performance.
[c] Based on analysis by anion exchange HPLC (AE-HPLC).

was followed for screening resins for both the cation-exchange and anion-exchange chromatography column. Four resins—Pharmacia SP Fast Flow, Pharmacia CM Fast Flow, Whatman S, and Bio-Rad Macro S—were screened for the first column. Nine resins—Pharmacia Q Fast Flow, Pharmacia Q High Performance, Pharmacia DEAE Fast Flow, Whatman QA52, Whatman Q, Whatman DE53, Bio-Rad High Q, Bio-Rad DEAE, and TosoHaas Q650M—were screened for the second column. Two parameters, product recovery and pool purity, were used to evaluate resin performance. Product recovery was defined as the sum of product peak areas (in mAU) in the pooled fractions per milliliter of injected sample. Pool purity was defined as the purity of the total pool formed by combining the fractions that meet the pooling criteria.

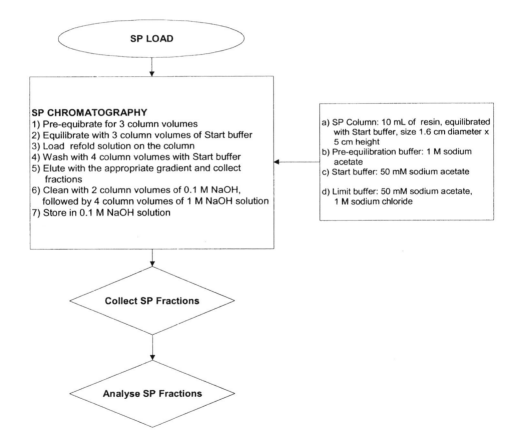

Figure 3 Operating procedure for the SP column.

Case Study

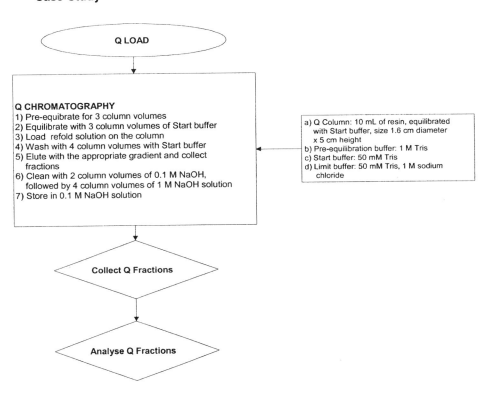

Figure 4 Operating procedure for the Q column.

Whereas the product recovery measures the quantity of product that can be recovered with designated purity, the pool purity describes the quality of the pool. After the product recovery and pool purity had been determined for each resin, these criteria along with economic considerations were utilized to reach a more accurate decision on the choice of the optimal resin. The results are presented in Tables 1 and 2. The optimal resins found for the CE and AE columns were the Pharmacia SP-Sepharose Fast Flow and Pharmacia Q-Sepharose High Performance, respectively.

B. Process Development

After the resin for each column had been chosen, we performed the process development experiments. The objective was to optimize operating parameters for the chromatography steps and also identify the parameters that require further characterization [9–11]. This was accomplished in four steps. First, parameters were identified for each unit operation for performing process development studies.

Second, effects of these parameters on the performance of that particular step were evaluated. Third, data were analyzed to obtain the optimum operating conditions. Fourth, based on the process data that were generated in these studies, a list of parameters was prepared for further characterization.

Experiments were performed to optimize the operating conditions for the two chromatography steps. The parameters that were studied included the pH, protein loading, flow velocity, and gradient slope. These parameters were chosen on the basis of our prior experiences with this and other recombinant protein products and published studies in the literature [5–7,9–12]. The operating procedures for the SP and Q chromatographic steps are illustrated in Figs. 3 and 4, respectively.

III. IN-PROCESS ANALYTICAL METHODS

Several analytical methods were used during development and characterization of the purification process. These included RP-HPLC, CE-HPLC, AE-HPLC, and UV absorbance by measurements at 280 nm (A_{280}).

A. Reversed-Phase HPLC of Reduced Samples

Reversed-phase HPLC of reduced samples gave a quantitative measure of the total amount of product present. The samples were reduced and denatured by treatment with a solubilizing solution (0.4 M DTT, 4% SDS, and 0.8 M Tris, 900 µL sample plus 100 µL solubilizing solution) and analyzed on a Vydac C4 column (#214TP54, 4.6 × 150 mM, 5 µM). A gradient of acetonitrile/water in the presence of 0.1% trifluoroacetic acid (TFA) was used for performing the separation.

B. Cation-Exchange HPLC

Cation-exchange HPLC was used to separate product from various impurities such as endotoxin, ROCP, aggregates, and product-related impurities. The assay used a Dionex ProPac WCX-10 column (4 mm × 250 mm) with 25 mM sodium acetate, with and without 250 mM NaCl, pH 5.25, as the mobile phase.

C. Anion-Exchange HPLC

Anion-exchange HPLC was used to separate product from various impurities such as endotoxin, ROCP, aggregates, and product-related impurities. The separation in this technique was based on charge differences. The method used a TosoHaas TSK-Q5PW column (7.5 mm × 75 mm) with 50 mM Tris buffer, pH 9.0, and bound protein was eluted with a linear NaCl gradient.

D. Ultraviolet Spectroscopy at A_{280}

Because proteins show significant absorbance at 280 nm, their concentration can be estimated on the basis of the UV absorbance at 280 nm in the absence of other A_{280} absorbing species. The extinction coefficient for the product was 0.98 $(mg/mL)^{-1}\,cm^{-1}$. If necessary, samples were diluted with 150 mM Tris, pH 8.5 to give absorbance readings within this range.

IV. RESULTS AND DISCUSSION

A. SP Chromatographic Column Development

As discussed above, the SP column is primarily involved in removal of endotoxin, misfolded monomer, and aggregates of the product. Refold solution was directly loaded on the SP column, and separations were performed following the procedure illustrated in Fig. 3. The resulting fractions were analyzed by CE-HPLC for purity, and the criterion for pooling the fractions was that the purity of a pooled fraction by CE-HPLC should be greater than 75%. The SP pool and the SP load samples were analyzed by RP-HPLC for measuring the quantity of the product, and thus the step yields were calculated.

A summary of results of the process development studies for the SP column is presented in Table 3.

1. *Effect of Buffer pH*

Separation in ion-exchange chromatography is based on interactions between the various charged species and the charged packing particles. Because buffer pH determines the nature and quantity of charge on both, it has a very significant impact on performance of an ion-exchange chromatography step. Experiments were performed at pH 5.0 and 6.5. At pH 6.5, approximately 73% of the protein by A_{280} was in the flow-through collected while loading the column, resulting in a step yield of ~17%. At pH 5.0, the step yield increased sharply to 80–90%, so pH 5.0 was chosen as the pH for this column.

Table 3 Optimization of Chromatographic Conditions for SP Column (Pharmacia SP FF resin)

	Flow velocity (cm/h)			Protein loading, (mg/mL)			Gradient slope (mM NaCl/CV)	
	50/50	100/50	100/100	5	10	20	0–400/20	0–400/10
Recovery, mAU/mL (by CE HPLC)	6.5	6.6	6.5	6.1	6.5	6.9	6.5	6.5
Pool purity, % (by CE HPLC)	80	78	79	80	79	80	79	76

2. Effect of Flow Velocity

Adequate binding of the different feed components to the resin particles requires a minimum residence time of the components in the column. Increasing the flow velocity helps reduce the process time and increase the productivity. However, if the flow velocity is too high, the molecules do not bind well to the resin and the resolution between the different components suffers. This leads to a loss in recovery and/or deterioration in pool purity for the chromatographic column. Experiments were conducted at loading/elution flow velocities of 50/50, 100/50, and 100/100 cm/h. Figure 5 illustrates the separation of the product from the early-eluting and late-eluting peaks (in the CE-HPLC chromatogram). It can be observed that both the recovery and the pool purity remain nearly unchanged over these variations in the loading and elution flow velocities. More experiments were performed to evaluate the performance at 200/100 and 200/200 cm/h. The results (not shown here) indicated a slight loss in resolution and pool purity at 200/200 cm/h. Hence, to minimize the processing time and get optimized separation, the loading and elution flow velocities of 200 cm/h and 100 cm/h, respectively, were chosen as the operating conditions.

3. Effect of Protein Loading

Protein loading is defined as the amount of protein that is loaded on a column per unit volume of the column resin (mg protein/mL resin). The maximum binding capacity of a resin is a function of the operating conditions (buffer conditions, temperature, etc.) as well as the resin and the protein under consideration. The maximum capacity is commonly obtained by loading a protein solution on a packed resin column and watching for the breakthrough (defined

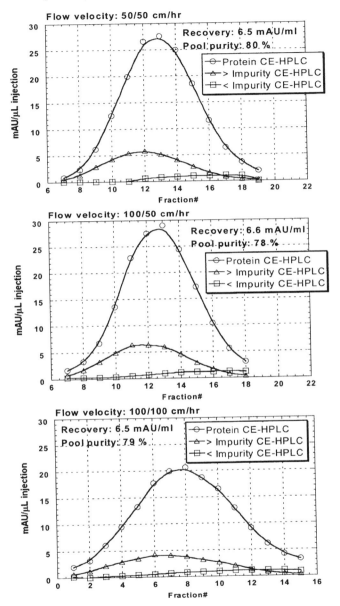

Figure 5 Optimization of SP column: flow velocity. (Gradient: 0–400 mM NaCl in 20 CV, 4 mL fractions.)

as a the point where the flow through from the column contains a certain percent of the protein concentration in the feed). The total amount of protein that has been loaded before breakthrough is divided by the total column volume to yield the maximum loading capacity for that protein on that resin. However, it is customary in large-scale chromatographic separations to operate at 70–90% of the highest protein loading, where the system can operate without compromising the quality of separation. Higher protein loading means that a smaller column will be required to process a fixed volume of feed stream and thus leads to reduced process times and increased productivity. Experiments were conducted at protein loadings of 5, 10, and 20 mg protein/mL resin. The results are illustrated in Fig. 6. It is observed that within the range of variation the protein loading does not seem to have any significant impact on either the recovery or the pool purity. Based on these results, the protein loading of 15 mg protein/mL resin was chosen as the optimum operating condition.

4. Effect of Gradient Slope

The slope of the gradient can potentially impact the resolution between the different components in the feed. A shallower gradient generally offers a better resolution. However, the peaks are broader, and thus the pool is more dilute. Experiments were conducted to compare gradients of 0–400 mM NaCl in the elution buffer over 10 and 20 column volumes. The results are illustrated in Fig. 7 and seem to indicate that there is a slight loss in pool purity for the case of the steeper gradient. As a result, the gradient of 0–400 mM NaCl over 20 CV was chosen as the final operating condition.

5. Effect of Column Length

The separation between the various feed components occurs gradually over the length of the chromatography column. Generally, a minimum bed height is required for adequate separation of the different species. The efficiency of separation, as measured by the number of plates a column offers, increases linearly with the length of the column. So, in an ideal case where dispersion or band broadening are absent, the column length can be increased proportionately with the difficulty of separation (as measured by selectivity), and any separation can be performed. In reality, however, band broadening occurs while the different peaks move across the column and increases linearly with the length of the column. The broadening of the peaks has a deleterious impact on the resolution and also causes dilution of the various components linearly

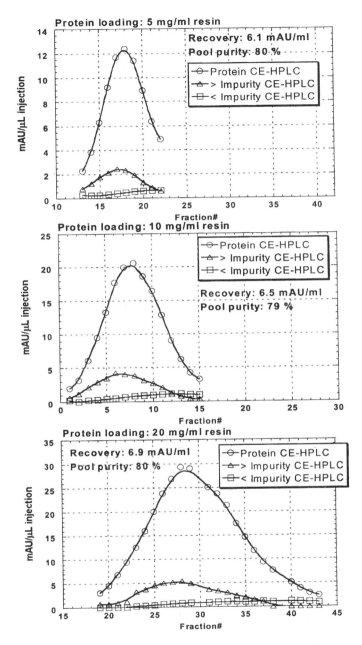

Figure 6 Optimization of SP column: protein loading. (Flow velocity: 100 cm/h. Gradient: 0–400 mM NaCl in 20 CV, 3 mL fractions.)

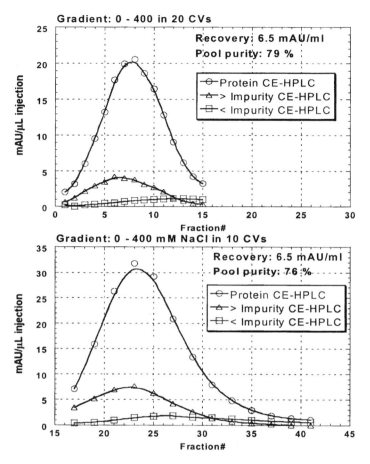

Figure 7 Optimization of SP column: gradient slope. (Flow velocity: 100 cm/h, 4 mL fractions.)

with the length of the column. Due to the presence of these counteracting phenomena during the separation process, there is an optimum column length for each separation. With this concept in mind, experiments were conducted to compare separations with column lengths of 5 and 10 cm. The results are illustrated in Fig. 8 and show an improvement in column performance with a 10 cm long column. Another experiment with a bed height of 20 cm (data not shown here) showed an incremental but smaller improvement in separation

Case Study

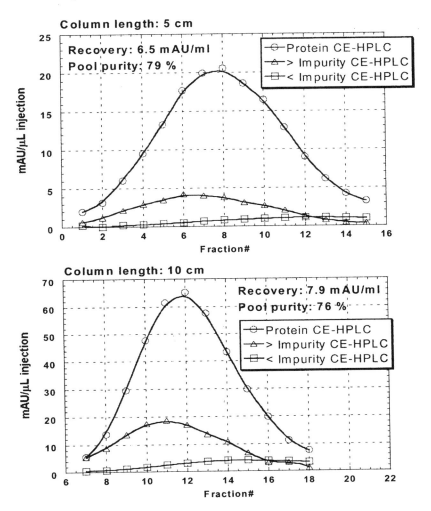

Figure 8 Optimization of SP column: column length. (Flow velocity: 100 cm/h. Gradient: 0–400 mM NaCl in 20 CV, 4 mL fractions.)

quality. As a result, a 20 cm column bed height was chosen for large-scale separations.

B. Q Chromatographic Column Development

As mentioned earlier, the Q column is primarily involved in removal of recombinant organism cell proteins, aggregates, and product-related impurities. Separations were performed following the procedure illustrated in Fig. 4. The resulting fractions were analyzed by CE-HPLC or AE-HPLC for purity, and the criterion for pooling the fractions was that the purity of a pooled fraction should be >85% by CE-HPLC or >95% by AE-HPLC. The Q pool and Q load samples were analyzed by RP-HPLC for measuring the quantity of the product, and thus the step yields were calculated.

A summary of results of the process development studies for the Q column is presented in Table 4.

1. Effect of Buffer pH

Experiments were performed at pH 8.0, 8.5, and 9.0. It was observed that the selectivity between the various components decreased at pH 8.0 in comparison to pH 8.5. Also, at pH 9.0 some precipitation of the product was observed, indicating that the protein was not stable at this pH. Based on these results, pH 8.5 was chosen as the pH for operating the Q column.

Table 4 Optimization of Chromatographic Conditions for Q Column (Pharmacia Q HP resin)

	Flow velocity (cm/h)			Protein loading (mg/mL)		Gradient (mM NaCl in 11CV)		
	50	100	200[a]	3	10	100–175	80–200	50–220
Recovery, mAU/mL (by CE HPLC)	54	49	0	49	44	100[b]	84[b]	71[b]
Pool purity, % (by CE HPLC)	91	87	0	87	87	98[b]	98[b]	98[b]

[a] None of the fractions met the pooling criteria.
[b] Analysis was done by AE-HPLC.

Case Study

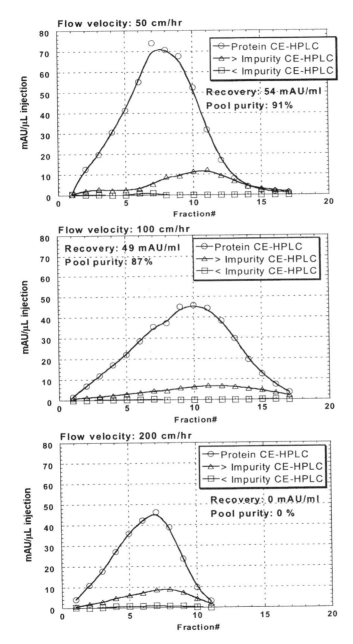

Figure 9 Optimization of Q column: flow velocity. (Gradient 100–175 mM NaCl in 11 CV, 1.5 mL fractions.)

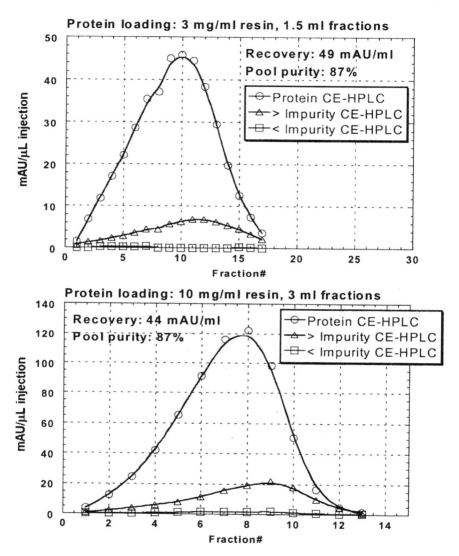

Figure 10 Optimization of Q column: protein loading. (Flow velocity: 100 cm/h. Gradient: 100–175 mM NaCl in 11 CV.)

2. Effect of Flow Velocity

Experiments were conducted at loading and elution flow velocities of 50, 100, and 200 cm/h, and the results are illustrated in Fig. 9. It was observed that unlike the SP column, recovery and pool purity of the Q column show a significant dependence on the flow velocity. This is perhaps due to the smaller bead size (~35 µm) of the Q High Performance resin in comparison to that of the SP Fast Flow resin (~90 µm). The smaller size of the HP resin makes it more critical for the molecules to spend more time on the column in order to properly bind to the column. As can be seen in Fig. 9, as the flow velocity was increased from 50 cm/h to 100 cm/h, the recovery fell from 54 mAu/mL to 49 mAu/mL and pool purity from 91% to 87%. A further increase in the flow velocity to 200 cm/h caused a sharp deterioration in the quality of separation, and none of the fractions met the pooling criteria. To minimize the processing time and get optimal separation, it was decided to perform the separation at loading and elution flow velocities of 100 cm/h.

3. Effect of Protein Loading

Experiments were conducted at protein loadings of 3 and 10 mg protein/mL resin, and the results are illustrated in Fig. 10. It is observed that as the protein loading is increased from 3 to 10 mg/mL, the recovery and pool purity dropped from 49 mAu/mL and 87% to 44 mAu/mL and 67%, respectively. Based on these results, the protein loading of 8 mg protein/mL resin was chosen as the optimum operating condition for operation at large scale.

4. Effect of Gradient Slope

Experiments were conducted to compare gradients of 100–175 mM NaCl over 11 CV, 80–200 mM NaCl over 11 CV, and 50–220 mM NaCl over 11 CV. The results are illustrated in Fig. 11 and indicate that even though the slope does not have a significant impact on the purity of the collected pool, the step yield decreases sharply with the increasing gradient slope. As a result, the gradient of 100–175 mM NaCl over 11 CV was chosen as the final operating condition.

V. CONCLUSIONS

Tables 3 and 4 summarize the results obtained during the process development of the SP and Q columns that are used in this purification process. The result

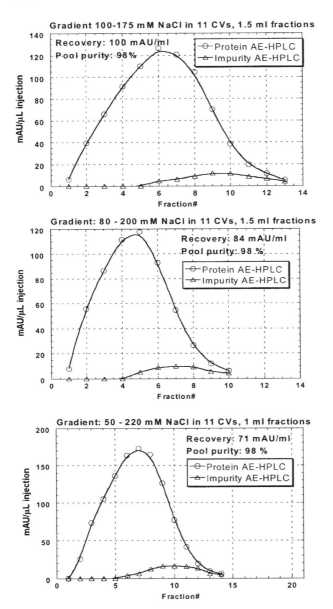

Figure 11 Optimization of Q column: gradient slope. (Flow velocity: 50 cm/h.)

of these studies was an optimized purification process that could manufacture the required amount of material of acceptable quality. Another objective of these studies was to identify parameters that could have a significant impact on the performance of the chromatography step and would require further characterization experiments to improve the robustness of the process. Buffer pH for the SP column and buffer pH, gradient slope, protein loading, and flow velocity for the Q column were identified as such parameters.

ACKNOWLEDGMENTS

I thank G. V. Johnson, R. A. Heeren, Joseph O. Polazzi, and Brian K. Matthews from Pharmacia Corp., Chesterfield, MO, for their help in providing the required analytical support, and M. E. Gustafson, D. E. Steinmeyer, and John D. Ludwig (also Pharmacia) for helpful suggestions.

REFERENCES

1. DO O'Keefe, P DePhillips, ML Will. Identification of an Escherichia coli protein impurity in preparations of a recombinant pharmaceutical. Pharm Res 10:975–979, 1993.
2. MJ Wilson, CL Haggart, SP Gallagher, D Walsh. Removal of tightly bound endotoxin from biological products. J Biotechnol 88:67–75, 2001.
3. J Briggs, PR Panfili. Quantitation of DNA and protein impurities in biopharmaceuticals. Anal Chem 63:850–859, 1991.
4. JE de Oliveira, CRJ Soares, CN Peroni, E Gimbo, IMC Camargo, L Morganti, MH Bellini, R Affonso, RR Arkaten, P Bartolini, MTCP Ribela. High-yield purification of biosynthetic human growth hormone secreted in the Escherichia coli periplasmic space. J Chromatogr A 852:441–450, 1999.
5. G Sofer, L Hagel. Handbook of Process Chromatography—A Guide to Optimization, Scale-up and Validation. New York: Academic Press, 1997, pp 27–113.
6. R Wisniewski, E Boschetti, A Jungbauer. Process design considerations for large-scale chromatography of biomolecules. In: KE Avis, VL Wu, eds. Biotechnology and Biopharmaceutical Manufacturing, Processing, and Preservation. Buffalo Grove, IL: Interpharm, 1996, pp 61–198.
7. G Sofer, C Mason. From R&R to production: designing a chromatographic purification scheme. Bio/Technology 5:239–244, 1987.
8. A Tsai, M Gallo, T Petterson, J Shiloach. Large-scale production and purification of clinical grade Pseudomonas aeruginosa exotoxin A from E. coli. Bioprocess Eng 12:115–118, 1995.

9. RL Fahrner, HV Iyer, GS Blank. The optimal flow rate and column length for maximum production rate of protein A affinity chromatography. Bioprocess Eng 21:287–292, 1999.
10. BH Chung, YJ Choi, SH Yoon, SY Lee, YIJ Lee. Process development for production of recombinant human insulin-like growth factor-I in Escherichia coli. Ind Microbiol Biotechnol 24:94–99, 2000.
11. GNM Ferreira, JMS Cabral, DMF Prazeres. Development of process flow sheets for the purification of supercoiled plasmids for gene therapy applications. Biotechnol Prog 15:725–731, 1999.
12. BD Kelley, P Jennings, R Wright, C Briasco. Demonstrating process robustness for chromatographic purification of a recombinant protein. BioPharm October 1997, pp 36–47.
13. AS Rathore. Resin screening to optimize chromatographic separations. LC-GC June 2001, pp 2–15.

Index

Adsorption isotherms
  linear, 2
  non-linear, 2, 81
Affinity chromatography, scale-up, 26
Alatrofloxacin purification, 303
  cGMP aspects, 314
  column cleaning, 309
  dimer reduction, 314
  method development, 304
  pilot scale demonstration, 312
  resin screening, 306, 313
  reversed-phase chromatography, 306
  scale-up, 308
Antibody fragment purification
  analytical assays, 233
  cation-exchange chromatography, 234
  column cleaning, 249
  column selection, 235
  conductivity, 239
  ionic strength, 245
  optimization, 238, 240
  pH, 238
  recovery process, 232

Batch adsorption
  effect of particle diameter, 57
  effect of column length, 58

Chromatographic method
  development, 3, 125
  optimization, 3, 10, 78, 82
Chromatography, model, 46, 63, 83, 88, 128
Column formats, 211
  adjustable volume pump pack axial flow, 222
  adjustable volume slurry pack axial flow column, 219
  fixed volume axial flow, 212
  fixed volume radial flow, 212
  fixed volume side pack axial flow column, 216
  packing densities, 226
Cycle time, 79, 95

Displacement chromatography, 94, 106
Distribution coefficient, 124

E. coli protein purification, 317
  analytical methods, 324
  column length, 328
  flow velocity, 326, 335
  gradient slope, 328, 335
  ion-exchange chromatography, 320, 332

[E. coli]
  method development, 325
  pH, 325, 332
  process design, 318
  process development, 323
  process flow diagram, 319
  protein loading, 326, 335
  resin screening, 320
Economics
  general considerations, 34
  simulated moving bed chromatography, 197
  size-exclusion chromatography, 267
Expanded bed chromatography
  plate height, 156
  scale-up, 23

Gradient elution, 103, 124, 140

Hen egg proteins separation
  adjustable volume pump pack axial flow, 222
  adjustable volume slurry pack axial flow column, 219
  anion-exchange chromatography, 207
  fixed volume axial flow column, 212
  fixed volume radial flow column, 212
  fixed volume side pack axial flow column, 216
  pressure and flow rate data, 216
  process flow diagram, 208
Hydrophobic interaction chromatography, scale-up, 24

IgG monoclonal antibody purification, 273
  anion-exchange chromatography, 276
  cation-exchange chromatography, 276
  clearance of cellular proteins, 278
  hydroxyapatite chromatography, 277
  pilot scale performance, 282
  polyethylene glycol precipitation, 276
  process flow diagram, 275
  pyrogen removal, 286
  scale-up, 279
  size-exclusion chromatography, 277
  virus removal, 286
Ion-Exchange chromatography
  antibody fragment, 234
  bed stability, 221
  column cleaning, 249
  column formats, 211
  column length, 328
  conductivity, 239
  dispersion, 150
  distribution coefficient, 124
  feed material, 206
  flow velocity, 326
  gradient elution, 140
  gradient slope, 328
  hen egg proteins, 207
  IgG monoclonal antibody, 276
  ionic strength, 138, 245
  mechanism, 123
  method development, 125, 206, 325
  model, 128, 140
  optimization, 238
  packing density, 226
  pH, 134, 238, 325
  plate height, 151
  pressure drops, 158
  productivity, 145
  protein loading, 326
  resin screening, 7, 237, 320
  scale-up, 24, 126, 280
  troubleshooting, 146
Isocratic elution, 98, 124

Light scattering, 255
Loading factor, 79

Mass transfer
  equations, 41
  heuristics, 38
  optimization, 41
  separator design, 39

Index

Membrane chromatography, flow distribution, 65
Mercury porosimetry, 257
Metal (chelate) chromatography, scale-up, 26
Modes of chromatography
 displacement mode, 94, 106
 gradient elution, 103, 124, 140
 interaction, 2, 23
 isocratic elution, 98, 124
 operation, 2, 22
Momentum transfer
 equations, 41
 optimization, 41

Normal phase chromatography
 chromatography method development, 292, 298
 equilibration volume, 296
 flow rate, 295
 mobile phase, 296
 sample load, 295
 scale-up, 296
 silica gel, 294
 solvents, 293
 synthetic pharmaceutical intermediate purification, 289

Optimization
 chromatographic separations, 3, 10, 78, 82, 115
 column length, 93
 cycle time, 95
 ion-exchange chromatography, 125, 238
 mass transfer, 41
 momentum transfer, 41
 simulated moving bed chromatography, 195

Phase ratio considerations, 18
Plasmid DNA purification, 251
 analytical assays, 256
 economic considerations, 267, 269
 light scattering, 255
 pore size of packing, 258
 process flow diagram, 255
 resin screening, 257
 scale-up, 266, 268
 reversed-phase chromatography, 254
 static capacity, 256
 dynamic capacity, 256, 264
Plate height
 contributions, 49, 153
 equation, 49, 80, 150
Pressure drop considerations, 158
Process design, 35, 318
Production rate, 79, 84
Purity, 79
Pyrogen removal, 286

Raw material costs, 267
Resin screening, 4, 237, 258, 306, 320
Reversed-phase chromatography
 alatrofloxacin purification, 303
 column capacity, 256
 column cleaning, 309
 pilot scale demonstration, 312
 plasmid DNA purification, 254
 pore sizes of packing particles, 257
 resin screening, 257, 306, 313
 scale-up, 25, 268

Scale-up
 affinity chromatography, 26
 bed stability, 19, 159, 221
 costing, 21
 expanded bed chromatography, 23
 flow distribution, 20, 60
 fraction collection, 21
 gradient separations, 20
 hydrophobic interaction chromatography, 24
 ion-exchange chromatography, 24, 126, 280
 media availability, 21
 metal (chelate) chromatography, 26
 model, 12, 139

[Scale-up]
  normal-phase chromatography, 296
  packing quality, 20
  practical guidelines, 19
  product loading, 20
  reversed-phase chromatography, 25, 308
  sample pretreatment, 22
  scale-down, 22
  simulated moving bed chromatography, 23
  size-exclusion chromatography, 24
  system design, 21
Separator design, guidelines, 39
Simulated moving bed chromatography
  column switching time, 186
  control, 195
  corrective responses, 194
  economics, 197
  mass transfer, 181
  operating flow rates, 187
  optimization, 196
  robustness, 190
  scale-up, 23
  single unit failures, 190
  theory, 175
  two unit failures, 190
Size-exclusion chromatography, scale-up, 24

Specific production, 80, 107
Stages of purification
  capture, 36, 206
  fractionation, 37
  polishing, 37
Synthetic pharmaceutical intermediate purification, 289
  chromatography method development, 292, 298
  equilibration volume, 296
  flow rate, 295
  mobile phase, 296
  sample load, 295
  scale-up, 296
  silica gel, 294
  solvents, 293

Thin layer chromatography
  method development, 292
  purification of synthetic pharmaceutical intermediate purification, 289

Van Deemter
  equation, 12, 49, 152
  plots, 54
Virus removal, 286

Yield, 79, 83

CIVILIZATIONS OF THE HOLY LAND

The hills stand about Jerusalem: even
so standeth the Lord about his people,
from this time for evermore.

Psalms 125:2

ENDPAPERS
A mosaic of loaves and fishes in the
Byzantine Church of the Multiplication.

ABOVE
A Philistine sarcophagus lid.

OVERLEAF
The wilderness of Sinai.

CIVILIZATIONS OF THE HOLY LAND

Paul Johnson

WEIDENFELD AND NICOLSON
LONDON

TO MY SISTER ELFRIDE

© Paul Johnson 1979
All rights reserved. No part of this publication
may be reproduced, stored in a retrieval system,
or transmitted, in any form, or by any means,
electronic, mechanical, photocopying, recording
or otherwise, without the prior permission of
the copyright owner.

House editor, Miranda Ferguson
Designed by Charles Elton
Picture research by Miranda Ferguson and Marcia Fenwick

ISBN 0 297 77610X

Filmset by Keyspools Ltd, Golborne, Lancs.
Printed in Gt Britain by Morrison & Gibb Ltd, Edinburgh

Contents

1 Canaanites and Phoenicians 7
2 The Civilization of the Bible 31
3 Greeks and Maccabees 87
4 The Age of Herod and Jesus 109
5 Byzantine Christianity 145
6 The Coming of Islam 169
7 Pilgrims, Crusaders and Saracens 185
   Bibliography 216
   Acknowledgments 217
   Index 218

1
Canaanites and Phoenicians

'CIVILIZATIONS OF THE HOLY LAND' is an expression not easily defined in either space or time. By the Holy Land, most of us mean the stretch of Near-Eastern territory, the nucleus of which is modern Palestine or Israel, intimately associated with the great 'Religions of the Book', Judaism, Christianity and Islam. Many of the events crucial to the origin and early development of these three faiths took place outside this geographical nucleus but cannot for that reason be ignored in this account. Equally, not all the cultures which have flourished in this region have been directly linked to the beliefs which, to us, make it holy but they are part of its history nonetheless, and must be brought into the story. The truth is that the history of this corner of the world is extremely complicated and does not easily accommodate itself to the straitjacket of a strictly systematic treatment. In telling it we shall sometimes find ourselves digressing both in chronology and geography before resuming the main thread of our narrative. In short, we shall be closer to the methods of Herodotus than those of Thucydides – with a dash of Pausanias and Strabo thrown in. No matter: what the tale loses in clarity it may gain in colour.

A glance at the map helps to explain why the history of the Holy Land has been so complex. It is small in itself, but fate placed it on the main highway of antiquity. It has always been a part of great events which it has rarely, if ever, been able to dominate. Somewhat unwillingly, and often helplessly, it has been close to the centre of the historical stage and has been exalted and battered by its dramas. Pre-historians are no longer quite so confident as they used to be that the origins of civilization are to be found in the plains formed and irrigated by the three great river systems of the Tigris-Euphrates, the Niles and the Indus. There is accumulating evidence from recently explored sites of innovatory early societies dwelling in the hills and mountains of Anatolia, Baluchistan and elsewhere in this huge region. All the same, the old theory holds good to the extent that we can still truthfully say that Mesopotamia produced the first unmistakable city-

The round stone watchtower of Jericho is Neolithic and dates from 7000 BC. It is part of a ring of fortifications excavated by Dame Kathleen Kenyon.

The clay tablet from Kuyunjik, excavated by George Smith, is inscribed with part of the Babylonian Legend of the Flood, which very closely resembles that in the book of Genesis.

civilization, in the second half of the fourth millennium BC; and that the first civilized State emerged, on the banks and delta of the Nile, about 3100 BC. The river systems were important, even if not all-important. They were areas of orderly government, agricultural wealth and progressive technology from the very beginnings of history, and as such formed the two horns of the Fertile Crescent, a concept of crucial importance in the Chalcolithic and Bronze Ages and which still has some significance today. Syria-Palestine stood at the centre of the crescent, and was therefore both the victim and the beneficiary of the movements of peoples and armies which flowed along it – Assyrians and Babylonians coming from the east and heading south; Egyptians coming from the south-east and heading north. Such major conquerors sought to control it mainly for strategic rather than economic reasons. Its intrinsic wealth was not enormous but it was sufficient to attract successive waves of desert-dwellers, and seaborne people from the rocky islands of the eastern Mediterranean. The history of the Holy Land is the history of ceaseless movement and of the interaction, often violent, of many different peoples.

Of these peoples, the most important, taking the history of the Holy Land as a whole, were to be the Semites. They came originally from the southern end of the Arabian desert, and during the fourth millennium BC the first waves of them spread into Mesopotamia, Syria-Palestine, and even into the Nile Valley and Delta. They mingled with, or displaced, the original Neolithic peoples of Palestine, termed by them variously 'the horrors', 'the ghosts', 'the howling people', 'the long-necked men.' A second wave of Semites, the Amorites, followed in the third millennium BC; and, towards the middle of the second millennium there was a third wave, the Arameans, who included the earliest Hebrew tribes. In addition, during the second millennium BC, large groups of other settlers moved in from the north and north-west, from Anatolia, Crete and the Aegean. These included the Hyksos, the Hittites and the Philistines, the last of whom gave Palestine its name. Hence for most of the Bronze Age and Iron Age the Holy Land was occupied by competing groups of peoples. It was also overshadowed by its bigger neighbours. For large tracts of time during the second millennium, the paramountcy was held by Egypt, either under its native dynasties of the Middle Kingdom and New Kingdom, or under the Hyksos military oligarchy. Between about 1150 and 850 BC, large parts of Syria-Palestine were virtually free from foreign control and it was during this period that both the Hebrew kingdoms and the independent city-states of Phoenicia flourished. Thereafter, the area formed successively part of the Assyrian, Babylonian, Persian, Greek, Roman, Byzantine and Islamic empires. Throughout, the Semites probably remained a majority, though by no means always a large majority, of the inhabitants.

Modern archaeology, however, is now beginning to uncover the pre-Semitic history of the Holy Land, and we must turn to this first. Palestine itself consists of a coastal strip, a mountain spine, and the deep cleft of the Jordan valley-Dead Sea rift, all three regions running north to south. To the east, beyond the rift, is the desert. To the conquerors of antiquity the important part of Palestine was the coastal strip along which they marched. The strip, however, ends well south of modern Haifa, where the mountains reach the sea. Instead, the coastal route turns inland at a pass guarded by the perennial fortress of Megiddo (or Armageddon), crosses the Plain of Esdraelon (or Jezreel), and then fords the Jordan on the way to Damascus, the great oasis on the eastern, or desert, side of the mountain barrier. The point where the coast route turns inland is thus the great hinge of the Fertile Crescent and it is no surprise that Megiddo and its strategic plain have been fought over fiercely from the time of the Egyptian Middle Kingdom at the beginning of the second millennium BC to Allenby's conquest of Palestine in 1918.

However, Megiddo is not the only strategic focus of this region. There is Jerusalem itself, on the spinal centre, 2,500 feet above sea level, thirty-five miles from the coast route on the west, and twenty miles from the bottom of the rift on the east, where the land sinks to 1,290 feet below sea level. There is a north-south route of sorts along the central spine, and Jerusalem controls a strategic gap along this inland route at a point where valley routes open up to both east and west: in effect, a crossroads. The eastern route descends from Jerusalem, some 2,500 feet in a mere fourteen miles, to the crossing of the Jordan where the Allenby Bridge now stands. And controlling this crossing, and the Jordan valley for many miles on either side is the oasis of Jericho. It is another crossroads, for it is not only a key link in the west-east route from Mediterranean to desert, but a stage in the north-south route down the valley bottom.

Jericho is Palestine's eastern gate into the desert. Equally, it is the desert's gate into Palestine. Hence Joshua, about to lead the Israelites into the Promised Land after their forty years in the wilderness, commanded his scouts to 'go view the land and Jericho.' Jericho is on the west side of the mountain walls which form the Jordan valley trough. Part of this west wall is the so-called Mount of Temptation, with its Greek church on top and its monastery halfway down. Opposite, on the eastern side of the trough, are the mountain walls of Moab and Gilead. The landscape is desolate, and must have been desolate even in prehistory, but Jericho is rich in the one resource which, in this part of the world, is essential to fertility: spring water. The trough descends lower as it proceeds south. The Sea of Galilee (otherwise known as the Sea of Tiberius or the Lake of Genessaret) is 650 feet below sea-level. Jericho is over 900 feet below sea-level and further south still, at the top of the Dead Sea, where salinity makes all water undrinkable, the level is 1,292 feet below. But Jericho has, and had throughout all known history, water of its own, a copious spring fed by an underground river which rises in the Judaean hills to the west. This spring produces 1,000 gallons of pure water a minute, and is reliable all the year round even in periods of acute drought.

The spring made Jericho and Jericho gives us our earliest glimpses of civilization, or proto-civilization, in the Holy Land. At Jericho, the excavations of Dame Kathleen Kenyon uncovered a Neolithic settlement which can variously be described as a large village, a town or a city. Its discovery came as a considerable surprise, for the earliest settlements previously investigated had been primitive villages of the fifth millennium BC. Neolithic Jericho went much further back, though quite how far back is still a matter of argument. The first datings using Carbon 14 put Jericho in the seventh millennium BC. This settlement was again carbon-dated in 1960, giving a date of 6850 BC, plus or minus 210 years. Tests in 1963 gave a date of 6935 BC, and more recent tests 7825 BC, plus or minus 110, or 8350 BC plus or minus 200 years. The results seem to vary from one laboratory to another and chiefly emphasize the severe limitations of this method of dating. Unfortunately there is no other in this case. The first inhabitants of Jericho could not make pottery, the key to pre-historical chronology. The great French archaeologist Father R. de Vaux, writing in the latest edition of the *Cambridge Ancient History*, dated the pre-pottery Neolithic settlement of Jericho, first phase, to around 7000 BC, and a figure of 8000 BC is now thought more likely.

Father de Vaux estimates its population at about 2,000. Does it rate as a city? Is Jericho the first example of civilized existence of which we have knowledge? De Vaux does not think so. It engaged in specialized agriculture. It traded, by exchanged bitumen, sulphur and salt for hardstones (for example obsidian) from Anatolia, but it does not seem to have produced a surplus of manufactured goods for trade. Of roughly contemporary sites which have also been uncovered by modern archaeology, Jarmo, in Iraq, was smaller; Catal Huyuk in Anatolia was much bigger and seems to have had a more advanced economy. Pre-pottery

A pre-pottery Neolithic 'B' plastered skull found at Jericho. This shows the beginning of portraiture; the features are modelled in plaster and cowrie shells inserted in place of the eyes.

Hacilar, in Anatolia, and the lowest level of Ras Shamra on the Syrian coast are other examples from the same period. But these were unwalled villages.

What distinguishes Jericho, and what gives it its claim to be the first civilized town, is that it is built on an urban not a rustic scale, and surrounded by massive walls. These defences, found in 1952, are of large undressed stones. They include a formidable round tower nearly thirty feet in diameter, with a stone-built staircase. The quality of the masonry is high, the concept is huge, indeed unique for the eighth to seventh millennia BC. Here, at any rate, is the world's first true example of architecture. Even before the walls were built there were houses. Some of these, with walls of sun-dried unbaked clay, were big, with a series of rooms linked by openings, and grouped around courtyards. The roofs were of reeds bound by mud, the floors of clay with a polished plaster surface, covered by rush mats. The citizens of Neolithic Jericho in its pre-pottery stage ate out of carved limestone dishes and bowls and used weapons and tools of flint. They had a religion and an art of sorts. One house had a family chapel, a small shrine-room with a cult stone of volcanic rock. Another, larger room may have been a communal temple. Miss Kenyon also found plastered and decorated skulls which she thinks belonged to ancestors whom these primitive town-dwellers wished to honour.

Jericho was probably not the only Neolithic city of the area. The walls are silent witness to competitors, probably urban ones. There is evidence that the original inhabitants, whom Miss Kenyon calls the hogback-brick people, were conquered by men from another civilized site whom she terms the plaster-floor people. There was, in fact, a form of urban society in existence, perhaps over quite a large area. We have here, possibly, the first example of a 'take-off' in history, brought into existence not only by human ingenuity but by a benevolent change in climate. The European ice-ages ended about 10,000 BC. Their equivalent in the Near East, the various pluvials, made possible phases of agricultural expansion. The last pluvial, coinciding with growing agricultural knowledge, could have produced a period of prosperity sufficient to sustain urban settlements of this kind. But the basis of the economy was narrow and fragile: two or three crops, a very marginal trade. Even much later, in the second millennium BC, the extinction of the civilized centres of Mycenaean Greece shows how easily over-specialization invited nemesis. Neolithic Jericho grew too big and grand for its own good, or even survival. As the last pluvial receded, increasing desiccation destroyed the surpluses which made urban Jericho viable. An early Dark Age supervened.

The extinction of this pre-pottery Neolithic town was followed by a complete break. Then new inhabitants arrived. They came with a fully-developed pottery culture. But in other respects there was regression. This fifth-millennium culture was simply a village one: no fine houses, no city wall. A primitive clay triad of man, woman and child, is the only evidence of religion. At the stages of these later levels, the *tell* of Jericho ceases to be unique and exotic – a Neolithic urban miracle out of its time – but it remains in miniature a history of the ebb and flow of Palestinian culture and the movements of peoples. First there are two Neolithic phases involving the use of pottery. Then in the fourth millennium came a long gap during the Chalcolithic (copper-and-stone) period when Jericho was unoccupied. Towards the end of the fourth millennium there was a big influx of Semitic nomads who gradually became urbanized as they moved into Palestine. There followed the Early Bronze Age (2900–2300 BC), in which a form of civilized town life again developed and spread over most of the territory; many of the most celebrated Biblical cities date their foundation from this time. Egypt and Mesopotamia already enshrined mature civilizations, with State unity in Egypt, law codes in Mesopotamia – and later the first empires – and writing in both. Palestine did not yet possess writing but it was coming within the proselytizing orbit of these older civilizations and it was developing its own system of petty city-states.

Their chief problem was security, against each other and against nomads. Like its Neolithic prototype, Early Bronze Age Jericho needed and had massive defences. Miss Kenyon found evidence in one trench she dug into the *tell* that the Early Bronze Age walls had been rebuilt no less than sixteen times. The walls were eroded by rainstorms, earthquakes (as many as four in one century), and enemies using fire. Constant repair was needed. On several occasions Jericho was overwhelmed. At the end of the third millennium BC there was a complete break, a new Dark Age, corresponding to the chaos of the First Intermediate Period in Egypt. There followed, around 1900 BC, the civilization of the Middle Bronze Age, moving in from the coast. For a time Jericho may have fallen under the control of the Hyksos military caste, as evidence of their very characteristic and sophisticated fortifications indicates. But this Middle Bronze Age culture was essentially the Canaanite-Phoenician civilization of the coast, based on a multiplicity of smallish, independent or semi-independent, and in either case strongly-fortified towns. Jericho gives us some precious glimpses of this urban culture, for its Middle Bronze Age tombs contained bits of furniture, mercifully preserved, perhaps (it has been suggested) by escaping gases which killed the micro-organisms which cause decay. The British Museum has thus been able to reconstruct what a room in Jericho looked like around 1600 BC, scientific archaeology providing substantiation for all the details, except the clothes. We cannot learn much from Jericho of the gods these people worshipped, but they evidently had a profound belief in life after death, as one would expect from those living on the periphery of Egyptian culture.

Thereafter the evidence from Jericho becomes less interesting and finally peters out. We turn instead to the coastal cities which are closer to the heart of the various Bronze Age cultures, and especially the Canaanite culture which lasted for the best part of the second millennium BC. In 1928–9, at a place now known as Ras Shamra, French archaeologists began to excavate the ancient city of Ugarit. This city is mentioned in a number of very ancient Near-Eastern texts, notably in

An imaginative reconstruction of how a room in prehistoric Jericho might have looked.

the letters of Eighteenth Dynasty pharaohs (fourteenth century BC) found at Tell El-Amarna on the Nile. The excavations showed it was of great antiquity. The first arrivals on the site were emigrants from the East, who seemed to have brought with them copper tools. About halfway through the third millennium, there was a fresh wave from Anatolia and around 2300 BC the city was destroyed by fire, possibly by the newly-settled Phoenicians. Around the beginning of the second millennium, Egyptian influence can be detected, followed by signs of the Hyksos supremacy (as at Jericho), and in turn the restoration of the Egyptian supremacy in the sixteenth century BC. Ugarit later slipped out of the Egyptian orbit, as did many Palestinian and Syrian city-states in the fourteenth and thirteenth centuries BC, and in the famous battle of Kadesh, fought by Ramesses II (Ozymandias) in 1286–5 BC, it sided with the Hittites. About 1200 BC Ugarit was destroyed by the 'Peoples of the Sea', the great coalition of barbarous sea-raiders which harried all the civilized states of the Eastern Mediterranean at this time and effectively destroyed the international culture of the Late Bronze Age. Like the Hittite Empire and the cities of Mycenaean Greece and Minoan Crete, Ugarit never really recovered from this assault.

What is interesting about the remains of Ugarit is that they give us perhaps our most revealing glimpse of Canaanite culture at its most flourishing. The term

The 'Weld-Blundell Prism' was discovered in Larsa, Iraq, and is inscribed in cuneiform script with the Sumerian King Lists.

Canaan derives from the fine purple dye, made from boiled snails, which is peculiar to this coastal strip, and which was exported all over the ancient world. It applies essentially to a culture, rather than a special ethnic group; indeed it was a product of the merger of the existing Early Bronze Age civilization with the Semitic intrusion of Amorite semi-nomads, who flocked in from the hills and deserts towards the end of the Early Bronze Age. The culture was shared by a medley of peoples, particularly on the coast but also in towns well inland. We pinpoint its development and phases from the evidence of Egyptian pottery found on Syrian-Palestine sites, which can be dated very accurately, and which links Canaanite history to the absolute dates of Egyptian dynastic chronology.

At Ugarit there survive a number of buildings from the fifteenth to fourteenth centuries BC. There are temples to Dagon and Baal; a house (and library) evidently belonging to a High Priest; a fortified palace, with administrative offices and courtyards; and a residential quarter. The palace and some of the houses have tomb-cellars lying beneath them, but these were rifled of their valuables in deep antiquity. Among things found are two gold vessels with hunting scenes. Much more valuable and exciting finds, however, were inscribed tables in the office of the palace and the High Priest's library. These writings are in alphabetic cuneiform – that is, written with a triangular wedge-shaped stick on wet clay, but using individual signs for each letter of the alphabet.

The introduction of alphabetic writing was the great achievement of Canaanite civilization. The earliest scripts, of which Sumerian and Akkadian were the most important, were based on pictograms which gradually developed phonetic qualities. But signs continued to represent whole words or syllables rather than individual letters. Egyptian hieroglyphs, another early writing system, remained a combination of pictograms and syllograms, though it also developed individual phonetic letters. In 1905, Sir Flinders Petrie discovered in the ancient mines of Sinai fragments of alphabetical writings which appear to have been developed from Egyptian scripts. A Semitic genius, working in or near the mines, seems to have seized on the alphabetic element in the Egyptian script and expanded it into a full alphabet for translating the sounds of his Semitic speech into signs. This development took place about 1500 BC and the new system moved slowly up the coastal route; though the script used in Ugarit was cuneiform. The Canaanite script was of the utmost importance for human history. Alphabetic scripts are much easier to learn and employ than the syllabic and combined scripts hitherto in use. They could be acquired by large numbers of people outside the narrow circle of the scribal class, and thus tended to make it easier for society to emancipate itself from the intellectual domination of court and temple. As used by the Phoenician traders and merchants, the alphabet was spread over large parts of the Near East and Mediterranean. The Canaanites used an underlying Semitic tongue, whose grammatical and syntactical structure make it unnecessary to indicate vowels by signs when it is written. The earliest alphabets therefore were solely of consonants. When the Canaanite alphabet was made the basis for written Hebrew, also a Semitic language, vowel signs were likewise considered unnecessary. But when the alphabet spread west and north into the non-Semitic Greek-speaking world, vowel-signs were needed. So the Greeks, in adopting the Phoenician-Canaanite alphabet, pounced on six signs for consonants which were alien to their tongue and transformed them into vowel-signs. This arrangement became the foundation for the first systems of alphabetic Greek, and later of Latin, and is perpetuated in all the written languages based on the European tongues.

One of the first alphabetic inscriptions we possess dates from the thirteenth century and is on the sarcophagus of a King Ahiran of Byblos, now in the Beirut Museum. It is no accident that the earliest alphabet should have been associated with Byblos, for this ancient settlement may well have been the world's first major port. Indeed, Philo, the first century AD Jewish philosopher from Alexandria, thought it 'the oldest city in the world'. It was the entrepôt and exporting centre for wood from the enormous Lebanese cedar forests, which throughout antiquity supplied the finest timber for palaces and temples. The first pharaohs of the United Kingdom of Egypt (which, apart from the sycamore, had virtually no native wood) were big customers, and Byblos was involved in the Egyptian timber-trade as early as the Old Kingdom, that is in the first half of the

Characters used in ancient and modern alphabets

This knife, made from gold and silver and dating from the eighteenth century BC, was found in the Temple of Obelisks at Byblos.

third millennium BC. In Egypt, in fact, all ocean-going vessels were known as 'Byblos Ships.'

The ancient site of Byblos is near a modern village. There is a twelfth-century Crusader castle, and part of the medieval walls. Excavations have revealed that Byblos was exporting copper as early as 3000 BC, and was even then a town of sorts with paved streets. They also uncovered early fortifications dating from the time of the Egyptian Old Kingdom. There are traces, too, of Hyksos-type defences, probably from the seventeenth century BC. Byblos, in fact, is a kind of reflection of Egyptian dynastic history, for it retained its close links with Egypt throughout the third and second millennia – interrupted by periods when Egyptian dynastic power (as under the Hyksos) was in retreat. There was a very early Egyptian-style temple at Byblos, and the inhabitants shared the Egyptians' passionate interest in life after death – the dead were placed in giant jars, and then enclosed in tombs cut into the rock. The people of Byblos had their own Semitic tongue and, as we have seen, used clay tablets inscribed in cuneiform rather than Egyptian scripts written with brush-pens on papyrus, but they were exposed over long periods to the influence of Egyptian literary forms and such Egyptian narratives as the *Story of Sinhue* or the *Tale of Wenamun* (part of which takes place in Byblos) circulated there.

The Egyptian 'connection' with this part of the world lasted intermittently for nearly 3,000 years – for as long as Egypt was a major power. Yet this was not really a case of imperialism. The Egyptians rarely conquered and settled foreign territories, for their all-pervasive religious beliefs made them unwilling to live abroad and they were terrified of dying and being buried away from the Nile. They preferred to exercise control in Canaan through their State trading operations and by maintaining a system of small client-states, petty kings who usually controlled only one important city each. Egypt paid them in gold (it was then the world's largest gold-producer) and sometimes provided small garrisons.

The system worked reasonably well so long as the Egyptian State was itself strong and united. If there was weakness in Egypt, the system tended to collapse. The petty states then acted independently or sought other protectors. This was what happened at the end of the Egyptian Middle Kingdom (nineteenth to eighteenth centuries BC). In the Berlin Museum, and in the museums in Cairo and Brussels, there are fragments of pottery bowls and figurines covered with the names of Asiatic towns and rulers who had severed their connection with Egypt and thereby offended the Egyptian State and its gods. Their names, and slogans cursing them, were written on these pots, which were then smashed in a solemn ritual. The Egyptians believed that by a process of 'sympathetic magic' the breaking of the vessels would destroy the power of these enemies. Thus these 'execration texts' can be regarded as documents on Egyptian foreign policy. One curses: 'The Ruler of Jerusalem, Yaqar-Ammu, and all his retainers, and the Ruler of Jerusalem, Setj-Anu, and all his retainers ... their strong men, their swift runners, their allies and associates ... who may rebel, who may plot, who may fight and who may talk of fighting, in this entire land.'

During the period of domination by the Hyksos, who for a time (the so-called Fifteenth Dynasty) ruled in Egypt also, Canaan was completely subjugated by this military caste. Their characteristic rampart-style defences, traces of which are found on many Syrian and Palestinian sites, as well as in Egypt, may have been designed to counter the battering ram, which was invented early in the second millennium BC. But these defences are virtually the only evidence of their rule: they seem to have been a feudal military aristocracy without a civil culture. The indigenous Canaanite culture continued uninterrupted under their domination. The Theban founders of the Egyptian New Kingdom eventually threw them out of Egypt, and chased them into Canaan, so re-imposing the Egyptian paramountcy.

The Egyptians usually left the local princely families in power. Sometimes they had their own garrison cities. One such was Beth She'an, situated where the valley of Esdraelon (Jesreel) meets the Jordan. At this important site, archaeologists have uncovered eighteen different levels, going back to the fourth millennium. The place, now a village, is famous for many things. There, the Philistines displayed the body of the fallen Saul, as we learn from 1 Samuel 31: 9–10. 'And they cut off his head and stripped off his armour, and sent into the land of the Philistines round about, to publish it in the house of their idols, and among the people. And they put his armour in the house of Ashtaroth: and they fastened his body to the wall of Beth'Shan.' Later it was an Israelite city, and from the time

of Pompey a Graeco-Roman settlement under the name of Scythopolis, a member of the league of ten cities (the Decapolis). Beth She'an is mentioned in the *Execration Texts*, its first historical appearance, and it figures again in the list of cities drawn up by the officials of the Pharaoh Tuthmosis III, shortly after 1468, to mark his conquests in Palestine. There are ample traces of Egyptian military occupation, though the city and its culture remained Canaanite, and no less than six Canaanite temples, dating probably from the thirteenth century BC, have been uncovered.

Much light on the geography and politics of Canaan around this time has been shed by the *Amarna Letters*, discovered in the late nineteenth century at Akhenaten's private city on the Nile. These diplomatic dispatches, from around 1389-1358 BC, are mainly appeals for help to the Pharaoh from his Palestinian and Syrian vassals. They are written in Akkadian cuneiform, the international language of the second millennium, but the scribes were evidently Canaanite, as can be seen from close analysis of the syntax and vocabulary. After examining

CANAANITES AND PHOENICIANS

LEFT The Roman theatre at Beth She'an, second century AD, with the ancient Tell el-Hosn in the background. Beth She'an was a member of the league of ten cities, the Decapolis.

RIGHT Amarna letter no. 61 is a letter from Labaya, prince of Shechem, explaining the capture of a certain town and refuting slanders which had reached the king. Labaya quotes a proverb: 'If ants are smitten, they do not acquiesce, but they bite the hand of the man who smites them to justify the action he has taken.'

them, the American archaeologist W.F. Albright concluded that there must have been a scribal school somewhere in the Megiddo area where scribes who worked at the courts of local rulers could learn the diplomatic jargon needed to correspond with foreign potentates (a fragment of the epic of Gilgamesh, written in a Phoenician hand of the fourteenth century, has been found there).

What the *Amarna Letters* tell us has been variously interpreted. They certainly suggest a period of declining Egyptian power, or inactivity, consistent with the self-imposed isolation of the 'heretic' pharaoh, Akhenaten. The chief Egyptian centres were Gaza and Joppa, and there were less important ones, like Beth She'an. We hear of Egyptian commanders, the *rabu* or 'great one.' The class structure of the little states is rigidly hierarchical. Their rulers sometimes have feudal relationships with each other, and sometimes are in conflict; they constantly accuse each other of corruption and disloyalty to the Egyptian suzerain. Most of them appeal for Egyptian help, usually against the Mitanni, expanding from the north. But the numbers they wanted were quite small (Albright estimates the total population of Palestine at this time to have been around 200,000). Megiddo asked Pharaoh for reinforcements of 100 soldiers; the princes of Gezer and Jerusalem wanted 50 each. The Prince of Byblos, very likely the biggest city, and unquestionably the most important to Egypt, asked for 200 to 600 soldiers and 20 to 30 chariots.

Egyptian paramountcy was restored under some of Akhenaten's successors of the Eighteenth and Nineteenth Dynasties, and for a time the Egyptians maintained the Mitanni against the expanding power of the Hittite empire as a kind of buffer state. But Egypt as an imperial power was never again alone in the field, and after the great irruptions of the Sea Peoples, in the twelfth century BC, the power of the Pharaohs stretched beyond Sinai only rarely and briefly. The Sea

Peoples and their allies in fact broke up the great empires and international configurations of the second millennium, and during the Dark Age that followed some city-states of Syria-Palestine did remarkably well for themselves, precisely because they were small and independent. These little states flourished mightily until the coming of the next great age of empires, introduced by the Assyrians in the eighth century BC.

Pre-eminent among these states were the Phoenician cities of the coast. As Ugarit and Byblos declined, so Sidon and Tyre rose. Sidon was by far the older; its people, whom Homer first called Phoenicians or 'dark red men' (probably a reference to their famous dye), had been on the site since about 3000 BC. In their purple-sailed ships, they were the first seamen we know of who dared to sail throughout the night. But even they could not sail to windward and therefore they always looked for harbour-sites with two entrances, either of which could be used according to the direction of the wind. They were also the first to build such harbours artificially where necessary. At Sidon, the old harbour works are now mostly under water. Some of their stones were used to build the thirteenth-century Crusader castle and the Crusader church, now a mosque. The town was rebuilt in the seventeenth century AD and still flourishes after a fashion, so it has not been possible fully to investigate its Canaanite and Phoenician under-pinnings, with the exception of the ancient cemetery, which once contained the proposed sarcophagus of Alexander (now in the Istanbul Museum). Sidon was certainly a great port in the last two centuries of the second millennium, possibly the busiest in the world, for the development of all-weather harbours and round-the-clock sailings transformed the Phoenician coastal towns into vast international trading centres and entrepôts. The Phoenicians gradually extended the range of their oversea ventures until they encompassed the whole Mediterranean. They were an important agent in the spread of metal cultures; they not only knew how to work iron but where to find it and buy it. They dealt in copper, tin, bronze, iron, both worked and unworked, as well as gold, silver and electrum. Their activities helped to pull the world out of the Dark Age of the late second millennium.

About 1000 AD, Sidon founded Tyre as a satellite. It soon became independent, richer and more important than its mother-city. Old Tyre was on the mainland. Then, under King Hiram, the friend and contemporary of Solomon, the offshore islands were joined up to the coast and a New Tyre built, with the characteristic twin harbours – the 'Sidonian' on the north and the 'Egyptian' on the south. A canal was dug across the island so that the ships could move from one to the other. Tyre became so big and so wealthy that its comparatively recent origins were forgotten. In Strabo's day it was thought to be the oldest of the Phoenician cities. To the writers of the Old Testament it was the city 'whose antiquity is of ancient days.' The Hebrews evidently saw Tyre as the archetypal citadel of materialism, where money and precious things, as well as abominable idols, were worshipped. Tyre had a monstrous temple to its god, Melkart, but it also had one

The temple at Tyre which was built to the god Melkart in 960 BC, and is largely Phoenician in character.

of the earliest mercantile forums. Isaiah called Tyre 'the crowning city, whose merchants are princes, whose traffickers are the honourable of the earth.' The Book of Ezekiel devotes three chapters, 26–8, to Tyre, and 27: 12–24 gives an impressive catalogue of the materials and products which were on sale in Tyre, and their provenance – though the purpose of this panegyric is merely to enhance the catastrophic nature of Tyre's fall, as a result of its worship of Mammon.

In fact Tyre did not fall easily. The phrase 'the crowning city' is a reference to its formidable battlements. It was one of the most strongly defended cities of the first millennium BC. It had a powerful navy, manned by seamen from Sidon and the island of Arvad, as well as Tyrians; and its garrison was reinforced, according to Ezekiel, by mercenaries: 'They of Persia, and of Lid and of Phut, were in thine army, thy men of war.' Sennacherib finally took Tyre in 701 BC – by which time the city's breakaway colony of Carthage was already in existence – but only after several attempts. The Babylonians invested it for thirteen years, then gave up the siege in despair. Even Alexander the Great, as we shall see, was at first baffled by Tyre's inventive defences. Not much of them, however, can be seen today – though the causeway Alexander built towards the island in order to get his siege-engines within range of its walls now forms an isthmus, so that it is an island no longer. Over the past thirty years, excavations of the southern part of New Tyre have uncovered the Byzantine and Roman levels which lay beneath the Crusader city-fortress, but beneath them is the Hellenistic level, and under that in turn are the Phoenician levels of the early and mid-first millennium.

In the fullness of time we shall undoubtedly know more about the culture of Syria-Palestine in the third and second millennia BC, which laid the foundations both of Phoenician mercantilism and of the Hebraic civilization. Some of the answers lie in the unexplored portions of the very earliest cities of the Holy Land. At present our fragmentary knowledge has to be supplemented by any cultural evidence – especially written evidence – which comes to hand, often from outside the area. But this is only natural: we are tracing a history of peoples in almost constant motion. So we look for clues not only in the territory we are surveying, but in Mesopotamia and Anatolia (as well as Egypt).

These early societies established their State framework as part of an all-pervasive religious cosmology. Law was a divine institution. The oldest law code we know of, a Sumerian pandect of about 2050 BC discovered in the Istanbul Museum, was the work of Ur-Nammu, 'king of Sumer and Akkad', of the Third Dynasty of Ur. It says that the god Nanna chose Ur Nammu to be king, and he dismissed all the corrupt officials and established a proper and just system of weights and measures. About 150 years later, some sixty property regulations were listed on two Akkadian tablets (now in the Iraq Museum) found near Baghdad in the ancient kingdom of Eshnunna, controlled by the law-giving god Tishpak. In the Louvre and in the University of Pennsylvania there are fragments of another code, of the nineteenth century BC, from the kingdom of Idi, further south in Mesopotamia. And finally, from the period 1728–1686 BC, there is the massive six-foot high piece of black diorite, found at Susa east of Babylon in 1901 and now in the Louvre, on which are inscribed in Akkadian the 300 provisions of the Code of Hammurabi. Hammurabi is entitled to be called the first great legislator of history, but even his work is invested with divine sanction. All these laws are said to have been written by the king concerned at the instruction of the local chief god.

Divinely inspired legislation is thus the fundamental background to all these early states. The written law-code seems to have come at the moment when the society (usually a city-state) achieved self-consciousness as an entity and its ruler acquired the luxury of scribes: at that point the chief traditional god emerged as a codifying law-giver, acting vicariously through the charismatic king.

There are also fragments of somewhat later law-codes, from the middle centuries of the second millennium. Far to the north, in Baghazkoy (Khattusas) in east-central Asia Minor, archaeology has uncovered Hittite laws covering such subjects as wills and covenants and oaths, as well as diplomatic treaties. And at Mari, the capital of an Amorite kingdom which was destroyed by Hammurabi of Babylon in about 1697 BC, French archaeologists in 1936 found about 20,000 clay tablets, of which some 5,000 are administrative documents. Some are intelligence reports, at one and the same time intensely obscure and unbearably suggestive: 'Yesterday I departed from Mari, and spent the night at Zuruban. All the Banu-Yamina raised fire-signals ... and so far I have not discovered the meaning of those signals. But I shall find the meaning, and I shall write to my lord whether it is so or not. Let the guard of the city of Mari be strengthened, and let my lord not go outside the city gates.' Other tablets, however, are perfectly clear, since they describe particular legal and property transactions, and so give to the law codes the extra dimension of case-law.

The evidence of these documents, taken together with the texts found at

ABOVE RIGHT The bronze and gold statuette of Hammurabi, King of Babylon, in prayer, from Larsa.
BELOW RIGHT The Monastery of St George is built on the slopes of the ravine of the Wadi Kelt.

CANAANITES AND PHOENICIANS

Ugarit, which of course have a more direct bearing on the institutions of the Syria-Palestine area, show that by the middle of the second millennium BC, the Canaanite culture was comparatively sophisticated. It had clear ideas about civic duties, a complicated code of land ownership and definite forms of taxation, which was collected in an orderly manner. It also possessed a literature, based on legends and epic cycles, as were its Akkadian and Sumerian predecessors, and it had evolved a religious system. The most revealing texts from Ugarit are written partly in Hurrian dialect, and partly in an ancient form of Canaanite which resembles the Hebrew of pre-Mosaic times. The chief god was 'El'; that is, 'God'. He is apparently depicted in a remarkable stele found at Ugarit with his hands stretched out in blessing over the city. He was a water-god who lived in 'the source of the two deeps' and he was also called 'the father of years' and 'the father of man.' His wife was Asheret. This couple gave birth to Baal (also Haddu, or Hatti in the Hittite pantheon), who gradually became the chief weather-god, author of storms, earthquakes, rain and who usurped El's central place in the pantheon – El eventually becoming a generic name for a supergod. Baal also appeared in the form of 'son of Dagon'. In addition there was Astarte (the name appears in a variety of forms), the fertility or mother-goddess.

Many of the details of religious worship are still obscure to us. There were city temples, which became the prototypes of the classical sanctuary, the earliest of which (dating from the fifth to fourth centuries BC) can be seen near Amrit on the coast. But there were also, in very large numbers, hill-altars and 'High Places', from which Baal the weather-god, the 'Rider of the Clouds', who hurled down forks of lightning and thunderbolts like his successors Zeus and Jupiter, could be appeased. Appeasement by sacrifice was in fact a central principle of Canaanite religion. The gods were not responsible and protective: they did not adopt the people of Ugarit, or anywhere else, and work lovingly and consistently for their welfare. They were unpredictable outsiders, to be approached with superstitious fear. We know that the Carthaginian settlers from Tyre, in the first millennium BC, practised human sacrifices. There is no evidence of this in the Ugarit texts (it was common in Mesopotamia, particularly in the third millennium BC), but indications that the gods were bloodthirsty and often destructive. They were also promiscuous. Astarte, in her various manifestations, was based upon the sexual principle; she had to be appeased not so much by sacrifice as by the enactment of sexual congress. This took place in her temple and on her sacred precincts on hills and was believed to set her in motion as a fertility source. Hence Canaanite temples, or some of them at least, sponsored ritual prostitution of both sexes and for reasons which are not yet entirely clear to us, some Canaanite temples also had homosexual guilds. The Ugarit tablets present to us fascinating glimpses of a Canaanite civilization which in many ways was subtle, complex and admirable, and undoubtedly advanced in its time, but whose cultic rites also contained elements of vehement sensuality and horrific barbarism. This was the society which faced the first Hebrew settlers of the Holy Land.

PREVIOUS PAGES The court of the Great Temple of the Sun at Palmyra, with the imposing colonnade on the right.
LEFT Mount Sinai, the traditional site where God revealed to Moses the stone tables of the Decalogue.

2
The Civilization of the Bible

THE ORIGINS AND EARLY HISTORY of the Hebrews and their religion raise problems of enormous complexity which have bred and continue to breed whole libraries of scholarly studies. Over the centuries there have been three different approaches. Until about AD 1800, the predominant view was 'fundamentalist': that is, the Book of Genesis and the other early books of the Old Testament (the Pentateuch) were accepted as a divinely inspired, literal and accurate account, and chronologies based on them were constructed to date the Creation, the Flood, the arrival of Abraham, the Exodus and other specific events. From the early nineteenth century, the so-called 'critical' approach, pioneered by German scholars, demolished the Old Testament as a strictly historical record and treated large parts of it as religious myth. It was now seen as a series of oral legends from a large number of different tribal sources which reached written form only in the post-Exilic period, during the second half of the first millennium BC, and were carefully edited by priests and scribes for their particular purposes – that is, to justify and give historical authority to their religious beliefs and practices.

This approach, which consisted in breaking up the Old Testament writings into what appeared to be their component parts, and then analysing the background to their composition, reached a high degree of sophistication in the work of Julius Wellhausen (1844–1918) and his school and followers, and still has some limited value. Unfortunately its methodology matured long before the age of scientific archaeology and many of its basic conceptions – still tenaciously maintained by a dwindling band of scholars – cannot accommodate the startling discoveries which have been, and continue to be made in the Near East. The exploration of ancient sites and the emergence of a growing number of written sources have tended over the past fifty years to restore the value of the Old Testament as a historical record. This third approach, which marries revised methods of textual criticism with other, non-Biblical, texts, and with archaeological science, offers the best prospect of discovering

A limestone relief from the Palace of Sennacherib (705–681 BC) at Nineveh, showing men of Lachish in the Assyrian Royal Guard.

what really happened, and it is the one I follow here.

The history of the Hebrews as a distinctive people with a special link to the universal deity begins with the migration of Abraham from Ur of the Chaldees to Canaan. The Old Testament is ambiguous about Abraham's theology. He is presented as the ancestor of a nation whom God had destined as his own, and in this sense a monotheist; but Genesis is contradicted by Joshua 24:2 which says he 'lived besides the Euphrates and worshipped other gods.' Wellhausen and his school deny the existence of a separate Hebrew history before the time of Moses, and treat Abraham, Isaac, Jacob and others of the patriarchal era either as completely mythical or as eponymous ancestors who gave their names to Hebrew tribal histories. In fact there is no good reason to doubt their historicity. In the 1920s, Sir Leonard Woolley discovered Ur itself, an important Sumerian city of the third and second millennium BC. It had a comparatively high level of culture. Woolley found a splendid helmet in the grave of, 'Meskalamdug, Hero of the Good Land', made in wig form from solid gold, with locks of hair hammered into relief. He found also a kind of religious standard, perhaps carried in processions, decorated with shells and lapis lazuli. Above all, he found an enormous ziggurat, the temple raised on a series of platforms which plainly inspired the story of the Tower of Babel. This was the work of Ur-Nammu of the Third Dynasty (2060–1950 BC), the lawgiver, who had himself portrayed on a stele (a fragment of which survives) as a workman carrying pick, trowel and compasses, with an angel flying overhead.

Abraham, therefore, came to Palestine not as a nomad but as a former city dweller, and the religious notions he brought with him must be seen accordingly. All the earliest societies of Egypt and the Near East seem to have begun with stationary gods, civic or local deities whose power was confined, originally at least, within narrow geographical limits. Nomads normally adopted the god or gods of the place where they took up their abode. Woolley maintained that at the time when Abraham left Ur the cult of city gods was for the first time being supplemented by worship of a private family or tutelary deity. The images and instruments of these private cults travelled with the family in trunks, so Abraham brought his religion with him into Palestine.

He also brought with him a number of notions which formed the bedrock of Hebrew law and history. As we have seen, Mesopotamian law-codes existed and had reached written form towards the end of the third millennium BC, before Abraham migrated. Abraham was thus familiar with the idea of a divinely inspired code, which he carried to Palestine. But it is very significant that early Hebrew legalists rejected or severely modified legal concepts which at the time or later became the outstanding characteristics of Mesopotamian law. Thus in Babylonian law-codes, especially Hammurabi's, the rights of property are paramount, whereas the Hebrews tended to emphasize the essential rights and obligations of a man and their laws were framed with deliberate respect for moral values. In 1903–14, German archaeologists unearthed at Qalat Shergat (ancient

The Assyrian poem of Gilgamesh in which Gilgamesh seeks a means to restore his dead friend Enkidu to life. This tablet bears the Legend of the Flood and is the eleventh of the series.

Ashur) clay tablets giving a Mid-Assyrian law code which may go back as far as the fifteenth century BC, and which provides for a horrific system of punishments, including impalement, facial mutilation, castration and what appears to have been flogging to death. It is notable that Hebrew codes, while strongly endorsing the idea of retribution, reject such savagery completely – from the start they embrace the idea that the human person is entitled to respect. Flogging was limited to forty strokes, and must be carried out 'before the face' of the judge, 'lest, if he should exceed, and beat him above these with many stripes, then thy brother should seem vile unto thee.' (Deut. 25:3).

Abraham also certainly brought to Palestine the ancient Mesopotamian creation and flood epics, which found their way into Genesis. It was clearly the devastating and sudden inundations of Mesopotamia – so unlike the regular Nile floods – that gave rise to the flood narrative which circulated in several versions during the third millennium BC. One, found on a clay tablet at Nippur, presented it as part of divine cosmogony. Thus: God created mankind and then regretted doing so and made up his mind to destroy it. However, Enki, the water-god, revealed the apocalyptic plan to a king-priest called Ziusudra, a Noah-figure, who built a boat and survived with it. Much closer to the Genesis story is the version in *Gilgamesh*, an epic built around a legendary Sumerian ruler of Uruk, in the fourth millennium BC, who figures in a number of episodes while searching for the 'plant of life.'

Woolley made strenuous efforts to find physical evidence of a spectacular flood in the earlier archaeological levels he investigated. At Shuruppak he came across impressive alluvial deposits at the Jemdet Nasr levels, and a similar stratum at Kish was of eighteen inches; he also found an alluvial deposit of eight feet at Ur – but the datings did not match and the evidence is inconclusive. What one can say is that belief in the great flood was very ancient and widespread outside Hebrew circles, though the interrelationship of the Babylonian and the Hebrew accounts has not been firmly established. One thing which is striking is that scholars of the textual criticism school, who dismiss the Old Testament flood story as late myth, find no difficulty in giving a very early date to the Gilgamesh flood, found – in the late version in which we have it – on the eleventh tablet of a series unearthed at Nineveh in the ruined library of Ashurbanipal, 669–627 BC. This demand for higher, or rather totally different, standards of proof for Old Testament assertions as opposed to non-Biblical ones is very characteristic of the difficulties in which scholars of the Wellhausen tradition now find themselves.

Another striking fact about the flood tales is that, whereas the Gilgamesh story is a collection of isolated episodes which lack a unifying moral and historical context, the Hebrew presentation in Genesis shows a definite concern for moral values and presents history as the working out of a providential design. It is the difference between secular and religious literature and it is also the difference betwen the writing of mere folklore and determinist history.

However, not all the episodes in Genesis are of transcendental importance: they simply illustrate the patriarchal way of life and its social and legal customs. Archaeological discovery provides a firm historical background to the patriarchal society. Such names as Abraham, Isaac and Jacob, far from being later eponyms, were in fact common in the region during the first half of the second millennium BC. The French excavations at the ancient palace of Mari, and still more the American excavations at Yorghan Tepe (ancient Nuzu) 100 miles north of Kirkuk, have produced an enormous number of cuneiform documents – over 20,000 clay tablets, dating from the fifteenth century BC in Nuzu alone – which illuminate the background to the patriarchal narratives. Many are from private archives, recording legal transactions. The proposal for the adoption of Eliezer as heir-presumptive to Abraham, the latter's negotiations with Sarah, the transfer of a birthright from Esau to Jacob and the binding power of a deathbed blessing and disposition of property, Rachel's theft of her father's teraphim (household gods), Jacob's legal relations with Laban – so baffling to us – all of these were in accordance with standard legal practice as illustrated repeatedly by the Mari and Nuzu tablets. To give only one instance, a tablet from Nuzu reads:

The adoption tablet of Nashwi, son of Arshenni. He adopted Wullu, son of Pohishenni ... When Nashwi dies, Wullu shall be heir. Should Nashwi beget a son, he shall divide equally with Wullu, but Nashwi's son shall take Nashwi's gods. But if there be no son of Nashwi's then Wullu shall take Nashwi's gods. And Nashwi has given his daughter Nuhuya as wife to Wullu. And if Wullu takes another wife he forfeits Nashwi's land and buildings.

In the light of these archives it is difficult not to believe that Abraham and his

One of the many cuneiform clay tablets found in Nuzu, dating from the fifteenth century BC.

progeny were real people. There are of course implausibilities, not to say impossibilities in the patriarchal tales, though not so many as were once thought. For instance, was the reference to camels anachronistic? The full domestication of camels on a large scale did not seemingly occur until the twelfth century BC. There are, however, references before this period. The Mari excavations turned up camel bones going back before the days of Sargon the Great, *c.* 2400 BC. Camels are mentioned in an eighteenth-century BC cuneiform list of fodder, from Alalakh in northern Syria and from roughly the same time we have a relief from Byblos depicting a kneeling camel.

As for the places named in Genesis, there is an unresolved controversy over the location of Sodom and Gomorrah. Sodom's name may have been preserved in Jebel Usdum, 'Mount of Sodom', which is on the west side of the southern part of the Dead Sea, a mount 700 feet high and several miles long, chiefly of chrystalline salt. It was once a pilgrimage site as is suggested by pottery found there which can be dated 2300–1900 BC. The end of the pilgrimages and of the potter sequences would mark the destruction of the cities. We are on firmer ground with the cities associated with the patriarchs. Beersheba means 'well of the oath' – the oath which Abraham swore with Abimelech (Gen. 21:27–31). It seems probable that there have always been wells on this site, where Abraham planted the tamarisk,

The impressive structure of Hebron, where God made with Abraham a covenant that he would be the father of a chosen people, and where now the tombs of the patriarchs rest.

Isaac built an altar and Jacob had his famous ladder-dream. The *tell* denoting the existence of an ancient city is actually three miles from the present town and has not yet been systematically excavated, but there is at present no good reason to reject the traditional association with Abraham. Much the same could be said of Hebron, the magnificent city of the patriarchs, 3,000 feet above sea-level, where Abraham's long migration ended. The Book of Numbers says it was founded seven years before the Egyptian city of Zoan (Tanis) in the Nile Delta, and the tradition that it contains the tombs of the patriarchs is so ancient and so strong that it cannot be discarded without evidence of falsity. Since Absalom set up his rebel standard there, it was evidently in King David's time still regarded as in some sense the capital. Both here and at the supposed site of the Oak of Mamre, a mile to the west, Herod the Great performed characteristic works of architectural piety towards the sacred places of Hebrew history, building enclosure walls, which at Hebron itself are over forty feet high and form a perimeter 600 feet in length. This massive structure was repeatedly repaired by Byzantine and Crusader Christians and by Moslems – for Moslems, like Jews and Christians, venerate the patriarchal tombs, and in the church-turned-mosque above them, the pulpit was the gift of Saladin.

If, then, Abraham and his progeny were real people and Hebron is the actual sepulchre of some of them, when did they live? In Genesis, antediluvian datings are of course schematic rather than actual. They vary somewhat as between the original or near-original Hebrew Massoretic text, the Septuagint and the Samaritan Pentateuch. Examinations of the lists of ante- and post-diluvian patriarchs show a structure of two groups with ten names on each. This corresponds roughly with parallels in non-Biblical literary records. The 'long' datings are similar to the lives of Sumerian kings before the flood at Shuruppak. The earliest king-list gives eight antediluvian kings, but a tradition recounted by Berossus, a Babylonian priest in Hellenistic times whose work corresponds to the historiography and chronology of Manetho in Hellenistic Egypt, provides ten kings, which fits the Genesis pattern.

All this indicates that in the earlier part of Genesis we are dealing with an attempted historical record rather than pure myth. After all, with the help of Manetho and other Egyptian king-lists, we can date Egyptian history with reasonable accuracy as far back as the First Dynasty, 3100 BC. Early Mesopotamian history is not so easily anchored in absolute time, but it now looks as though Sargon and the Old Akkadian period can be dated roughly 2360–2180; the law-giver Ur-Nammu and the Third Dynasty of Ur to c. 2060–1950, and the Old Babylonian period to 2060 and after, with Hammurabi, who is an authenticated statesman, lawgiver and human being, being dated with some confidence to 1728–1686. How do we link the Genesis patriarchal narratives to secular chronology? The evidence all suggests that they belong in the Middle Bronze Age, that is between 2100 and roughly 1550 BC. They cannot easily be fitted into the Late Bronze Age, because that would place them in the Palestine of the Egyptian Empire of the New Kingdom, and it is notable that there is no Egyptian presence in patriarchal Palestine. Some day it may be possible to place Abraham reasonably precisely within Bronze Age chronology. At present he wanders in time, as he was once a nomad in space. Albright changed his mind repeatedly about Abraham's dating, pushing him backwards and forwards between the twentieth and the seventeenth centuries BC. His final conclusion, that Abraham cannot have lived before the twentieth century or after the nineteenth is probably sound. From this it would follow that Jacob and his family went into Egypt during the period of alien Hyksos rule (Fifteenth Dynasty) when large numbers of Asians migrated to the Nile Delta.

A further problem is to fit the patriarchal records into what we know of a distinctive Hebrew people. Unfortunately, this is not much, and what there is remains open to argument. In Mesopotamia there were a group of people referred to in tablets and inscriptions by the ideogram SA.GAZ, or as Hapiri, Habiru. Egyptian sources, in the Late Bronze Age, also speak of Abiru or Habiru. By this they did not mean bedouin, for whom they had a different expression. 'Habiru' seems to have been an abusive term, attached to non-city dwellers who could not be classified as regular tribes moving according to the cycle of seasons. Their culture was superior to that of most desert tribes. Sometimes they held jobs as mercenaries, government employees, servants, or worked as tinkers and pedlars; at other times they travelled with donkeys in caravan, as merchants. But they were also outlaws and bandits. The authorities found them a problem and somewhat of a mystery. From this group, almost certainly, the first Hebrew settlers of Palestine came, and they continued to come throughout the Middle Bronze Age and the early part of the Late Bronze Age.

Some of these people moved to Egypt in the time of the Hyksos, others remained behind in Palestine. Their way of life is described in the Song of Deborah, one of the first of the Old Testament books to achieve written form, perhaps as early as 1200 BC, and which bears little trace of subsequent editing. As these Hebrews settled and multiplied, their activities began to be reflected, albeit faintly, in the secular records. The *Amarna Letters* were written at a time when (I assume) the Hebrews had settled in Egypt but before the Exodus. Some are by and others refer to a chief called Labaya, or Lion Man, who controlled the wild hill-country of central Palestine. He was almost certainly a Hebrew. He was not a very civilized man – some of his letters lapse into almost pure Canaanite – and he was a thorn in the side of the Egyptians and of his neighbours. But he seems to have carved out a petty kingdom for himself which included the city of Shechem and this would accord with the tradition of the capture of Shechem in chapter fourteen of Genesis. He met a violent death during the reign of Akhenaten, but his sons carried on his work, and it seems very likely that this group of Hebrews or their descendants were still in occupation of Shechem at the time of the conquest of Israel by Joshua. At any rate, there is no statement that Joshua took the city: instead, as soon as the Israelites got into the hills north of Jerusalem, they held at Shechem the ceremony of the Covenant which officially brought into existence the Hebrew tribal league (Josh. 8:30–35).

Shechem is near the modern city of Nablus, a name derived from the Neapolis or New City built in AD 72 under Vespasian. There can be no doubt about the identification of the site, for Shechem was near the holy Mount Gerizim (Judg. 9:7) and Josephus, writing about AD 90, says the city was located between Mounts Ebal and Gerizim, further evidence being provided by Eusebius in his *Onomasticon* (before AD 340), who says it is in the suburbs of Neapolis near Jacob's well. It is, in fact, the first city in Palestine to be mentioned in the Old Testament (Gen. 12:6–7). Abraham got the divine promise there. Jacob on his deathbed told Joseph: 'I have given to you rather than to your brother one Shechem, which I took from the hand of the Amorites with my sword and with my bow.' How this fits with Labaya's conquest is not clear. Shechem was already a city in the nineteenth century BC, since it is mentioned in a document of Sesostris III (1878–1843), and later, in the eighteenth to seventeeth centuries BC, it acquired cyclopean walls – which may still have been in existence when Labaya took it in the fourteenth century BC. The Biblical archaeologist G.E. Wright

The surviving stonework of the eastern double gateway to the Ancient City at Shechem, located between Mounts Ebal and Gerizim.

believes that it then remained in Hebrew hands until Joshua's conquest, the returning Israelites recognizing its inhabitants as related to them by race, culture and religion, and the nearby hills as places of peculiar sanctity. This would establish continuity between the patriarchal settlement and the post-Exodus conquest. Shechem is thus in a sense the original capital of Hebrew Palestine; certainly geography made it a natural one.

Labaya took Shechem about 1360 BC. When did the Hebrews of the Exodus rejoin this remnant of the original patriarchal settlement in Palestine? If, as is most likely, the original Hebrew migration to Egypt took place under the Hyksos, then they moved to Egypt some time during the seventeenth century BC. A case can be made for their return to Palestine during the fifteenth century, on the grounds that the *Amarna Letters* show them already established and active in the area. But such an early date conflicts with the evidence of the Old Testament. This points clearly to the time of the Nineteenth Dynasty pharaoh Ramesses II, 1290–24. Exodus says that, prior to their revolt and departure, the Hebrews worked as conscript labour on the two pharaonic cities of Tanis (Avaris) and Pithom (Tell el-Retabeh). Both these cities were built, or rebuilt, by Ramesses II – Tanis in fact was known as the House of Ramesses. There is also the evidence of the Meneptah stele *c*. 1224–16, discovered by Sir Flinders Petrie in 1895 in the temple of Meneptah in Thebes. This inscription is a boastful recitation of the king's victories over various foreigners and it contains the passage: 'Israel is

39

destroyed. It has no seed-corn. It has become a widow for Egypt.' This is the first surviving mention of Israel or the Israelites in history and it implies that they were by this date a recognizable and sizeable group established in Palestine – the Exodus had evidently taken place some time before.

Accepting, then, that the migration from Egypt to Palestine took place in the first half of the thirteenth century BC, we must appreciate that this particular movement of Habiru or Hebrews was only one aspect of the settlement of Palestine, which took place perhaps over as long a period as two centuries, involving several Hebrew groups and other, non-Hebrew settlers. We ought to see the events recorded in Genesis and Exodus as a selection from a series of oral tribal histories. Very likely the Hebrews brought out of Egypt by Moses were only one tribe, but the fact that their particular history eventually subsumed that of the other migrating groups indicates the important role they played in the achievement of a successful migration and settlement and, not least, the greatness of their leader and lawgiver Moses and his militant successor Joshua.

Moses is rightly regarded as the crucial figure in Hebrew history. Before his time we have isolated episodes and anecdotage, on which any real pattern is superimposed anachronistically by hindsight. By the time of his death, the Hebrews are a recognizable people with a religion, an aim, State laws and institutions, a philosophy of destiny. Moses was very likely a Levite and it seems to me that he must have been literate – as would have been almost inevitable if, as Exodus states, he was brought up at the pharaonic court. All the early societies of the ancient Near East matured at the point that their rulers and priests acquired writing and first employed it for public purposes. Moses drew up the earliest State documents of the Hebrew people, even if they have not come down to us in precisely the form he gave them.

Was Moses a monotheist? Assuming that the patriarchs were not, and that Moses himself did not inherit a monotheistic tradition, he must be seen as a religious innovator. He was not a polytheist, though his background in Egypt was certainly polytheistical. He might be termed a henotheist – that is, a believer in the god of one's people. Most of the tribal societies which moved into Palestine at this period were henotheistical. Thus the Moabites had Chemosh as their national god and the Ammonites had Milkom. The Hebrew god was Yahweh, but he was not always referred to by this name. There are references to Shaddai and El Elyon, who may once have been separate deities. It is not clear that the god who left Egypt with Moses and the Hebrews was the same deity to emerge with them from their desert wanderings. It has been argued that Moses was a priest who had a special connection with Yahweh, who was the god of Qadesh, the Sinai mountain-god who lived near the watering place of Ain Qadais, ninety miles south of Jerusalem. The Hebrews of the Exodus conquered the oasis from the Amelekites and joined forces with other related tribes who pitched their tents there. It was the 'centre' of a desert religion, and the feeling of solidarity, and later nationhood, which grew up among the Israelite tribes sprang from their

THE CIVILIZATION OF THE BIBLE

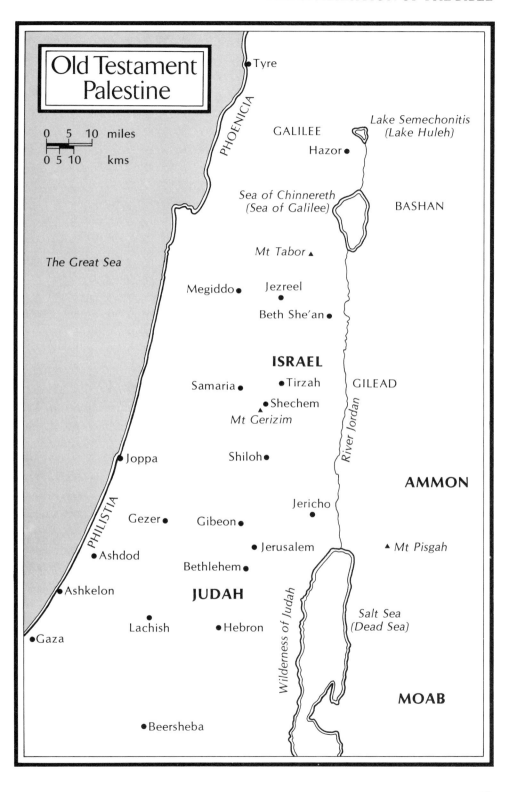

attachment to the cult of Yahweh at Qadesh, which for them was a pilgrimage centre as well as an oasis. There comes a point in the development of a henotheistic cult when by a process of cultural and political self-assertion, it matures into monotheism – the god is the supreme god not merely of a people, but of the entire world. This moment came under Moses's leadership.

Certainly, it is difficult to see Moses the teacher and lawgiver except in terms of monotheism. His god was the creator of everything, the sole source of justice, without sexuality or mythology, who cannot be seen by human eye and cannot be represented in any form. There is no parallel to such a deity in the Late Bronze Age, not even Akhenaten's ephemeral sun-disc. The Wellhausen school argued that Mosaic monotheism evolved slowly out of a polytheistic background but there is no evidence for such a process in the materials we have. That is why the great Jewish historian, Yehezkel Kaufman, insisted that Israelite religion was monotheistic from the beginning and that it had no antecedents.

There were, however, both antecedents and contemporary Late Bronze Age parallels for the Mosaic law-codes. Hence it is difficult nowadays to accept the Wellhausen school theory that the legal sections of the Pentateuch were composed many hundreds of years after the time of Moses and foisted on him by the editors who put together the Pentateuch in post-Exilic times. Leviticus, which gives the enactments, largely ritual, constituting the legal basis for organized civil and religious life among the Chosen People, is consistent with what we know of the political history of the Hebrews at this time, that is the thirteenth and twelfth centuries BC. So is Deuteronomy, which by contrast is a popular exposition of the more technical priestly writings, especially the Covenant obligations, for a wider audience. The kind of material contained in these two books – cultic, professional, medical, juridical, scientific, dietary – is absolutely consistent with similar non-Biblical material circulating in the Near East in the Late Bronze Age. Detailed cultic writings go back to the *Pyramid Texts* of the Fifth and Sixth Egyptian Dynasties, inscribed at Sakkara in the middle of the third millennium; Sumerian religious texts are also third millennium BC.

The Mosaic Pentateuch is rich in medical material. The famous passage dealing with leprosy, outlining the duties of a particular branch of the priesthood in diagnosis and therapy, is unique. But written medical lore, enshrined in law-codes or presented as professional material, is not rare in the ancient Near East, even in the second millennium BC. The Egyptian tradition in fact goes back to Imhotep, *c.* 2650 BC, and four of the most important Egyptian medical papyri, even in the copies we possess, are all later than or contemporary with the Mosaic codes – the Edwin Smith Medical Papyrus, *c.* 1600, the Ebers Papyrus and the Hearst Papyrus *c.* 1500, the Berlin Papyrus 3038 *c.* 1290. We find similar medical empiricism in the Code of Hammurabi, written half a millennium before the Mosaic era, and in other medical enactments from Mesopotamia and Egypt.

Mosaic dietary regulations also fit in, though with significant differences, to the pattern of government elsewhere in the Late Bronze Age. Like the Egyptians, the Hebrews were forbidden to eat sea creatures without fins and scales. Pious Egyptians were not supposed to eat fish of any kind but they consumed quantities of water-fowl, forbidden to Jews (though they could eat pigeons, turtle-doves, goose, quail, partridges and domestic fowl). The regulations seem to divide the clean and the unclean on the basis of some kind of proto-scientific empiricism rather than pagan diabolism and superstition. Thus edible quadrupeds, cloven-footed or with parted hoof had to chew the cud; 'clean' animals were, on the whole, exclusively vegetarian. Permissible wild animals were fallow-deer, gazelle, ibex, antelope, roebuck and mouflon. Predatory and carnivorous animals were prohibited, and so were camels, swines, hares and coneys. This was because the camel was valuable and a source of milk, while swine harboured parasitic organisms and were particularly dangerous when eaten raw or partly cooked. Rapacious and predatory birds were also excluded. Hebrew laws on hygiene were not dissimilar to contemporary practice in Egypt. But on sex Leviticus is much stricter than Egyptian custom (there were no laws as such in Egypt), which regarded consanguinity as fairly unimportant, and still more so than the customs depicted in the Ras Shamra tablets, which treat adultery, fornication, bestiality and incest as permissible in certain circumstances. In the Late Bronze Age the Hittites also permitted forms of bestiality (though not incest). The Hebrews were certainly the first people we know of to compile a list of forbidden degrees of marriage, based on affinity as well as consanguinity.

Hence already in the time of Moses, the Hebrews were emerging as a people with a strict and strictly defined moral sense, and with a unified moral view of the universe illuminated by Yahweh, a sole god on the verge of becoming a universal one. It is possible that, during Moses's lifetime, Yahweh was in some sense still tied to his particular mountain. One theory is that even after the Hebrews settled in Canaan, the Sinai mountain was kept up as a place of pilgrimage. If so the tradition gradually lapsed, the pilgrimages ceased and eventually the locality of Sinai was forgotten. The present Mount Sinai has always been a Christian site; its identification goes back certainly to the fourth century AD and probably to about 200 years before, but even that was a whole millennium since the original Sinai had ceased to be a Hebrew pilgrimage centre and it is highly improbable that Mount Sinai is the sanctuary of Moses.

Tracing the Exodus and the wanderings is not a very fruitful occupation, though it has been done many times. Very likely the Hebrews took the southerly route out of Egypt, crossing a vast papyrus marsh, the Yam Suph or Reed Sea, between the Bitter Lakes and Zilu (Thiel). Within the Sinai peninsula, tentative identifications have been made of such sites as Elim (Wadi Gharandel) and Dophkah (Serabit el-Khadem). The Hebrews kept close to the oases and to the grazing lands south-west of the Dead Sea. They seem to have spent thirty-seven years in the region of Qadesh (Ain Qudeirat-Ain Qadais).

With the actual conquest of Canaan, however, there is the beginning of a long continuity and the archaeological evidence becomes much more solid. In recent

St Catherine's Monastery on Mount Sinai. The present site has always been Christian, and it is unlikely that it was ever the sanctuary of Moses.

decades the Book of Joshua has re-emerged as a historical account, albeit full of mysteries and *lacunae*. It would probably be a mistake to see the conquest under Joshua as a sudden invasion. It was more a process of gradual infiltration, punctuated by skirmishes and sieges. While the Hebrews were settling in Palestine, other desert and tribal peoples were moving into Syria, Moab, Ammon, Edom and similar territories to the north and west of the wilderness. It was one of the great migrations of history.

In the case of the Hebrews it was necessarily a long process. Materially, the Canaanites had a higher civilization and they were almost certainly better armed. It was one thing for the Hebrews to move into the hill country running down the spine of Palestine, quite another to take over the strongly defended cities of the plains and the sea-coast. Of the two mountain cities, Shechem they did not need to take, since it was already held by people they recognized as belonging to their culture, and Jerusalem they bypassed – not taking it until the reign of David, 200 years later. Most of the fortress cities of Canaan – Megiddo, Gaza, Hazor, Beth She'an, Taanach, Dothan, Yenoam – had originally been Hyksos strongholds and they were still organized on feudal lines, as is indicated in the Ugarit tablets, with social privilege largely dependent on military status. These were tough nuts for the Hebrews to crack.

Their first success was at Jericho. Indeed, the crossing of the Jordan from the east, preparatory to the siege of Jericho, marks the point at which the wilderness wanderings ceased and the settlement began. The Hebrews camped at Gilgal, where they were circumcised with knives of flint. The manna ceased and they ate their Passover cakes of unleavened bread for the first time in the Promised Land 'in the lowlands of Jericho.' Then follows one of the most striking passages in the whole of the Old Testament: the week of marching round the city, sounding the trumpets of rams' horns, and the spectacular shout on the Seventh Day, when the walls fell down. All the inhabitants were massacred except Rahab the prostitute and her family who had sheltered Joshua's scouts. Jericho was burnt and Joshua cursed anyone who should rebuild it. Kathleen Kenyon's excavations do not confirm this narrative, though they are not inconsistent with it either. Erosion has removed all traces of the walls of the city taken by the Hebrews, but it is clear that the site remained unoccupied for many years after the walled city was taken, in accordance with Joshua's threat: 'Cursed be the man before the Lord, that riseth up and buildeth this city, Jericho.' Indeed, the siege by Joshua effectively ended Jericho's grandeur as a city. Herod the Great built an enormous Roman-style villa there, but nearly two miles to the south-west, and, centuries later, the Arab prince Hisham, built a splendid palace at Khirbet Mefjer, but this also is a mile or two away. The city itself never recovered.

At other sites archaeology confirms the hypothesis that the Hebrew conquest under Joshua took place in the thirteenth century BC. Bethel, excavated by Albright in 1934, was destroyed by a tremendous fire during this century, thus fitting the Bible record. Lachish, excavated by J.L. Starkey from 1933 on, was also destroyed in the thirteenth century BC. Perhaps the most important of the early Hebrew conquests was the great city of Hazor, which was thoroughly excavated by Yigael Yadin. Hazor figures in the *Execration Texts*. We learn from the Mari archives that it was important enough for Hammurabi to have resident ambassadors there, and its king, Ibni-Adad, is recorded as receiving three shipments of tin for bronze making. Nearly every New Kingdom pharaoh mentions Hazor, and from the *Amarna Letters* we learn that its ruler was a king and had authority over other towns.

The city revealed by Yadin's work was huge: a lower town of 200 acres, and the actual *tell*, or upper town, of 24 acres. Together they accommodated over 50,000 people. The fortifications were impressive and the gates massive. There were a number of temples within, one of them, in three parts, a prototype of the later Temple of Solomon in Jerusalem. Among the debris Yadin came across a potter's workshop, with a clay mask, a hematite cylinder seal, two cherubs, a bronze plaque of a Canaanite official, a carved stone lion, a clay model of a cow's liver used in divination, and most important of all a deliberately mutilated temple stele of the moon-god Baal Hamman, with the uplifted hands which were the symbol of his wife, Tanit or Tinnith.

There is one inconsistency in the Old Testament record as it stands, for Hazor appears to have been taken twice (Josh. 11:1-5, and Judg. 4:23-4). On balance it seems more likely that the campaign described in Joshua corresponds to the historical facts. As one would have expected from its previous history, Hazor under its king, Jabin, formed a coalition of all the northern cities of Canaan against the encroaching Hebrews. The Canaanites were defeated near

The victory of Joshua by the 'waters of Merom' marks, according to the Bible, a decisive phase in the conquest of northern Canaan. The consequent taking of Hazor by Joshua resulted in the burning of the strategically placed city.

THE CIVILIZATION OF THE BIBLE

the 'waters of Merom'. The narrative continues:

And Joshua turned back at that time, and took Hazor, and smote its king with the sword; for Hazor was formerly the head of all those kingdoms. And they put to the sword all who were in it, utterly destroying them; there was none left that breathed, and he burned Hazor with fire ... But none of the cities that stood on mounds did Israel burn, except Hazor only; that Joshua burned.

The identification of the site excavated by Yadin is established by a cuneiform clay tablet, recording litigation over land at Hazor in the presence of the king. The discovery on the site of immense quantities of Mycenaean imported pottery, in a series which can be closely dated becuase it is also found at Amarna in Egypt, for which we have virtually absolute dates, permits us to say that the city was taken and destroyed in 1250–30 BC, which fits the Joshua narrative. The destruction was immense – the enormous lower city was never re-occupied – and evidently religious in fervour.

In some cases the Hebrews did not take cities but allied themselves with them or accepted their submission. At Gibeon, a few miles north of Jerusalem, Joshua sentenced its citizens to be 'the hewers of wood and the drawers of water.' Near the city Joshua won a great victory over the five Amorite kings, 'having marched up all night from Gilgal'. There followed one of the most spectacular episodes in the Bible: 'Then spoke Joshua to the Lord in the day when the Lord gave the Amorites over to the men of Israel; and he said in the sight of Israel "Sun, stand thou still at Gibeon and thou Moon in the valley of Aijalon." And the sun stood still, and the moon stayed, until the nation took vengeance on their enemies.' But it was a long time before Gibeon was integrated with the Hebrew community. Saul violated the original peace-treaty with its people by killing some of them and so in the time of David, who was frightened by a famine, the Israelites agreed to hang in expiation two sons and five grandsons of Saul 'before the Lord at Gibeon.' Chapter twenty-one of the Second Book of Samuel gives us a moving picture of Saul's widow Rizpah, who kept watch by the gallows, 'and took sackcloth, and spread it for herself on the rock, from the beginning of harvest until rain fell upon them from the heavens; and she did not allow the birds of the air to come upon them by day, or the beasts of the field by night.' At that period the Gibeonites were considered to be 'of the remnant of the Amorites', and it was not until the time of Solomon that they united themselves with the Israelites.

There are altogether forty-five references to Gibeon in the Bible (there is also a mention in a hieroglyphic text on the south wall of the Temple of Amon at Karnak, dating from the reign of Sheshonk [Shishak], 945–924 BC), but its site at El-Jib was not finally established until the post-war excavations of James Pritchard. There, as at the sites of Lachish and Beth-She'an, the actual name of the ancient city was found among the detritus – in the case of Gibeon on no less than twenty-five handles of wine-jars. Gibeon was evidently a wine-producing centre, for the excavators found a whole series of underground cellars for storing wine in

The enormous well with its spiral staircase leading to the pool of Gibeon, a few miles north of Jerusalem.

nine-gallon vats. But the most spectacular discovery which illuminates Old Testament references to Gibeon's water-supply and its 'pool', was of an enormous well with a spiral staircase inside it, linked to a complicated tunnel system and water-cistern, fed from a spring under the walls. This fine piece of Early Iron Age engineering, which gave the citizens access to a fresh water supply even when Gibeon was under siege, is a late example of a type found in antiquity in the second millennium BC, at Gezer, Megiddo and Ibleam (Khirbet Bel'ameh, south of Jennin) in the Holy Land, and also at Mycenae and Athens in Greece and Susa in Persia.

Both from the Old Testament narrative, and from the finds of archaeology, it is clear that Joshua won a series of spectacular victories in the latter part of the thirteenth century BC. He conquered the south and then conquered the north – already the two parts of the Promised Land are taking shape. But from the second part of the Book of Joshua and the first chapter of Judges, it is evident that the consolidation of the settlement was not a simple one. The tribes tended to act independently and at cross purposes and even in opposition to each other. After Joshua's death there was a revival of Canaanite power and the final conquest took a long time. We are dealing here with a period of about 200 years, from 1200 to 1000 BC, when the Kingdom of Israel came into existence. This period is covered by the narratives in the Book of Judges, but if we are to get the dating right it is important to realize that some of the judges were contemporaries, and that the original compiler of the book has made a selection from among a mass of material contained in oral tribal histories.

There is only scanty evidence from non-Biblical sources of Israelite history at this time. A limestone plaque found in Solomon's dowry city of Gezer, sets out the annual rhythm of this profoundly rural society in blank verse:

> Two months of ingathering;
> Two months of sowing;
> Two months of late sowing;
> A month of hoeing up of flax;
> A month of barley harvest;
> A month of harvest, then festivity;
> Two months of vine-dressing;
> A month of summer fruit.

The Israelites were organized as a tribal confederation, the outline of which they brought with them from the wilderness. Those scholars who are most sceptical of the early Old Testament record have advanced the theory that the Hebrew community devised its organization only after the settlement in Canaan, and that the twelve-tribe system, with its cult of Yahweh at a common sanctuary inaugurated by Joshua at Shechem near Mount Gerizim, was based on Canaanite ideas, of Mediterranean origin. Such sacred leagues or amphictyonies, usually of twelve groups of tribes, were common in Italy and Greece, especially in the first millennium BC. But this is a very complicated and clumsy method of discrediting the Biblical evidence. It is far more likely that the Israelite structure, at all events a very loose one until the time of Saul, was rooted in the Semitic patterns of the second millennium BC.

It should be noted that, until the Peoples of the Sea and other barbarians destroyed the international societies of the Late Bronze Age in the twelfth century BC, there was no clear division between east and west – between the Greek and the Oriental and Semitic worlds. That abyss was the consequence of the Dark Age which followed. Before it, there were many parallels in social development in Mycenaean Greece, Minoan Crete, pharaonic Egypt, Canaanite Syria-Palestine, and the Hittite and Mitanni states. One characteristic of Mycenaean society of the twelfth century BC for which there are clear parallels in the Israel of the Judges period is the idea of heroic kingship or leadership.

Such charismatic kings had to be of aristocratic or even royal blood, but the throne was not decisively determined simply by linear succession. Rulers emerged from the battlefield or as a result of some spectacular feat in the public interest. 'Judge' in Old Hebrew means 'deliverer' as well as ruler and lawgiver. Most of the Israelite judges came to the fore during a period of crisis, when they were invested charismatically with the special power needed to save the nation; then they continued to rule after the fighting was over, though not to found dynasties. Such judges were local as well as national heroes, and several might be active at once, though as a rule one would hold the paramountcy.

This system fitted well with the tribal parochialism and democratic instincts of early Israelite society, which at that time had little more to fear than the emnity of petty kingdoms. But the national need changed with the emergence of Philistine power. The Philistines, or *Pulesti*, were one of the groups forming, or linked to,

A plaster cast of the head of a Philistine soldier from a temple wall relief at Thebes, Egypt, twelfth century BC.

the Peoples of the Sea. When they were beaten back from Egypt by Ramesses III, in the tremendous sea and land battles of which we have a magnificent pictorial record at Karnak, they turned north-east and settled on the coast which still bears their name. Their top echelons were organized as a feudal aristocracy under 'princes'; they were aggressive, disciplined, skilled in war, and they already possessed iron weapons. Their five great cities in Palestine, Gaza, Gath, Ascalon, Ashdod and Ekron, have not been systematically or fully excavated and there is still a lot to be learnt about their culture. Their pottery seems to have been Aegean. Their original occupation of the Palestine coast, early in the twelfth century BC, took place at the expense of the indigenous Canaanite occupants, not the Israelites. But they slowly expanded inland and by the end of the eleventh century they were a formidable and growing threat to the Israelite tribes of the interior. It was Philistine aggression which effectively created the Kingdom of Israel.

To meet the Philistines, with their iron weapons and their high level of military organization, it was necessary for the Hebrews to change from a tribal confederacy into a centralized kingdom. The changes can be followed in the two Books of Samuel. The first stage occurred when the Israelite tribes agreed to act together and brought the Ark of the Covenant from its shrine at Shiloh. After its disastrous capture by the Philistines, the second stage took place, power being concentrated in the person of Saul. Initially, Saul 'emerged' like one of the charismatic judges, being annointed by the prophet Samuel, 'the man of God.' But after his massive defeat of the common enemy, the Philistines, 'the whole people' opted for a constitutional change at Gilgal near Jericho, and 'made Saul king before Yahweh.' This, it should be noted, was a conditional not an absolute kingship. The people made a covenant with Saul, as they had made one with Yahweh; there were obligations on both sides. Saul was king as long as he supplied resolute military leadership and civil justice. If he failed in either the bargain lapsed and there was no suggestion that his post was hereditary. The bargain was also with God; if the king was seen to be out of harmony with

Yahweh, his powers likewise lapsed. Indeed the presumption was that failure signified God's anger at irreligious behaviour on the part of king, people, or both.

Being a free tribal society, the Israelites approached the institution of monarchy with much fear and hesitation. Their worries were powerfully voiced by Samuel himself, before he agreed to annoint Saul. A king had the right to call out all the tribal levies, a right hitherto exercised only by the tribes individually. Thus his mere existence in a sense dissolved the democracy of the tribal system. Samuel foresaw high taxation, royal store-cities, mercenary troops and the whole apparatus of autocracy. But he himself, and his like, constituted a control on regal behaviour. In Hebrew society the order of prophets was an older institution than kingship. Indeed it went back to the very beginning. Abraham was a prophet; so, above all, was Moses. The origins of the word, *nabhi*, are not clear: it may have meant 'to bubble forth'; more likely 'one who is called.' 1 Samuel 9:9 says: 'He who is now called a *nabhi* was previously called a *roeh*' (seer). Seers were often attached to shrines, as was Samuel at Ramah, but they might wander about the country, or reside in the temple, or even have an honoured place at court. Some of them may have been official priests, too, and it is wrong to suppose they were in permanent and institutional conflict with the priesthood and the official cult. Of course they all believed with Samuel (1 Sam. 15:22), 'Behold, to obey is better than sacrifice, and to hearken than the fat of rams.' They represented the puritanical and the fundamentalist principle in Hebrew religion, as opposed to the ceremonial and mechanistic tradition of the priesthood. Thus they tended to accuse the priests of misleading the Hebrews into believing that provided they attended the festivals and paid their temple offerings they did not need the spiritual impulse. The prophets urged that both were essential. This was the message of Moses, the archetypical priest-prophet, and his teaching was most clearly upheld by Ezekiel who demanded absolute ritual purity as the only sure protection against idolatry, but moral sincerity also.

Prophets, then, were notable for their insistence on the fundamental moral content of religion, and their harking back to the meaning of the Covenant. But they were also judged on their ability to predict. As such they were part of a great international movement in antiquity. Prophecy and oracles formed a central strand running through Egyptian history from the third millennium BC, and the *Tale of Wenamun* indicates it was also part of the Phoenician tradition at least as early as 1100 BC. From thence it spread to the Greeks. Plato, in the *Phaedrus*, says that those possessed by the gods could make assertions as mere agents of a deity without relying on human reasoning. The gift, or affliction, was called 'enthusiasm' or divine madness. In Hebrew practice the prophet might deliver himself of an oracle after a process of consultation or self communion. Or, like Elisha, he might ask for musical inspiration (2 Kgs. 3:15): 'But now bring me a minstrel. And it came to pass, when the minstrel played, that the hand of the Lord came upon him.' There is even the suggestion, in Isaiah 28:7, that intoxication was sometimes a factor: 'The priest and the prophet have erred through strong drink, they have swallowed up of wine, they are out of the way through strong drink; they err in vision, they stumble in judgment.'

The importance of the prophet in Israel's political history is that once such a man established himself in the public estimation as a seer who felt 'the hand of the Lord', he was an alternative source of divine guidance to the monarch and therefore a check on the monarch's activities. He presented the democratic, or rather the populist, tradition. Of course the kings tried to win over the prophets by giving them an honoured place at court and by putting up with their contradictions and verbal chastisements. Even so, prophets of any status were not easily neutralized by royal patronage and they remained angular figures even in the Presence Chamber. They were a reminder that the king, like everyone else, was under the Law. In this respect Israel can always be correctly described as a state under the rule of law, with the prophets forming an unofficial constitutional court.

The Hebrews' first experiment in monarchy, under Saul, must be described as a failure. Within a year or so of his coronation, the Philistines came up through the Plain of Esdraelon and defeated the royal army at Mount Gilboa, killing Saul and his son Jonathan. Saul is strikingly portrayed in the Books of Samuel as a man hovering on the brink of madness and sometimes slipping over it, gifted and brave, alternating between sudden generosity and angry violence, an unpredictable oriental potentate. But his failure was due to rather more than his faults of character. It is clear that he lacked the political skill to unite the North and the South into a unitary kingdom. And his personal heroism and energy were in the end not enough to compensate for the superior military organization of the Philistines.

David was a different kind of figure altogether. Although as a youth he came to prominence as Saul's armour-bearer, his real military and political training was with the Philistines. He became an expert professional warrior and commander and was awarded a feudal fief under Achish of Gath. He might indeed have identified himself wholly with the Philistines. As it was, he built up a group of professional knights and soldiers attached to him personally and who expected to be rewarded by him with land. It was around this nucleus that he created his victorious Israelite army and his personal regiments of guards remained faithful to him even during the populist conspiracy of his son Absalom. If David brought Israel fully into the Iron Age in a military sense, he also imposed upon it the more sophisticated political concepts of the coastal cities. We can write of David with some confidence because we are entering a period rich in historical annals, some of them manifestly based on official court records. There are, of course, many lost books of the Old Testament, to which we have tantalizing references – 'The Book of the Acts of Solomon', 'The Book of the Chronicles of the Kings of Israel', 'The Book of the Chronicles of the Kings of Judah', as well as at least one lost 'prophetic' source. Nevertheless, the two Books of Samuel, especially the second, are admirable records, compiled by a man or men intimately acquainted with the

A portrait of David as psalmist from *The Psalter of the Duc de Berry*, early fifteenth century AD. David's love of music was inherited by his son, Solomon, who made it a central feature of his temple cult.

events described. They rank as one of the masterpieces of historiography in deep antiquity. During the monarchy, moreover, we begin to get a number of synchronisms with extra-Biblical sources such as the Assyrian *limmu* or eponym lists, which can be cross-referenced to Hebrew king-lists; and with the Egyptian canon of Ptolemy to hand we can make accurate datings. During the early monarchy the dates may be ten or so years out, but no more, and in the later monarchical period of the two kingdoms we get virtually absolute dates. We can say with some confidence that Saul was killed *c.* 1005, David reigned until *c.* 966 and that Solomon died in 926 or 925 BC.

During David's forty-year reign, he created and maintained a large mercenary army and with it he penned the Philistines into a narrow coastal strip. He then set about making Israel the paramount state in the whole area. As we have noted, the Israelite tribal confederation was only one of a number which moved in from the desert. Each had its narrow, particularist religion and national god, and each fought holy wars against its neighbours, envenomed by *odium theologicum* as well as land-hunger – hence the merciless savagery to which the Old Testament bears abundant witness. David effectively cleared up this situation. Within the area of primary Hebrew settlement he eliminated the non-Hebrew pockets or forcibly converted them; he got himself accepted as king first by the southern tribes, with their capital at Hebron, then by the northern, with their sanctuary at Shechem. He cleared and occupied the coast both of Ashdod and as far as the frontiers of Tyre in the southern Lebanon, and inland to the north he made

Damascus a dependency with a governor. To the east and south he turned the petty kingdoms of Edom, Moab and Ammon – enemies of the Israelites and their god for generations – into vassal states. This was not conquest for conquest's sake. He was in the process of turning Israel into a viable, centralized and 'modern' state, and such a state needed a source of income other than taxes on its fields and flocks. David thus secured sea-routes by his occupation of the Mediterranean coast and by his conquest which brought him to the head of the Gulf of Aqaba on the Red Sea. Equally important, the territories he now controlled to the east enabled him to tax the inland desert caravan route, which went from south to north, from Arabia (and ultimately India), up through Edom, Moab and Ammon, to Damascus, and so to the Phoenician coast.

David's master-stroke was the conquest of Jerusalem. This was the most important city of the interior and also the most heavily defended. Indeed, for over 200 years the Israelites had been unable to take it: 'As for the Jebusites, the inhabitants of Jerusalem, the children of Judah could not drive them out: but the Jebusites dwell with the children of Judah unto this day.' This was still true when David became king. He had been invested by covenant at both Hebron and Shechem, but his kingship of both the southern and the northern tribes was conditional and he wanted a central capital city which was his by absolute right of conquest. Jerusalem was the obvious, indeed the only, place for it. The Second Book of Samuel, combined with the excavations of Kathleen Kenyon, tell us what happened. Jerusalem is built on the eastern side of the main mountain spine of Palestine. The Old City as we know it is spread over three valleys: Kedron on the east, Hinnom on the west, and a central valley which Josephus called the Tyropoeon – all three eventually converging into the Brook of Kedron to the south. The earliest, pre-Hebrew city was built on the eastern ridge because this was the only one which had a constant water-supply, from the Spring Gihon. As at Megiddo, Gibeon and Gezer (and, as we have seen, elsewhere in antiquity in the second millennium BC), a secret tunnel connected the interior of the city to the spring, so that water could be obtained even when under siege.

The Jebusites were sure they could hold the city against David – hence their sneer about the blind and the lame (2 Sam. 5:6). The discovery of the secret water conduit illuminates David's command: 'And David said on that day, Whosoever getteth up to the gutter, and smiteth the Jebusites, and the lame and the blind, that are hated of David's soul, he shall be chief and captain.' It was Joab who performed this exploit, climbing up the conduit and so getting inside the walls (a French soldier in the early thirteenth century AD likewise helped Philip Augustus to take Château Gaillard by climbing up a latrine-shute). Kathleen Kenyon also found part of the original walls, dating from about 1800 BC, which were still in use in David's time. From this it was possible to map the Jebusite and Davidic city on the east ridge, south of the great Herodian wall. It was a small place, 150 yards wide and 400 yards long, enclosing less than 11 acres. The houses within were built on a series of terraces, at an angle of 45 degrees; and the terraces or *millo* (a word originally meaning 'filling') had to be constantly maintained to prevent the city sliding down the hill in heavy rain; hence 2 Samuel 5:9. 'And David built round about from Millo and inward.'

The archaeological evidence suggests that while David kept the city in good repair he did not develop and expand it, no doubt because he was too busy enlarging and consolidating his kingdom. The rationalization give in the First Book of Chronicles is that Yahweh told him not to build a temple for the Ark because he was a man of war who had shed blood. Nevertheless, David did the politically decisive things – he brought the Ark, which was the symbol of Hebrew unity, to his own city of Jerusalem, which was his not by election or covenant, but by right of conquest, and which he could bequeath by will. Hence he moved a further stage towards autocratic monarchy by identifying the throne with his own house. The house of David became the royal line and the kingship hereditary, though not by primogeniture. In fact three of the Davidic princes who might have succeeded, Absalom, Amnon and Adonijah, met violent ends, and David exercised the absolute right to designate his successor, even in his dotage. Solomon did not have to take the field to secure his throne – unlike all his predecessors he never seems to have been an active general – though he behaved with great ruthlessness to put down potential opponents. In all essentials he was a king by divine right.

The reign of Solomon marked the culminating point of the Jewish state and kingdom, though not its greatest extent, for Solomon never exercised power over Damascus as his father did. He preferred to win glory through trade and diplomacy rather than war. His greatest coup was to 'make affinity with Pharaoh king of Egypt' by marrying his daughter, the greatest matrimonial prize of the day, who brought as dowry the city of Gezer. The Old Testament (1 Kgs. 11:1) says Solomon 'loved many strange women together with the daughter of Pharaoh, women of the Moabites, Ammonites, Edomites, Zidonians and Hittites', and this almost certainly refers to a series of marriage alliances with neighbouring and vassal states. Solomon also formed an alliance with Hiram, the great king of Tyre. In all things he was anxious to assume the panoplies and ceremonials of established monarchy. He imitated Egyptian coronation ritual, and it is doubtless from his time that we can date the only title list of the kingdom of Israel that has survived (Isaiah 9:6): 'And his name shall be called Wonderful, Counsellor, the Mighty God, the Everlasting Father, the Prince of Peace.' In his day, Solomon was famous for his appreciation of beauty, his love of luxury and his wisdom. The last manifested itself in his judgments in hard cases, which he tried personally, and in his witty exchanges with friendly monarchs. Josephus says that Solomon would hold riddle-contests with Hiram of Tyre – a not uncommon form of diplomatic exchange in the Near East of the Early Iron Age – with large sums of money or even cities as the forfeit. Solomon also turned his hand to hymns, poetry and proverbs.

The Old Testament, confirmed by a good deal of archaeological evidence, shows that Solomon held his kingdom together, and administered and taxed it by building and fortifying a series of royal cities which included Hazor, Megiddo and Gezer. The evidence of previous Israelite building, which is to be found at some of the smaller cities, Beth She'an, Lachish, Beit Mersim and Bethshemesh, is undistinguished, a mere continuation of Canaanite models. Solomon's royal cities were something quite different, and almost certainly the work of specially recruited foreign masons, probably from the Phoenician coast, but quite possibly from Egypt also. At Hazor, Gezer and Megiddo, Solomon built royal quarters, that is a primitive form of acropolis or upper town, with a fortress-residence for himself or his governor, a sacred area, and barracks and storage facilities, the whole walled and separated from the rest of the city. At Hazor, Yigael Yadin found that Solomon restored the deserted town, burned by Joshua 200 years before, and made the high *tell* itself into his royal enclave, surrounding it with casemated walls and a magnificent high gateway. At Megiddo there were two palaces, one for the king, one for his deputy or governor; and there is evidence of the same casemate-type walling and a grand gate with two towers, as at Hazor. At Gezer, Solomon again found the opportunity to build afresh. 1 Kings 9:16 says: 'For Pharaoh, King of Egypt, has gone up and taken Gezer, and burnt it with fire, and slain the Canaanites that dwelt in the city, and given it for a present

Phoenician merchants bringing monkeys as presents to the Assyrian king Ashurnasirpal II. These merchants were 'servants of Hiram', king of Tyre, with whom Solomon formed an alliance.

THE CIVILIZATION OF THE BIBLE

unto his daughter, Solomon's wife.' Here again the new royal quarters was magnificently walled, and equipped with the characteristic towering gateway. We still possess a drawing of a relief (which was subsequently lost) showing what Gezer looked like when it fell to Tiglath-Pileser III, the Assyrian king. The First Book of Kings calls these places 'chariot cities', and it is evident that Solomon had a large mercenary army equipped with a massive chariot force which was based on these cities and on Jerusalem. But the stables found at Megiddo and elsewhere are, in fact, from later reigns, when the numbers of chariots were further increased. Solomon built the basic structure of royal power, to which his successors added.

His greatest work of course was at Jerusalem, the chief of his royal fortress-cities, where he made good his father's omission to build a temple to the Lord. The Book of Kings says the whole building programme was financed by a capital levy: 'And this is the reason of the levy which King Solomon raised: for to build a house of the Lord, and his own house, and Millo, and the wall of Jerusalem, and Hazor, and Megiddo, and Gezer.' Millo, of course, was the old Jebusite and Davidic city on the southern terraces; 'his own house' was the royal palace inside its fortified royal quarter; a new wall was built round the city as a whole, and 'the house of the Lord' was the first temple to be built in Jerusalem. Nothing

The remains of the city gate at Gezer, where Solomon did much rebuilding.

now survives, or at any rate is visible, of Solomon's temple. The site was the threshing-floor of Araunah the Jebusite, bought by David and on which he had erected an altar; this must have been outside the original city wall. Kathleen Kenyon concluded, almost certainly rightly, that Solomon's temple must have been beneath the site of its two successors, built by Zerubbabel after the return from exile, c. 515 BC, and by Herod the Great in the first century BC. The northern boundary of the old Jebusite-Davidic city is separated by a gap of 200 yards from the south wall of the present Haram temple-platform, the work of Herod the Great, which survives to a height of 128 feet from its foundations. In the 1960s, excavations revealed a 'join' in the platform in the south-east corner, where Herod appears to have enlarged the area by about 100 feet. His very characteristic masonry merges with masonry from two earlier periods. The earlier masonry has a Persian look about it, and resembles Persian work at Byblos and the great Temple of Eshmoun near Sidon, which is late sixth, early fifth century BC. This could, then, be the work of Zerubbabel, which almost certainly followed the line of Solomon's temple, with the actual Solomonic masonry underneath. As for Solomon's wall, it looks as though it lay inside the line of Herod's giant wall, which of course includes the Wailing Wall, and archaeologists have been able to plot the route it followed. But nothing of this survives, because it was made into a quarry under the Romans, and the rock on which Solomon's work stood was literally quarried away. The rest of his wall seems to have collapsed and disappeared completely. So Solomon's Jerusalem has vanished, except the portions which are forever hidden under Herod's massive construction.

It is clear, however, that Solomon built his temple and palace about 250 yards from the old city he inherited. For the rest we must rely on the literary evidence, given in chapters six to seven of the First Book of Kings. We know that the craftsmen were imported from the coast. Solomon had written to Hiram: 'Now therefore command thou that they hew me cedar trees out of Lebanon; and my servants shall be with thy servants according to all that thou shalt appoint: for thou knowest that there is not among us any that have skill to hew timber like unto the Sidonians.' In fact the King of Tyre sent craftsmen of all sorts, including another Tyrian called Hiram, a superb artist in bronze. They naturally built in the Phoenician tradition, and Solomon's temple was thus akin to the most sophisticated Canaanite temples of the period. Useful surviving comparisons are the Late Bronze Age Canaanite temples at Lachish and Beth She'an, the two temples in the Canaanite city of Hazor, destroyed by Joshua in the thirteenth century BC, and the ninth century BC temple at Tel Tainet in Syria. The last three temples were built on the three-room principle, along the same axis, a form which originally derived from Egypt, via Byblos. Solomon's temple was small: it had three rooms, all thirty-three feet wide, one in front of the other – the Ulam or porch, sixteen feet long, the Hekal or holy place, sixty-six feet long, and the Holy of Holies, which was thirty-three feet square. The last room was totally dark, as in the inner sanctums of Egyptian temples: 1 Kings 8:12 says: 'Then spake

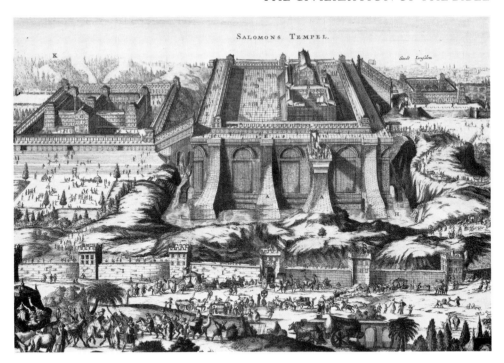

The Temple of Solomon: a representation of the fanciful seventeenth century model built by a rabbi in Hamburg.

Solomon, The Lord said that he would dwell in the thick darkness.'

In the temple, the stone was perfectly dressed, which implies the use of Phoenician masons. There was a cedar floor, and cedar was also used to line the walls, being carved with birds and flowers and veneered with gold. Between the Hekal or main hall, and the Holy of Holies, where the Ark of the Covenant stood, there was a screen of hanging gold chains. The Ark was protected by two winged cherubim, fifteen feet high – cherubim being originally winged lions, or other powerful beasts, a Mesopotamian, not a Hebrew concept. There was a gold altar, with ten candlesticks of pure gold, and two enormous bronze pillars, called Jachin and Boaz, each nearly forty feet high. We do not know what these pillars symbolized, unless the high standing stones of the native Canaanite religion, another case of Hebrew adoption. There was a 'basin on wheels', that is a bronze stand for a laver, similar, one presumes, to the remains of sacral objects found in Cyprus and in Megiddo. And there was the great 'molten sea', containing 2,000 baths of water, and supported on 12 bronze oxen, in which the priests performed their ritual lustrations after making sacrifices. Talmudic tradition says that the Holy of Holies also contained a stone called Even Shetiyah, or Foundation Stone, said to be the central core from which the whole earth had grown (not unlike the Greek *omphalos* at Delphi). A fourth century AD rabbinical legend locates it thus. 'The land of Israel is the middle of the earth. Jerusalem is the middle of Israel. The

Temple is the middle of Jerusalem. The Holy of Holies is the middle of the Temple. The Holy Ark is the middle of the Holy of Holies. And the Stone of Foundation is in front of the Holy of Holies.' In fact according to the Talmud, the Holy of Holies was a relic-sanctuary which contained the pillow on which Jacob slept when he had his ladder-dream; the manna-jar; the staff of Moses, and Aaron's rod. All these, together with the Ark, were the venerated cult-objects of Yahweh-worship, but all had disappeared before 587 BC, the date of the fall of Jerusalem. The Talmud says they were buried for safety and will be restored by Elijah when the Messiah comes.

It may be asked, what were these humble reminders of the desert origin of Yahweh-monotheism doing in Solomon's magnificent Phoenician-style temple? Solomon was clearly taking over a lot of the magnificent apparatus of Canaanite-Phoenician religion, which to the Hebrews was mere image-worship. He was seemingly aware of the acute religious dilemmas which his creation of a temple-shrine posed. Thus, although his own palace was part of the temple complex, built with a vast hypostyle hall of forty-five cedarwood pillars, in the Egyptian fashion – 'the house of the forest of Lebanon' – the separate palace he built for his Egyptian wife, who retained her old religion, was outside the sacred area: 'My wife shall not dwell in the house of David King of Israel, because the places are holy, whereunto the Ark of the Lord hath come.' The implication was that Yahweh actually dwelt in the new temple: 'I have surely built thee an house to dwell in, a settled place for thee to abide in for ever.' Yet in the same chapter eight of the First Book of Kings, Solomon contradicted himself: 'But will God indeed dwell on the earth? Behold the heaven and heaven of heavens cannot contain thee; how much less this house which I have built?' He seems to have resolved the difficulty by supposing a kind of symbolic presence: 'That thine eyes may be open towards this house night and day, even towards the place of which thou hast said, My name shall be there.' But Yahweh's presence, if only symbolic, was real enough to produce a powerful *shekinah*, or divine radiation, which went outwards from the Holy of Holies, in concentric circles of diminishing intensity.

The temple, of course, replaced the desert tent which, in the days of Moses and Joshua, had contained the Ark, just as Jerusalem, the new royal city of David, had replaced Sinai or Gerizim as the stationary site for the Lord's abiding. Scholars of the Wellhausen school reject this explanation. They regard the building of the temple as the virtual adoption by Solomon of Canaanite religion and they maintain that the tent was invented by the post-Exilic priests who (according to their theory) compiled the Pentateuch, to provide a primitive prototype for the temple arrangements, and so explain away Solomon's manifest apostasy. Other scholars think the tent and the Ark were rallying-points of two different theological parties among the Israelites.

Such explanations are almost perversely complicated and far-fetched. Yet it does seem that the early Hebrews were not clear in their own minds whether Yahweh was universal and ubiquitous, or whether he did concentrate his

THE CIVILIZATION OF THE BIBLE

presence in one particular place. The Ark itself was a chest of acacia, four feet long, two feet six inches deep, and the same in breadth. It was carried by poles passed through rings on either side, both chest and poles being overlaid with gold. Inside it were the tablets of the Law. In one half of their minds, the Hebrews saw the Ark simply as a repository of God's commandments, not as a cult-object to be worshipped. But in another half, the Ark was seen, originally at least, as a kind of throne, in which or on which Yahweh dwelt. The stories of the Ark's adventures in the Books of Samuel imply a very close relationship between the Ark and Yahweh, and in Numbers 10:35-6 there is likewise the assumption that the Ark was a kind of throne: 'And it came to pass, when the Ark set forward, that Moses said, Rise up, Lord, and let thine enemies be scattered; and let them that hate thee flee before thee. And when it rested, he said: Return, O Lord, unto the many thousands of Israel.'

This Mosaic view of the Ark, a kind of dog-kennel from which the Lord issued and to which he returned, certainly resembles the portable shrines of Egypt, of which we have many representations. Such a notion of the Ark as throne-repository would also explain the winged cherubim, which in the Near East nearly always implied the near-presence of a god. An Assyrian rock-relief at Maltaia shows a pantheon of gods standing on winged creatures, and Ishtar is often shown supported by cherubim – hence the Old Testament expression 'he sitteth

One of the earliest representations of the Ark of the Covenant in stone bas-relief from the second century AD synagogue at Capernaum.

enthroned upon the cherubim.' As Yahweh, like the Egyptian supergod Amun, was ubiquitous and invisible, the box did not need to contain anything except air – and, of course, the tablets of the Law.

What is clear, of course, is that Solomon was not a strict Jew, at any rate by later definitions. Not only did he graft on to Hebrew practices much of the adorational apparatus of Near Eastern absolutist religion in his new temple – it is even possible that the first High Priest of Jerusalem was a non-Hebrew associated with the Jebusite cult-centre there – but the Old Testament admits that he maintained some of the Canaanite 'high places'. His religion, in fact, was syncretistic, as perhaps befitted an absolute monarch ruling a racially and religiously mixed society, which he held together not with the sword, like his father, but by statecraft. Yet however much he took over some of the forms of non-Hebrew cults he does not seem to have changed its fundamental characteristic, which distinguished it from all other religions throughout antiquity – its rejection of images. Jewish hostility towards images is curious, since Yahweh himself was a personal deity who spoke and did things; there is plenty of anthropomorphism in the Old Testament. Moreover, it was not true, as the Hebrews supposed, that the devotees of the religions they condemned regarded images as gods; on the contrary, as Egyptian theology for instance makes abundantly clear, they recognized that gods were spiritual in essence and in no sense bound by the material of which images were made. In at any rate the more sophisticated cults of the ancient Near East, the image was a means by which the worshipper communicated with and drew nearer to god – as it still is in Roman Catholicism, for instance.

In the view of the Hebrews such a visual aid diminished God. They were not merely monotheistic, they attributed to God infinite power. God was bigger than the world or the universe. He was not one of the forces which kept the cosmos going and in order, or even the sum totality of all such forces – he was their creator. The Hebrews rejected completely the notion that devils and divinities roamed through creation doing things. Everything that happened, military victories and disasters, famines and earthquakes, plague and plenty, good and bad, was the work of one unique, all-powerful, indivisible and ubiquitous Providence. The essence of the Hebrew faith, what set it apart from all the religions of antiquity, was this rock-like belief in a single causative factor who could not be identified with any of his creations and was therefore unrepresentable. Indeed, any image of the Creator, however magnificent, was an act of lese-majesty. Hence the Jews did not merely reject images but viewed them with implacable hostility. It was this, above all, which aroused resentment against them in antiquity, especially in the cosmopolitan Graeco-Roman world, where the essence of international *politesse*, the ecumenical spirit as it were, was a friendly genuflection towards another man's gods. The Israelites were guilty of the cardinal sin of shunning intercourse: and, as Tacitus put it, of insulting what others held sacred: *Profana illis omnia quae apud nos sacra*

The inscription on the rock tomb in Siloam runs: 'The tomb chamber in the side of the rock ... This is [the sepulchre of Shebna] yahu, who is over the House (steward). There is no silver and gold here but (his bones) and the bones of his slave-wife with him ... cursed be the man who will open this.'

(They hold profane everything that is sacred to us).

The rejection of images is one reason why the material survivals of the culture of the Israelite kingdom are so meagre. The household gods and votive figures, the sacred animals and scarabs which antiquity left behind in prodigious quantities, can be found in the Hebrew levels of ancient Palestinian cities. But another reason was that the Hebrews were not much interested in death – as the Book of Job puts it, 'Thou prevailest for ever against man, and he passes; thou changest his countenance, and sendest him away' – and certainly never made an artistic cult of it. Israelite tombs hold few secrets and little treasure. We have found only one dating from the days of the Hebrew monarchy which bears an inscription, a rock tomb in the village of Silwan (Siloam) across the Kidron. Even this has been defaced, perhaps at the time it was built or shortly after. We learn from it that it was the tomb of a royal chamberlain, whose name ended in '... iah', probably Shabaniah. This is a variant of Shebna. Isaiah 22:15ff tells us that Isaiah castigated Shebna, the royal chamberlain, for his presumption in providing himself with an elaborate tomb in the rock.

What, then remains of Solomon's glory? No coins, of course: it is too early for that. But we have weights in polished stone and metal. Some are perhaps fraudulent: hollowed out in their flat base, or with pieces of metal inserted, which bear out the claim of Amos that shopkeepers kept two different sets of weights for buying and selling, 'making the shekel great and the ephah small.' It may be that we owe to Solomon the saying recorded in Proverbs 11:1: 'A false balance is abomination to the Lord: but a just weight is his delight.'

What we do know is that Solomon inherited from his father a passionate love of music and psalmody and made it a central feature of his temple cult. Music was something the Canaanites shared with their Hebrew incursors. The Beni-Hassan tablets, which have been dated *c.* 1900 BC, show Palestine nomads walking behind their animals in Egypt, accompanied by a lyre; and a lyre also features on an inlaid ivory plaque found at Megiddo, which shows court musicians entertaining a king who sits drinking on his throne. Of course the most common Hebrew instrument was the *shophar*, or horn made from a ram (and later the ibex and antelope). But Genesis 4:21 says: 'And his brother's name was Jubal: he was the father of all such as handle the harp and the organ.' David, we know, shook the sistrum; this was of great antiquity, found in Egypt and in the Sumerian levels of Ur and Kish; and Albright found a Canaanite one at Bethel in 1934. There were also pipes of various kinds. A silver single pipe was found at Ur, distant ancestor perhaps of the pipe mentioned in 1 Samuel 10:5; and double pipes feature on Akkadian tablets. Some instruments were definitely secular. Thus the *toph*, which occurs seventeen times in the Old Testament, was not allowed in the Temple precincts. But the *kinnor* or lyre, which had between three and twelve strings and corresponded to the cithara of the Graeco-Roman world, was both sacred and secular. David played it and so did the Levites, who were regular members of the temple orchestra; and this too was very old since it features on the Beni-Hassan tablet. Temple music was mostly stringed, but cymbals, trumpets and timbrels were also used. These pre-Exilic Temple musical rituals accompanied the sacrifices in Solomon's sanctuary. Some of the psalms were also sung to music, as part of the Temple liturgy (others being devotional poetry), and there is no good reason why we should reject the tradition that they were composed and perhaps set to music too by David himself, Solomon providing the grandiose architectural and decorative setting to his father's artistry.

The impression we get from the Books of Kings is that this setting was created, at any rate in the first instance, by imported craftsmen and that the material culture of the Israelites was something they acquired from the more advanced Canaanite civilization among whom they had settled, rather than something they brought with them or created for themselves. It is notable, for instance, that they learned the craft of dyeing from the Phoenicians and in time Shechem, Lydda and above all Jerusalem became leading centres of dyeing, especially the jacinth or hyacinth purples, from the *bucinum* or *murex* snail, which was regarded as the best. Though the Jews were later to become notoriously conservative in their dress, they initially adopted the habitual garments of the age and place. In Ur, they wore the long tailored shirt, later called the *kekoneth*, and a long role of cloth, worn as an outer garment or carried slung over the shoulder. This was the *simlah*, the equivalent of the Greek *himation* or the Roman *pallium*. The oldest object on which Israelites are definitely represented is the Black Obelisk of Shalmaneser III, dated 842 BC, now in the British Museum. The second of its five registers shows Jehu King of Israel and his followers paying tribute: they all wear the *simlah* over the shoulder. On its corners were tassels, in accordance with Numbers 15:3. 'Speak unto the children of Israel and bid them to make a tassel on the corners of their garments throughout their generations and to put on it a twined cord of blue.' Though this was to become the Jewish law, it was not

A detail of the 'Jehu stele', showing Jehu and his followers making their submission before the Assyrian king.

originally a Hebrew custom, for in the Jordan Archaeological Museum there is a statuette of an Aramaean deity wearing a *simlah* with tassels on the corners. The side-locks or *peoth* were not worn by Jews in antiquity but were characteristic of Libyans, Cretans and some Syrians. But gradually certain marks of appearance did become identified with Jews: the tassels or *tsitsith*, and the *tefillin*, the small leather boxes with leather thongs, each with four passages from the Pentateuch, worn on the forehead and left arm, and apparently a derivation from esoteric tattoo-marks or amulets. Jewish clothing rituals became a mark of exemplary piety – hence Jesus's remark in Matthew 23:5. 'They make broad their philacteries and enlarge the borders of their garments.' These Jewish fashions were also notable among some of the early Christians, who continued to put tassels on their garments and wore *tefillin* in church.

The highest standard of sartorial religiosity was, of course, set by the High Priest in Jerusalem, who acquired his regalia in Solomon's day. It was originally based on regal costume and did not change over the centuries of the first millennium: the dress of Assyrian kings, as for instance the relief, *c.* 650 BC, of Ashurbanipal in the British Museum, matches up well with the description of the High Priest we have from Josephus, who was a member of a priestly clan and witnessed the traditional ceremonies in the Temple before it was finally destroyed in AD 70. There was a very elaborate multi-coloured sash over an embroidered *ephod* or tunic, with a bejewelled breastplate. The jewels on the breastplate, according to Midrash tradition, corresponded to the sons of Jacob: Reuben (red), Simeon (green), Levi (white, black and red), Judah (azure), Saachar (black), Zebulun (white), Dan (sapphire), Gad (grey), Naphtali (rose), Asher (beryl), Joseph (black) and Benjamin (all twelve colours). The most complicated item was the headgear – a linen coronet, with a muslin headdress over it, then a blue headdress embroidered in gold, then a gold crown with a gold plate with the name of God, which was fixed to the High Priest's forehead.

Solomon's transformation of a tribal confederation into a centralized monarchical temple-state was a costly business, which he financed by trade and industry. He controlled the desert trade routes, and the visit of the Queen of Sheba, a South-Arabian princess who dealt in gold, spices and jewels, was undoubtedly connected with commerce. Solomon himself, like many Oriental autocrats of antiquity, himself bought, sold and manufactured on a big scale – his chief commerce being in horses and chariots. He was the arms supplier for a wide area. Much of this business was carried out in conjunction with his Tyrian allies who supplied craftsmen and mariners for his ships. At the head of the Gulf of Aqaba, the American archaeologist Nelson Glueck found the copper refinery built by Solomon and his Phoenician partners. It was on the island of Hirbet el-Kheleifeh, half a mile from the shore, where high prevailing winds were caught by the flues of the blast-furnaces. A special town, Ezion-Geber (1 Kgs. 9:26), which combined a refinery with a shipyard and port, was built for these industrial and trading activities, and like Megiddo and Hazor had a great double gateway.

Despite these efforts, Solomon's courtly magnificence and his immense building programme made it inevitable that he would oppress the people, as Samuel had warned was an inseparable consequence of Oriental monarchy. One of the functions of the royal cities Solomon built was to store the king's share of the harvest, and we possess ostraca, written in ink, recording later deliveries to the Crown of produce for taxes, recorded under regnal years (these date from about 800 BC). Even more resented was the forced labour, or corvée, an Egyptian institution which stirred up angry Hebrew memories of the sufferings before the Exodus. These exactions of the united monarchy created strains which needed all Solomon's diplomacy to smooth over. As we have seen, the Hebrew settlement was in two parts right from the beginning and the united kingdom was always to some degree an artificial creation. It is significant that when Solomon died, his heir, Rehoboam, having established his title in Judah, found it necessary to go to Shechem, the northern shrine-city, to be crowned as king of Israel – so Jerusalem was not accepted as a national capital. There, in answer to the protests of the northerners, who not only resented Solomon's impositions but no doubt regarded the loss of Damascus as neglect of their interests, Rehoboam, in one arrogant sentence, undid all his father's careful diplomatic work: 'My father chastised you with whips, but I will chastise you with scorpions.'

The break up of the kingdom which followed was never healed. The two kingdoms of Israel in the north, and Judah in the south, went to their historical doom separately. For a time, the northern kingdom flourished. King Omri, who gets only brief treatment in the Old Testament, was in fact a formidable monarch, who evidently gained control of northern Moab and was exacting tribute. This is made clear by the Moabite Stone, discovered in 1866: 'I am Mesha, son of Chemosh ... King of Moab, the Dibonite ... Omri King of Israel ... oppressed Moab many days because Chemosh was angry with his land. And his son succeeded him, and he also said I will oppress Moab.' Chemosh was, of

The site of 'King Solomon's pillars' near Eilat, which was believed to be the slave-camp for the copper mines.

The west gate of the city of Samaria, later known as Sebaste, showing the round Herodian tower built on the remains of a square Greek tower.

course, the pagan god for whom, as we learn from 1 Kings 11:6–7, Solomon in his diplomatic way, had made provision: 'Then did Solomon build an high place for Chemosh, the abomination of Moab, in the hill that is before Jerusalem.' Omri linked his inland kingdom with the sea by marrying his son Ahab to the King of Sidon's daughter Jezebel, and a new capital of the northern kingdom was built on the hill at Samaria, from which the sea can be seen twenty miles away.

Samaria is one of the most fascinating sites in Palestine, for it was later settled by Macedonian veterans in Hellenistic times and then rebuilt by Herod the Great as Sebaste, a Roman-style city where much of his work survives. But its Israelite

A Phoenician ivory panel carved with cherubim back to back, found in the North-West Palace at Nimrud.

origins, which lie beneath, are also of great interest, for it is one of the earliest eastern cities of which we have an approximate foundation-date, c. 875. Like Solomon's royal cities, it had a walled acropolis on the hilltop, and a fortified palace quarter. It was the scene of the most ferocious and prolonged spiritual battles of the Old Testament; here Ahab deserted Yahweh for his wife's worship of Baal: 'But there was none like unto Ahab, which did sell himself to work wickedness in the sight of the Lord, whom Jezebel his wife stirred up. And he did very abominably in following idols ...' (1 Kgs. 21:25–6). Ahab reigned very successfully for twenty-five years, nonetheless, and twice defeated the King of Damascus, until he was killed in his chariot by 'a certain man' who 'drew a bow at a venture.' He left behind him 'the ivory house he had made', that is to say one of the magnificent ivory-decorated palace rooms which were a feature of royal splendour in the Near East of the Iron Age. In the excavations at Samaria in 1931–5, a deposit of decorated ivory was discovered, and specimens of this work are now in the Rockefeller and National Museums in Jerusalem. More illuminating, however, are the ivory plaques in low relief discovered by Sir Max Mallowan at the Assyrian capital of Nimrud. The Assyrian King, Sargon II, destroyed and looted Samaria in 722–1 BC, and his store-room at Nimrud, which was full of precious ivories, may have included some of the finer portions of Ahab's 'house of ivory'. The Nimrud plaques, which were chiefly fixed to bedsteads and other royal articles of furniture, give some impression of the lost splendours of Israel and Judah in the age of monarchy.

Equally illuminating are the changes and improvements which such kings as Ahab and Jehu made in Solomon's chariot cities where they built immense stables. Ahab's army of chariots was the largest in the coalition of small states from Syria and Palestine which fought the Assyrians at the Battle of Karkar in the Orontes valley in north Syria in 853 BC. The records of King Shalmaneser III say that Ahab contributed 2,000 chariots and 10,000 infantry. Jehu also was a great chariot man. When he descended on Jezreel to seize the Kingdom of Israel from Ahab's son Joram, the watchman reported: 'And the driving is like the driving of Jehu, the son of Nimshi; for he driveth furiously.' The first batch of stables excavated at Megiddo, and evidently built by Jehu, housed 450 horses: with three horses per chariot, this would mean a regiment of three squadrons, with 50 chariots each.

There is no doubt that the military effort of the Northern Kingdom delayed the advent of Mesopotamian power, represented by the warlike autocracy of Assyria, and its successor, the Neo-Babylonian kingdom. But the triumph of force was probably inevitable in the long run. The independent kingdoms of the Hebrews, and for that matter the self-governing city-states of the Phoenicians, flourished in the gap between two great ages of empires – the Egyptians and Hittites in the second millennium, and in the first millennium, the Assyrians and Babylonians, followed by the Persians, Greeks and Romans. The gap was probably made possible by the coming of the Early Iron Age, bringing with it a revolution in military technology. For a time at least the new iron weapons, manufactured in quantity by such enterprising small states as Solomon's Israel and Hiram's Tyre, kept the great empires at bay. But in time their size and organized military power made the extinction of such city- and temple-states inevitable. Israel and Judah, too, were weakened by their division and by internal conflicts over religious and political policy. The royal acropolis built by Omri and Ahab at Samaria carried the physical separation between the palace quarter and the people, which Solomon had first established in his cities, a noticeable stage further. The citadel dominated and oppressed the town. This was not a part of the Hebrew tradition and was never accepted by all the people, who remained attached to residual concepts of tribal democracy.

The social breach was envenomed by religious conflict. In the north, Canaanite and Phoenician cults gained a stranglehold over the palace culture, but prophets such as Elijah and Elisha sprang up to remind the people of Mosaic fundamentalism, and of the subordination, monarchy included, to the law of the Pentateuch, which was already in some kind of written form. In the southern kingdom of Judah, which was materially poorer but which had the Ark and the Temple, there was a similar conflict between Canaanite idolatry and the periodic revivals of the pristine monotheism based on the Covenant. Isaiah sharpened and systematized the ethical approach to life which was implicit in the Mosaic teaching, and Jeremiah taught that, even without the cultic apparatus of the Ark and the Temple sacrifices, the Covenant with Yahweh could be fulfilled by personal morality and observance of the law – thus preparing Judah for its spiritual survival in Babylon without a Temple. In a real sense, Hebrew religion

A limestone relief from the South-West Palace at Nimrud depicting Tiglath-Pileser III capturing Ashtaroth-Karnaim (Astartu) in Gilead. *Above* the city, standing on a mound and surrounded by a double wall with towers. *Below* the king standing in his state chariot beneath a parasol.

The clay prism containing the final edition of the Assyrian king Sennacherib's 'Annals' in cuneiform script, 691 BC. The inscription talks of 'Hezekiah the Jew' as a 'caged bird'.

was maturing during the late-monarchical period, but the new concepts were difficult to grasp and imperfectly understood – they were, after all, entirely new in the history of world religion – and so Judah, no less than Israel, faced the imperialist threat divided and doubtful.

We have noted that, even in the ninth century BC, the Black Obelisk of Shalmaneser shows that Jehu had been forced to pay tribute to Assyria. In 740 BC the Assyrian annals of Tiglath-Pileser record a further stage in the subjugation of the Northern Kingdom under King Menahem of Israel: 'As for Menahem, terror overwhelmed him ... he fled and submitted to me ... silver, coloured woollen garments, linen garments ... I received as his tribute.' The end came in 722, under Sargon II, as the *Khorsabad Annals* note: 'I besieged and captured Samaria, carrying off 27,290 of the people who dwelt therein. Fifty chariots I gathered from among them ...' In accordance with Assyrian imperial practice, the city and the surrounding area was resettled by aliens forcibly moved from another part of the empire:

So Israel was carried out of their own land to Assyria unto this day ... And the King of Assyria brought men from Babylon, and from Cuthat, and from Ava, and from Hamath, and from Sepharvaim, and placed them in the cities of Samaria instead of the children of Israel: and they possessed Samaria and dwelt in the cities thereof.

All traces of Israelite royalty were systematically erased. At Samaria, the royal quarter was completely destroyed. At Shechem and Tirzah the towns disappeared. Megiddo was smashed, and Assyrian-type buildings erected on the rubble. Hazor lost its walls. Some Israelite peasants and artisans remained and intermarried with the new settlers, and they kept up the cult of Yahweh at the shrine of Mount Gerizim, near their original capital of Shechem. But the worship of Yahweh in the north had always been suspect in the eyes of the Judaeans or Jews of the south. Such suspicions reflected the bifurcation of Hebrew settlement and history which had taken place even before the Egyptian period and the Exodus. They were increased by the Northern acceptance of Canaanite paganism. The circumstances of the fall of the northern kingdom, and the intermarriage of the remnant with aliens was used as an excuse to deny to the Samaritans their Hebrew heritage – from now on their claim to be part of the true community of Israel was never again fully recognized by the Jews.

In any case, the religious forms practised in the south began to change as the kingdom of Judah, spurred on by the shock of seeing its northern sister destroyed, took heed of prophetical warnings and reformed itself. The changes introduced by Josiah, following the discovery of ancient law-books in the Temple – probably the original text of the Pentateuch – centralized all sacrifices in Jerusalem: 'And he brought all the priests out of the cities of Judah, and defiled the high places where the priests had burned incense, from Geba to Beersheba ...' (2 Kgs. 23:8–9). These religious reforms had been preceded by more practical preparations in Jerusalem, under Hezekiah, to meet the Assyrian threat. The Second Book of Chronicles says that Hezekiah enforced the pure worship of

Hezekiah's inscription from the Siloam Tunnel. The Tunnel conducted water from the Spring Gihon to a rock-cut cistern within the walls of Jerusalem so that the people could obtain fresh water when under siege.

THE CIVILIZATION OF THE BIBLE

Yahweh, but he also 'set to work resolutely and built up all the wall that was broken down, and raised towers upon it, and outside it he built another wall.'

One remarkable aspect of Hezekiah's defensive work was brought to light in 1867, when the Siloam Tunnel was investigated. This conducted the water from the spring Gihon to a rock-cut cistern from which a channel carried the overflow down into the Kedron Brook. From within the town there was access to the huge cistern, and the strategic beauty of the arrangement was not merely that the garrison could obtain fresh water while Jerusalem was under siege, but that the besiegers were quite unaware of the arrangement. In 1870, the American family of Vesters, who founded the celebrated American Colony Hotel in Jerusalem, discovered in the tunnel a remarkable Hebrew inscription recording its construction:

This is the story of the boring through: while [the tunnellers lifted] the pick, each towards his fellows, and whilst three cubits [yet remained] to be bored through, [there was heard] the voice of a man calling his fellow, for there was a split in the rock on the right hand and on [the left hand]. And on the day of the boring through, the tunnellers struck, each in the direction of his fellow, pick against pick. And the water started to flow from the source to the pool, twelve hundred cubits.

By such means Jerusalem remained independent for another century. In fact the Assyrians never took it. In 612 their empire suddenly collapsed to the new power of Babylon. Seven years later the Babylonians won a great victory at Karchemish in Syria, destroying the Egyptian army, the 'broken reed' on which Judah had relied for support. In 597 Jerusalem fell to the army of Nebuchadnezzar. This was the end of the royal city. A section of the *Babylonian Chronicle* dealing with the years 626–539 BC, now in the British Museum, notes laconically:

In the seventh year, in the month of Kislev, the Babylonian king mustered his troops, and having marched to the land of Hatti, besieged the city of Judah, and on the second day of the month of Adar took the city and captured the king. He appointed therein a king of his own choice, received its heavy tribute and sent [them] to Babylon.

We have here the exact date of the fall: 16 March 597 BC. The Old Testament (2 Kgs. 24:14) fills in the story. The king, Jehoiakim, was taken to Babylon with 'all Jerusalem, and all the princes, and all the mighty men of valour, ten thousand captives, and all the craftsmen and all the smiths; none remained, except the poorest people of the land.' The treasures of the Temple were carried off, the gold being 'cut into pieces.'

Under its Hebrew governor, Zedekiah, the city revolted and was again besieged. In 1935 J.L. Starkey found in the excavated gatehouse of Lachish a group of inscribed ostraca now known as the *Lachish Letters*. Dating from the autumn of 589 and covering the last phase of Jerusalem as a free Israelite city, they are despatches from a military outpost to a staff officer at Lachish. One has a tantalizing reference to a 'prophet', perhaps Jeremiah. Another reads: 'When the army of the king of Babylon was fighting against Jerusalem and against all the cities of Judah that were left, Lachish and Azequah, for these were the only fortified cities of Judah that remained.' Starvation eventually compelled Jerusalem to surrender in 587. Zedekiah was blinded and sent to Babylon. The Temple was wrecked, the fine houses demolished, the walls pulled down and the terracing collapsed, all the old town of Millo sliding into the ravine.

Between 734 and 581 BC there were six distinct deportations of the Israelites, and others fled voluntarily to Egypt and elsewhere. The Edomites moved in from the south-east, pushed by the insurgent Arab tribe of the Nabateans. In the first years of the Babylonian exile, the central shrine of Judah shifted from the Temple to Bethel in the hills of southern Ephraim, and the political centre moved to Mizpah in the hills north of Jerusalem. But the Jews were comparatively well treated in Babylon. Tablets found near the Ishtar Gate of the ancient city list the rations allotted to captives, including 'Yauchin, king of the land of Yahud' (595–570 BC). This, of course, is Jehoiakim. Some of the exiles became wealthy merchants, and the first success-stories of the Diaspora were written. During the exile, the old Hebrew faith gradually transformed itself into Judaism. Obliged to develop in Babylon a non-sacrificial type of worship, the Jews, under the teaching of Ezekiel, now conformed to a rigid monotheism with high standards of personal and social morality, and a stress on fasting, confession, prayer, law-reading and study. From this phase we can identify the rise of the scribal class as custodians and commentators on holy writ; they took over the leadership and founded the Pharasaic Judaism which was to survive the destruction of the Temple by the Romans. In exile, the Jewish faithful tried to differentiate themselves as much as possible from the pagan Babylonians: indeed, the period seemed to make Yahweh more powerful and his relationship with the Jews more exclusive, for Yahweh was behind everything that had happened in the great convulsion of

The clay cylinder inscribed in cuneiform with a text of Cyrus the Great, king of Persia.

Near Eastern history, and had used the mighty Assyrian and Babylonian empires as the rod of his anger, to execute his judgment against his chosen people. As Yahweh became more powerful in Jewish eyes, so he became more attached to his people, and the Israelites struggled to preserve their special identity in a hostile world striving to annihilate them. For this purpose the Jews developed the three great institutions of circumcision, the sabbath and the law. In these developments the heroes were Ezekiel the priest-prophet, Ezra the scribe, and Nehemiah the builder and statesman.

In fact the Babylonian empire survived for only fifty years after the final fall of Judah. It was replaced by the alliance of the Medes and Persians created by Cyrus the Great. The faith of the new ruling Persians was universalistic and ethical, and therefore tolerant, rather than narrow and nationalistic. Inscriptions show that Darius I, 522–486, was definitely Zoroastrian but even in Cyrus's day, the influence of Zoroaster was strong and growing. Hence the Persians developed a completely different imperialistic policy to the Assyrians and Babylonians. Like the Romans who eventually replaced them, they tended to support the religious beliefs of their subject peoples, and they seem to have been anxious to reverse the monstrous deportations and persecutions of their predecessors. The story of Cyrus's conquests is given in the *Babylonian Chronicle* and in his own foundation-cylinder, from the temple of the Moon-god at Ur, which he restored. On this Cyrus says he was appointed to punish Nabonidus's insults to Marduk, the God of Babylon. Marduk 'looked through all the countries, searching for a righteous ruler. ... He called the name of Cyrus, King of Anshan, and pronounced his name to be ruler of the world.' This is echoed by Isaiah 45:1–2. 'Thus said the Lord to his annointed, Cyrus, whose right hand I have holden, to subdue nations before him ...' The cylinder, now in the British Museum, says Cyrus, 'resettled upon the command of Marduk, the great Lord, all the gods of Sumer and Akkad whom Nabonidus had brought into Babylon

... in their former chapels, the places which made them happy.'

In this general restoration of peoples and cultic objects, the Jews shared. Cyrus gave back the Temple vessels looted by Nebuchadnezzar – now more important than ever, for the Jews had come to believe during the Exile that Yahweh could only be worshipped properly in Jerusalem. Ezra 1:1–4, gives Cyrus's speech to the Jews:

> The Lord, the God of heaven, has given me all the kingdoms of the earth, and he has charged me to build him a house at Jerusalem, which is in Judah. Whoever is among you of all his people, may his God be with him, and let him go up to Jerusalem, which is in Judah, and rebuild the house of the Lord, the God of Israel – he is the God, who is in Jerusalem.

The first Jews returned in 538 BC, under Sheshbazzar. More followed in 520, under Zerubbabel, a man of royal Davidic blood, who was appointed Governor of Judah. This group probably included Haggai and Zechariah. The Temple was rebuilt, in a more humble fashion than its Solomonic prototype, as Haggai 2:3 makes clear, though cedars of Lebanon were again used, and very likely Phoenician masons. Some portions of this building can be seen. The rebuilding of the Temple marked the official split with the Samaritans who asked to be allowed to help in the work but were refused on the grounds of their racial contamination: 'You have nothing to do with us' (Ezra 4:1ff). They complained to 'Ahasuerus' (Xerxes) but to no avail.

The Temple was finished in 515 BC, under Darius, and fifty or sixty years later work was begun on rebuilding the walls of the city. This was undertaken by Nehemiah, a Jewish official at the court of Artaxerxes I, who made him Governor of Judea, probably in 445 BC. His autobiography is told, and very vividly too, in the first six chapters of the Book of Nehemiah, written in the first person. Here is Jewish history at its best: the first survey of the ruined walls in secret, by night; the lists of those who helped and what they built; the attempts of the Arabs, Horonites and Ammonites to stop the work; its continuation under armed guard – 'For the builders, everyone had his sword girded by his side, and so builded' (Neh. 4:18), a method the Jews were to resurrect during the settlement of modern Israel between the wars. The builders went back into the city at night, but 'none of us put off our clothes, saving that everyone put them off for washing.' Nehemiah says the walls were completed in fifty-two days, no doubt in a rough and ready fashion, using old ashlars. The ruinous terraces of Millo on the east side were abandoned as beyond rescue and the new eastern wall was built along the crest of the ridge. The resulting city was smaller than Solomon's and of course it was impoverished – 'The city is wide and large but the people within it were few and no houses had been built' – but it was gradually repopulated by people from all over Judah, chosen by lot.

The next two centuries are the lost years of Jewish history. Nothing eventful seems to have happened; or, if it did, we have no record of it. Israel was a temple-

76 The south-east corner of the walls of Jerusalem. The lower right section is the original Solomnic wall, which was built upon, probably by Phoenician masons, in the fifth century BC.

THE CIVILIZATION OF THE BIBLE

state within the Persian Empire, free to practise its religion. The Jews were also free to move throughout the wide-ranging satrapies. Large Jewish communities developed and prospered, not only in Mesopotamia but in Egypt and elsewhere. Of all their conquerors, the Jews liked the Persians best: when the Egyptians rose against Persian rule, Jewish mercenaries helped to restore order.

But if, for the Jews, the centuries of Persian rule were silent and uneventful, they were also fruitful. They saw the compilation, more or less in its present

form, of the Old Testament. Before 400 BC we hear nothing of a great body of Jewish religious writing; after 200 BC, it is there. As we have seen, the civilization of the Hebrews was not innovatory or even accomplished, in the plastic arts. It has left us no great pottery or sculpture or painting; no architectural heritage. But for most of the second and first millennia BC, the Hebrews were accumulating, first stored in their memories, later in written form, a historical and religious record which, when it finally assumed a permanent shape in the last centuries before the birth of Christ, constitutes one of the greatest literatures of antiquity, or indeed of any age.

Individual authorship is claimed, specifically or by tradition, for most of the books of the Old Testament; but the Jews themselves saw their literature as essentially collective, performed by official historians in the name of God and the people, and subject to social control. Josephus, in his apology for Judaism, *Contra Apionem*, put it like this:

With us it is not open to everybody to write the records ... the prophets alone had this privilege, obtaining the knowledge of the most remote and ancient history through the inspiration which they owed to God, and committing to writing a clear account of the events of their own time just as they occurred. ... We do not possess myriads of inconsistent books, conflicting with one another. Our books, those which are justly accredited, are 22 in number, and contain the record of all time.

The phrase 'justly accredited' implies a canon. The word 'canon' was originally Sumerian, meaning a reed; hence straight, upright. The Greeks use it in the sense of rule, standard or boundary. But the concept of canonical writing is undoubtedly Jewish, and is very ancient. It means divinely inspired pronouncements of unquestioned authority, and as the divine spirit operated through prophets, each book or work had to have a prophet as its accredited author. Hence the idea of the canon probably emerged when the first five or Mosaic books, the Pentateuch or Torah, came into existence in written form, which may have been as early as the period of Samuel. It expanded enormously in the immediate pre-Exilic period, when readings of the Torah, and possibly other books, became a regular part of Temple services, and it received a further impulse during the Exile, and in the continuing Diaspora, when the non-sacrificial service and the synagogue became a permanent part of the Judaic cult. But of course the canon varied. The Samaritans recognized only the Pentateuch, the basic law-code, which ended, like the earlier laws of Hammurabi, on which to some extent it was modelled, with an imprecation against anyone who added to, or took away from, its contents. Having been excluded from the rebuilding of Jerusalem and further alienated by the reforms of Ezra and the whole scribal movement, the Samaritans took no part in the canonization of the post-Mosaic writings and therefore did not recognize them; indeed they eventually (Josephus says in 330 BC) built their own Temple on Mount Gerizim. They claimed that their text of the Pentateuch went back to Abishua, the great-grandson of Aaron. It undoubtedly represents one of the oldest traditions and it is remarkably incorrupt, though deliberately changed in places to suit Samaritan sectarian ends.

The Septuagint, the Greek version of the Old Testament compiled among the Jewish Diaspora of Alexandria in Hellenistic times, included all the books plus the Apochrypha and pseudepigraphs, rejected by the stricter Jews because they tended to assimilate Mesopotamian and Greek myths. By contrast, the Hebrew canon comprises twenty-four books in three divisions: the Law, the Prophets and the Writings. The Law, of course, is the Pentateuch. The Prophets are usually divided into two: 'former prophets', that is the books of Joshua, Judges, Samuel and Kings (Samuel being traditionally named as the author of Judges, and Jeremiah of Kings); and 'latter prophets', that is Isaiah, Jeremiah, Ezekiel and the twelve minor prophets. The third division, the Writings or Hagiographa, comprised books dealing with various aspects of Hebrew devotion, public and private. Thus they include psalms, wisdom writing and proverbs, plus historical and prophetic writings written from a particular theological angle, and attributed to Daniel, Ezra (who is also credited with Chronicles) and Nehemiah. The Torah, which originally consisted of the Pentateuch alone, was gradually extended to include the second and third divisions.

The canon may have been complete as early as 300 BC. As it was based upon divine inspiration rather than orthodoxy, a great deal of it is severely critical of the Jews but of course it is precisely this lack of censorship which adds to its value as history and literature. Some books – Esther, Ecclesiastes, Canticles – were only admitted reluctantly, because they were in popular use on Jewish feast days. Most of the books that were rejected have vanished without trace, though in a few cases we know their titles. The Old Testament, in fact, is only a small selection of what was once a vast literature, for the ancient Hebrews were industrious and prolific writers. But in antiquity unless a manuscript were constantly recopied the work would disappear within a generation or two. The Old Testament in its canonical form survived because the professional body of scribes existed to ensure its uniform and accurate transmission from one generation to the next. As one of the most famous of the early Rabbis, Aqiba (died *c*. AD 132) put it, 'The accurate transmission (*massoreth*) of the text is a protective fence for the Torah.'

Transmitting the canon thus became a hereditary occupation for many generations of Jewish scholars, first termed scribes and then (AD 500–1000) *massoretes*. From their work the so-called Massoretic text eventually emerged, though the massoretes did not always agree and there will never be any such thing as a perfect text. Thus, the Samaritan Pentateuch differs from the Massoretic text in 6,000 instances, and of these it agrees with the Septuagint, as opposed to the Massoretic, in 1,900. And Massoretic texts vary. In the early tenth century AD, there were two leading Massoretic families in Palestine, the Ben Ashers and the Ben Naphtali. Five generations of Ashers worked on the text. Ben Asher himself copied a codex, or bound book, of the prophets in AD 895; it is now in the Qaraite Synagogue, Cairo. About 1010, one Samuel Ben Jacob made a copy of the

ABOVE The fourth century AD Codex Vaticanus is one of the earliest, and finest, of the Christian manuscripts.

BELOW The incomplete Codex Sinaiticus is of the fourth century AD.

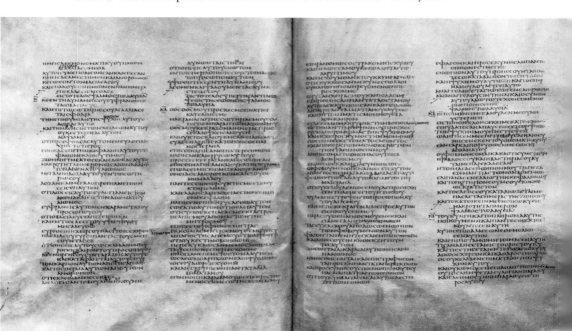

complete Ben Asher text, which is now in Leningrad. No complete Ben Naphtali text survives but their work can be seen in the Reuchlin Codex, dated 1105, now in Karlsruhe. Until the discovery of the Dead Sea Scrolls, these were the earliest surviving Jewish texts of the Old Testament, though various fragments of the Septuagint have been found in Egypt. The earliest Christian manuscripts are the fourth century Codex Vaticanus, the finest and still, of course, in the Vatican; the incomplete fourth century Codex Sinaiticus and the fifth century Codex Alexandrinus, both in the British Museum. There is also a Syriac version, the oldest manuscript of which goes back to AD 464. Since the Dead Sea Scrolls date from about 250 BC to AD 70, they antedate the earliest Hebrew texts by a whole millennium and illuminate almost every aspect of Old Testament study.

The problems which cluster around the dating and composition of the books of the Old Testament are infinite and will continue to occupy scholars until the end of time. The old 'liberal' view that all the written Old Testament is post-Exilic can no longer be maintained. It now seems probable that some books achieved written form as early as 1200–1100 BC, and that the work of composition stretched over 1,000 years. In its original and earliest form, it was a law-code, and should thus be related to the other legal texts of the ancient Near East – the Laws of Ur-Nammu, the Lipir-Ishtar Lawcode, the Code of Hammurabi – which mark the beginnings of written jurisprudence in the third and second millennia BC, and are the antecedents of the great classical codes, from the fifth century BC Code of Gortyn, in Crete, to the great compilations of Justinian in late antiquity.

The Old Testament, like the other early law-codes, begins by placing its commands in a formal religious context. What makes it unique is that the context, or setting, in religion and history, continues to expand and acquires its own momentum. The law-code remains the core, but in a sense it becomes an excuse for writing the history of a people, a civilization and its divine patron. The Hebrews were great lawyers but they were superlative historians. As an antique law-code, the Old Testament is only one, albeit the best, of many; as a work of history it is unsurpassed, for even the achievements of the classical Greeks are fragmentary by comparison. The sense of history is an important stage in the development of civilization. The ancient Egyptians and Mesopotamians did not possess it. They confined their reflections on the past to mere antiquarianism. They were incapable of creating a wide, comprehensive view of the past from individual events and so they could not present the history of their nations in unified and orderly form. There is much sense in the view that it was the Hittites who were the first to write proper history. They transmitted their skill to the Hebrews, with whom they were in contact for a great part of the second millennium BC, and then in turn to the Greeks, whose first master-historian, Herodotus, lived in Halicarnassus in Western Anatolia, where he imbibed the Hittite historical tradition. But of course the Hebrew historical tradition was already centuries old by the time Herodotus was born.

The Jews seem to have developed a kind of compulsion to find out, and then reflect upon, the origins of their nation. It began with aetiology, that is the assignment of causes: why was there a heap of stones before the city gate at Ai? (Josh. 8:29). What was the meaning of the twelve stones at Gilgal? (Josh. 4:20). This then broadened into the desire for a general explanation of the present in terms of the past. What was the Jewish people? How did it come into existence? Where was it heading and why? No nation in antiquity was so self-conscious, so well-informed about its origin and progress, so anxious to amass documentary evidence of its past. Indeed, the desire for a general explanation giving a gigantic answer to the big question of life, but capable also of supplying satisfactory answers to all the little ones, is central to Hebrew monotheism and the concept of a god with limitless power. 'Does evil befall a city unless Yahweh wills it?' asked Amos. Of course not: God is active in all things, ubiquitous, tireless, relentless, unchanging.

Thus the Hebrew sense of history helped to create Hebrew theology and the theology, in turn, gives to Hebrew history its marvellous unity and sense of purpose. But added to this was the undeniable Hebrew gift for presenting a narrative with clarity, economy and drive. The Hebrews could tell a story better than any other people of antiquity and the Old Testament can be seen as a collection of brilliant short stories. It is arguable that they acquired this gift from the Egyptians. Certainly, the Egyptians were the first to write stories – the *Story of the Eloquent Peasant*, *Sinhue*, *Wenamun* and so on, some of which in their original form go back to the third millennium BC. But the Hebrews placed such episodes within the framework of a general, purposeful narrative, something which was quite beyond Egyptian skill; moreover, each story, as it is told, is made to seem momentous and to carry moral weight. The inconsequential triviality of much of Egyptian literature, its sheer lack of gravity, is in striking contrast to the Old Testament, which conveys to the reader on almost every page a sense of great events inexorably unfolding.

The story of King David and the accession of Solomon, written evidently by a man who was an eye-witness at court, is a sustained masterpiece of historical literature, working into its main theme, without ever losing sight of it, innumerable subplots, and handling with dazzling skill a huge cast of sharply-differentiated characters. The portrait of David himself, a great man torn by conflicting emotions and urges, rivals any biography of antiquity, and contrasts notably with the presentation of Solomon, the cool and calculating sybarite. The narrator is Thucydidean; he stands apart, rarely obtruding with his judgments, allowing the actions to speak for themselves; God, likewise, rarely appears, though his controlling hand is felt throughout. And there is a masterly theme, a presentation in microcosm of the theme which runs through all the Bible, and indeed throughout Jewish history: sin redeemed through suffering.

Of course much of the Old Testament cannot rise to such heights, but it rarely falls below the importance of the events it describes. The Jews seemed to have turned to history to express their genius because they were conscious of a particular destiny. As the great German scholar Gerhard von Rad put it: 'Only a political state which makes history can write it.' No doubt the Greeks felt this same compulsion. And of course Hebrew history was essentially didactic: history, indeed, is the supreme form of didacticism. But the Jewish love of the past was so strong that their scribes crammed their narratives with information even when the didactic purpose was obscure or non-existent. The gifted narrator of the Second Book of Samuel cannot resist telling us about Benaiah the son of Jehoiada,

who had done many acts. He slew two lionlike men of Moab: he went down also and slew a lion in the midst of a pit in time of snow. And he slew an Egyptian, a goodly man: and the Egyptian had a spear in his hand; but he went down to him with a staff, and plucked the spear out of the Egyptian's hand, and slew him with his own spear.

These lines are not much to the purpose of the text; but they are true, vivid, and interesting.

Within this legal and historical framework the Hebrew writers contrived to place all the worthwhile notions which they could express in writing. Thus prayer and poetry jostle with morality, practical wisdom and the deepest philosophy. The poetry has the inestimable merit that it survives in translation, partly of course because it had no rhyme but largely because it has a flexible structure based on sense. Josephus says Hebrew poetry was written in trimeters, pentameters and hexameters, but he was rather desperately trying to apply western classical concepts for the benefit of his Roman readers. In fact there is little evidence of metre and certainly no strict metrical form. The stanzas and strophes follow the logic of the thoughts the poetry is expressing: form, in short, is determined by content. Hence we can, or at least we feel we can get as much out of the psalms, for instance, as the congregations for which David and others originally composed them.

Perhaps the most remarkable aspect of the poetry of the Old Testament is the inclusion of the Song of Solomon, which, however interpreted – as allegory, liturgy, moral-didacticism, marriage hymns or religious drama – is also, and inescapably, an anthology of love poems. The Song is the first of the five Megilloth, or Canticles, read at the Hebrew feasts: Canticles (Passover), Ruth (Pentecost), Ecclesiastes (Tabernacles), Esther (Purim) and Lamentations (Destruction of Jerusalem). Hence it was popular, and its popularity, plus the association, real or spurious, with Solomon, are the only reasons why it got into the canon. Jewish tradition says that at the so-called Council of Jamnia, when the canon was settled finally, Rabbi Aqiba said: 'For in all the world there is nothing to equal the day on which the Song of Songs was given to Israel, for all the writings are holy, but the Song of Songs is the Holy of Holies.' He qualified this by adding: 'He who, for the sake of entertainment, sings the song as though it were a profane song, will have no share in the future world.' That is as may be, but whichever way the song is sung, it is magnificent poetry.

A papyrus fragment of the ancient Egyptian classic *The Wisdom of Amenope* which was written in the hieratic script and clearly influenced Hebrew wisdom literature.

In many cases it is hard to draw a distinction between Hebrew poetry and wisdom literature. Indeed, Solomon is credited with both. Wise sayings and proverbs were treasured and written down in the ancient Near East from the third millennium onwards. Early writers, and indeed actual vocabularies, often made no clear distinction between wisdom and skill as used in art and medicine for instance – both being derived from the gods. 'I will praise the Lord of Wisdom' begins the great Akkadian didactic poem, the *Ludlul Bel Nemeqi*. Early Hebrew writers were probably familiar with such texts, and with more locally produced didactic literature, of which the Ugaritic tablets found at Ras Shamra provide abundant evidence. They certainly knew one at least of the ancient Egyptian classics, *The Wisdom of Amenope*, since there is unmistakable evidence of direct borrowing in the Book of Proverbs, 22:17 to 23:11. But Hebrew wisdom literature is incomparably superior to any of the Near-Eastern patterns from which it was derived. Indeed, it produced two works which have been recognized as masterpieces in all ages, Ecclesiastes and Job.

The merits of Ecclesiastes are too obvious to need commentary, though there is unresolved argument over its date and provenance, some scholars even claiming to detect traces of Greek and especially Stoic influence, which would mean a very late dating. Job, however, is a far more mysterious document. This great essay in theodicy and the problem of evil has fascinated and baffled mankind. From a literary point of view it has been perhaps the most influential of all the Old Testament texts, and many writers have hailed it as the supreme masterpiece of Biblical literature. But what does it mean? Indeed, what it is? The obscurities of Job raised difficulties for all the ancient scribes and translators. There are over 100 words in Job which are not found in any other text or inscription. The language is genuine Hebrew but it clearly comes from the desert fringe of the Hebrew world. Its background may be Edomite; alas, we know little of the language of Edom. Or it may come from the Hauran, south of Damascus. One suggestion is that Job was a Hebrew settler in the Hijaz Oasis during the period about 550 BC when there was a Jewish occupation of the northern Hijaz. The fifth century AD theologian Theodore of Mopsuestia suggested that Job was a derivative of Greek drama, a view still held by some. Or it can be seen as an epic poem. Another possibility is that it used the form of an Arab poetry contest of the kind in which tribal poets competed at oasis-fairs, with Yahweh acting as the umpire. But equally it can be considered as a scholarly natural history in poetic form; like similar though far inferior Egyptian texts used as scribal manuals, it seems to be presenting an encyclopaedic or proto-scientific catalogue of cosmic and meteorological phenomena.

No one who reads Job can be unaware of the richness of its poetry or of the manner in which the author is stretching his mind to the limits of its potentiality. It might be described as an ultimately unsuccessful attempt to solve the mystery of the universe. But theodicy baffles everyone. The author of Job cannot solve the problem posed by his hero's sufferings, but he does indicate the way in which it should be approached. In chapter twenty-eight he gives a remarkable description of mining in the ancient world, a passage, unfortunately, which is largely misunderstood in the Authorized Version. Through this image he presents a view of the limitless technological potential of the human race and then contrasts it with man's pathetically weak moral reserves and his pitifully limited grasp of true wisdom. In the end, he says, to understand and master the physical order of the world is meaningless unless we come to accept and abide by the moral order. The author is trying to show that it is hopeless to try to answer the questions of theodicy in the way man poses them and he points to the right way. Hebrew philosophy supposed there was a secret to the universe but of an altogether different order to the secrets of, say, mining technology. The secret was personified as Wisdom, as in some of the Hebrew proverbs, and wisdom came to man only through obedience, the true foundation of the moral order: 'And he said to man: "Behold, the fear of the Lord, That is wisdom: and to depart from evil is understanding."' The point is enlarged in the Apocrypha, in chapter twenty-four of Ben Sirach's wisdom-poem, the Ecclesiasticus, where he says that after the Fall of Man, Wisdom failed to find a dwelling-place among the nations until God selected his Chosen Race; and so men in turn find wisdom by choosing God. It is to Job, again, that St Paul refers when he says, in his first Epistle to the Corinthians, that divine folly is greater than man's wisdom, and that divine weakness overcomes the strength of this world. Indeed the Book of Job illuminates all that central area of religious philosophy which contrasts the moral with the physical order; and it serves to remind us that the heritage of Hebrew civilization is not primarily in stones and artefacts but in the imperishable thoughts it has implanted in the minds of all succeeding ages.

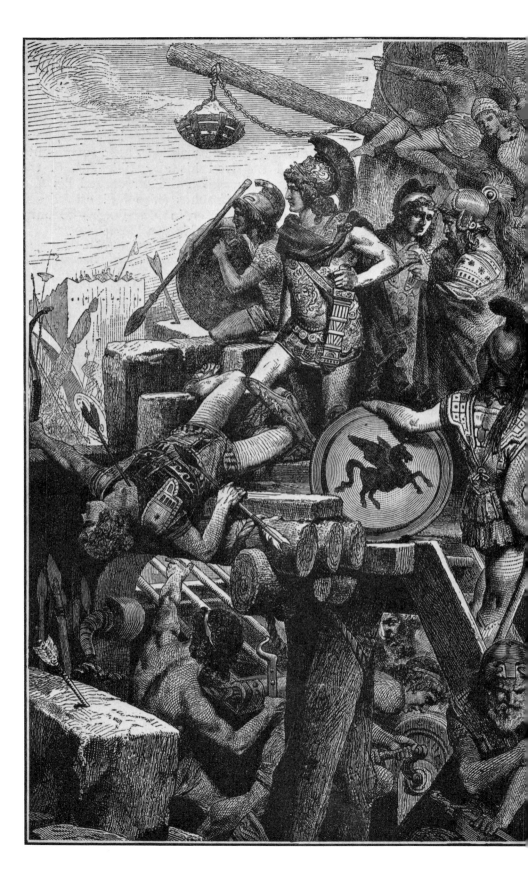

3

Greeks and Maccabees

WHILE THE EGYPTIAN, Assyrian, neo-Babylonian and finally the Persian empires moved across the stage of the Holy Land, a new civilization had been growing up in the northern Aegean. In the Bronze Age of the second millennium BC it was not possible to speak of East and West, of Europe and Asia, in cultural terms. The petty kingdoms of Mycenaean Greece, of Minoan Crete, of Canaanite Palestine and Phoenician Syria, of Hittite Anatolia and the Egyptian Empire of the New Kingdom to some extent shared a common international culture, just as many of them corresponded with each other in Akkadian cuneiform. There was no Continental cleavage. Then followed the barbarian anarchy of the twelfth and eleventh centuries BC, and the Dark Ages. When the world re-emerged into the fulness of the Iron Age, the great East-West division began to appear. By the middle of the first millennium BC, Greece had created a civilization of its own, centred round the free institutions of the city state or *polis*, which stood in sharp distinction to the autocratic courts and theocracies of the Orient. The rivalry of cultures soon took the form of a political and military struggle between the Greeks of the Aegean and the great Persian Empire.

During these centuries, the Greeks bred a continuing surplus population and created a great maritime commerce. They pushed into the central Mediterranean and founded colonies all over southern Italy and Sicily – the cultural empire of Magna Graecia. Then, early in the fourth century BC, under the leadership of Macedonia and with the personal inspiration of its brilliant young King Alexander, they turned east, and brought the sprawling Persian Empire down in ruins. In their crimson and scarlet coats, gilt armour and brightly plumed helmets, the Greek soldiers scattered the multitudinous armies of the East. Their instrument of terror was the Macedonian phalanx. It was sixteen, sometimes thirty-two, ranks deep, each man equipped with the *sarissa*, or twenty-one-foot-long spear, carried in both hands, the first five ranks holding theirs horizontally, so that the front of the phalanx was a glittering forest of metal points. As all the

Alexander the Great at the Siege of Tyre, one of the most ferocious sieges in history. Although Alexander was well-equipped with catapults and battering rams, Tyre was a walled island and able to resist for seven months.

other ranks, from eleven up to twenty-seven, were pushing behind the first five, the impetus of a charge was irresistible. The phalanx could not easily turn once it had started and its flanks were vulnerable unless closely protected by cavalry. It was eventually mastered by the all-purpose Roman infantrymen at the Battle of Magnesia near Pergamum in Western Anatolia in 189 BC. But until that date it did not fail in open battle and the Greeks were masters of the Near East.

Not that all Alexander's conquests were easy. At Tyre, the biggest, richest, most independent-minded and best defended of the Phoenician city-states, he was obliged to conduct one of the most memorable and ferocious sieges in history, the lines of which can still be traced in this much-battered town. The siege was a striking early example of technological warfare. Alexander had a stone-throwing catapult, powered by springs of twisted sinew. The Tyrians had borrowed from their Carthaginian colonists the secret of the arrow-firing catapult, first developed in Greek Syracuse. But they did not have the torsion-spring, which had been invented in the workshops of Polyeidus the Thessalian, chief engineer to Alexander's father, Philip of Macedon. Alexander had first used it to take Halicarnassus. It could kill people at 400 yards, and at 150 it could inflict significant damage to a city wall. He also had immense siege-towers, stretching up to 180 feet, from which men could be deployed at up to 20 different levels, to shoot and operate battering-rams. He had a boring-machine on wheels, for use on the mud-brick stretches of city walls, or for when the outer layer of ashlars had been smashed or removed. His battering-rams were 'tortoises', 48-feet square, with ropes and rollers to work the huge ram. Each was protected by a three-storey tower, the first and second storeys carrying water for putting out fire, the top a catapult.

But Tyre was a walled island, and when Alexander began the siege his fleet had not arrived, so the Tyrians had sea-supremacy. To get his engines into range, he started to build a mole out from the mainland, pulling down Old Tyre and scouring the Lebanese cedar forests for materials. It is not quite clear from the contemporary accounts whether he carried his moles actually to the foot of the island walls, for 100 yards from their foot the water grew suddenly deeper. But at this point his engines were within range, and began their assault. Later he was joined by the ships of Rhodes, Cyprus and Tyre's Phoenician rivals, and on these he mounted rams and catapults, and attacked by sea. The Tyrians retaliated with fire-ships, anti-personnel harpoons, whirling marble wheels which deflected the blows of missiles, armoured ships to cut cables, underwater divers, skins stuffed with seaweed to cushion the blows of the Greeks' giant arrows, and they poured down loads of red-hot sand on the besiegers, which got between the chinks of their armour and caused agony.

Alexander was exasperated by the seven-month siege. He won in the end by overkill – an attack in great strength from all points of the compass, by land and sea, using all his array of war-technology and leading the main assault in person. Inevitably he took his revenge on the fallen city. Some 8,000 Tyrians were massacred in the assault, 2,000 were later crucified along the shore, and 30,000 sold into slavery. There is a lasting memorial to this bitter siege: Alexander's mole began a silting process which now links island Tyre to the mainland by an isthmus.

The Greeks came to Asia to stay. They saw it as a case of 'manifest destiny.' They immediately began to settle veterans in agricultural colonies, to build cities, of which Alexandria in Egypt was the prototype, and to invite all men of Greek culture to join them. There followed an enormous Greek Diaspora. Of course Alexander did not live to see it. When he died, his generals, the Diadochi, divided the spoils and carved out the so-called Hellenistic monarchies for themselves – Ptolemy in Egypt, Seleucus in Syria and Mesopotamia, eventually Attalus in Anatolia. The Greek adventure into imperialism meant big changes for them as well as for their new subjects. Their Bronze Age states had been charismatic tribal monarchies. Their Iron Age *polis* was democratic in concept. Even the new Macedonian monarchy was semi-constitutional, since the army arbitrated a disputed succession and the throne was balanced by the council of the King's Friends or Companions. Such institutions had to adapt themselves to the immense territories of the Orient, and the autocracy of Hellenistic kingship developed. The King's Friends were kept but they ceased to be an aristocracy sprung from the soil and became one by royal appointment. Ptolemies and Seleucids acquired the apparatus of despotism: vast Crown lands, State monopolies, mercenary armies, protective tariffs. Alexander had created his empire as an ideal; he 'ordered all men to regard the world as their country ... good men as their kin, bad men as foreigners.' He wanted to fuse races and he said that Greek institutions should adapt themselves to local conditions. But his successor-kings argued that autocracy was inevitable since only absolute monarchs, standing above Greek and Oriental alike, could unite East and West. From this it was only a step to the deified monarch.

The concept seems absurd to us; for the Greeks it was utilitarian. They came to it by a process of practical constitutional reasoning, not, like the Egyptians of the Old Kingdom, through theological development. The Greeks always tended to elevate the human and to lower the divine. They did not see the difference between the two planes as enormous: gods were not much more than revered and potent ancestors. All men sprang from the gods in a way. Among the aristocracy the link was more direct, and with a great man, a man of destiny, the link might be immediate. Aristotle argued in the *Politics*:

If there exists in a state an individual so pre-eminent in virtue that neither the virtue nor the political capacity of all the other citizens is comparable with his ... he should not be regarded as a member of the state at all. For he will be wrong if treated as an equal when he is thus unequal in virtue and political capacity. Such a man should be rated as a god among men.

Hence Greek political theory tended to obscure the distinction between human and divine, or made it seem narrow. Apotheosis was a comparatively small step.

All founders of city states were deified, *a fortiori* founders of kingdoms. Three leading oracles said Alexander had sprung direct from Zeus. By virtue of this descent Alexander was accepted into the circle of the deities of the Greek city states in 324 BC. This was a constitutional device – king-worship, in fact, was essentially a political, not a religious, mechanism – for a king could not legally enter or deal with a city-state whereas a god could. Again, if such a state accepted a mere king as ruler it lost its liberty but if the king were a god its liberty was preserved. Moreover, the practical Greeks observed that a god-king was not only constitutionally more acceptable than a mere king; he was in practical terms more powerful than a mere god. Gods could not confer salvation anyway, and they often failed when appealed to. When the sacred island of Delos could not collect the money owed by the city-states for the upkeep of its temple of Apollo, it beseeched the god, who did nothing; it then appealed to Ptolemy of Egypt, who sent his fleet – and, lo! the money was there. There was an Athenian song about Demetrius: 'The other gods are not, or are far away. They hear not or give no heed. But thou art here, and we see thee not in wood or stone but in very truth.'

Hence both Ptolemies and Seleucids deified themselves. Ptolemy I began it when he instituted a State cult of Alexander in 289 BC. Ptolemy II and his sister-wife, Arsinoe, began to share a temple with Alexander, as living brother-sister gods, about 270 BC. Antiochus II was the first Seleucid to encourage a cult of himself in his own lifetime. He claimed descent from Apollo. The cult names these kings chose emphasized the practical side of their divine virtues or their family ties: Soter (saviour), Euergetes (benefactor), Philadelphus, Philopater, Philometor. Antiochus IV went a stage further when he chose the name Epiphanes, the god made manifest – an Egyptian notion. He was the first Seleucid to use his divine title on his coins, and it may be he actually believed he was a god; if so, he was the only Seleucid to do so and it is more likely he stressed his godhead to restore unity to his kingdom during a time of particular trouble.

Divinity enabled the Hellenistic kings to reconcile monarchy with the free *polis*, and so Greek cities sprang up all over the Near East. In the new cities and often in many of the old ones, Greeks or Greek-speaking citizens were in the majority and the cities were the nails which held the structure of the Hellenistic empire together. In the cities, the Greeks guarded the formation of their youth, all over the imperial Diaspora, through the educational and cultural institution of the gymnasium. And the kings used the wealth of Asia and Egypt to endow the cities with all the apparatus of State-subsidized Greek culture: temples, universities, libraries. Some of these cities were enormous. Antioch, on the lower Orontes, a new city to which the Seleucid capital was shifted in 300 BC, eventually had a population of 500,000. It had magnificent public buildings, adorned by Eutychides, pupil of the sculptor Lysippus. Antioch was one of the greatest cities of antiquity, housing a multitude of races, in time a centre of Judaism and early Christianity (St Peter lived there), as well as Hellenistic and Roman culture, a Byzantine city re-founded by Justinian, and in its last great phase a Crusader

The remains of the walls of the ancient city of Antioch, where Greek culture and learning thrived in the third century BC.

stronghold. In AD 1268 Baibars and the Mamluks destroyed it, and the present Turkish town is only about a tenth of the city once enclosed by the walls of Theodosius. There is an old Roman bridge and a Byzantine citadel, but savagery, floods and earthquakes have swept away virtually everything from its splendid past. However, much remains beneath the surface to be excavated, and still more, perhaps, at Apamea, the Seleucid military headquarters not far away. This, too, was a great city, with a population of 120,000. It supplied the stud for the cavalry and quartered the 600 war-elephants which constituted the glory of the Seleucid army. These splendidly trained beasts came from India. They were dressed in red housings and carried four men in their turrets. They stood firm in the noise and horror of battle, unlike the African elephants employed by the Ptolemies, which were bigger and wilder but often turned tail and trampled their own infantry. The Seleucids awarded their elephants campaign medals.

Even Apamea, which stank of military dung and was noisy with arms-forges, had the full apparatus of Greek culture; indeed, it possessed a famous philosophy school. Such establishments, lavishly financed by the Hellenistic kings, were

ABOVE Indian war-elephants were the glory of the Seleucid army. The Ptolemies used African elephants which were less easy to control.

LEFT Some of the carvings on the cliffs near the Dog River (Nahr el Kelb) in the Lebanon.

ABOVE RIGHT The six-century mosaic m of Jerusalem found at Madaba, near the Dead Sea.

BELOW RIGHT The Temple of Ed Deir a Petra, Jordan.

OVERLEAF The domes and minarets of t Old City of Jerusalem.

GREEKS AND MACCABEES

needed not only to keep the Greek settlers Greek but to attract new colonists from the Greek heartlands. Syria and Palestine, held first by the Ptolemies, then after the victory of Antiochus III at Panion (Banias) in 200 BC, by the Seleucids, were areas of massive Greek settlement and of rapid Hellenization of their non-Hebrew inhabitants. The coast was completely Hellenized. Such cities as Tyre and Sidon had their councils, magistrates and popular assemblies like other *polites*. The Hellenistic kings were generous in awarding Greek-style cities freedoms and privileges though they soon tended to fall into the hands of local dynasts or tyrants: such were Gaza, Straton's Tower, Byblos, Tripoli. The Greek coastal cities, in turn, set up colonies in the interior, or Hellenized existing towns. There was a Sidonian colony at Shechem which asked the king for permission to turn the sanctuary of Mount Gerizim into a temple of Zeus. There was another Sidonian colony at Marissa in southern Palestine, which enjoyed self-government, and where painted Greek tombs of about 200 BC have been found. Such inland centres as Gamala and Philadelphia (Amman) were major Greek cities. Of course both Ptolemies and Seleucids kept inland garrisons and built castles to house them – at least one of them, Tobias, built in the late third century BC, survives as a ruin. But the real impress of Hellenism was provided by the cities, swarming with Greeks and semi-Greeks, whose loyalty was not so much to the monarchy, as to the culture it promoted and defended.

Gradually, in Ptolemaic times, a ring of such Greek cities came to surround Hebrew Samaria and Judaea. The countryside and the hills tended to remain Oriental, and within the Seleucid empire there were a number of such 'temple states', which the Greeks regarded as an anachronism, soon to be swept away by the creeping spread of Hellenic ideas and institutions. Not, of course, that the Greeks wished to stamp out Oriental religions. On the contrary, they absorbed and adapted them by the syncretistic process at which they were so adept. As polytheists they did not wish to extend imperialism to religion. In the Greek Diaspora, fidelity to the original Greek gods was seen as a matter of individual conscience. What the authorities did was to maintain order and encourage temple sharing, and to promote this they grouped deities together. Thus Apollo-Helios-Hermes was one mutant. Asclepios, the Greek god of healing, was merged with Imhotep, his deified Egyptian equivalent. The Greek rites of Dionysos were blended with the Egyptian Isis-cult. Gods cannibalized each other or migrated. The Greeks were natural ecumenists; indeed, they really saw no intermediate stage between the *polis* and the *ekumene*, the world. They saw a temple to the highest god, Zeus to them, as an ecumenical place of worship where men of varying races could, according to taste, cultivate Zeus, or the Persian Ahuramazda, or the Egyptian Ammon, or – why not – the Yahweh of the Jews.

Such a notion was, to the Jews, teeth-grinding anathema, 'the abomination of desolation.' They first fell into the Greek sphere of influence in 333, when Alexander destroyed the Persian army at the Battle of Issus; thereafter, until the Seleucids took over in 198 BC, they were ruled from Egypt. Ptolemaic rule was

LEFT A view of the Citadel courtyard with the Tower of David in the background.

The ruins of the citadel on Citadel Hill at Philadelphia (Amman) which was a major Greek city.

tolerant; the historian Polybius says the common people of Palestine favoured it. But there seems to have been a pro-Seleucid party among aristocratic Jews, whose point of view was put by Ben Sirach, the author of Ecclasiasticus, who says that under the Ptolemies the land was full of tears, the dead were happier than the living, and even the birds acted as spies for the authorities. The Hellenistic monarchy's State support of city culture was expensive and taxation was heavy, but some Jews did well out of the tax system. In his *Antiquities of the Jews*, Josephus tells the story of Joseph, son of Tobias, who went to the auction of the tax-farmers at Alexandria: 'Now it happened that at this time all the principal men and rulers went up out of the cities of Syria and Phoenicia, to bid for their taxes; for every year the king sold them to the men of the greatest power in every city.' Joseph got the contract by accusing the other bidders for the taxes of Coelesyria, Phoenicia, Judaea and Samaria, of forming a secret cartel and putting the price too low. He held it for twenty-two years 'and brought the Jews out of a state of poverty and meanness to one that was more splendid.' Of course he was well-connected: his mother was the high-priest's sister and taxes were normally paid through the high-priest and the Temple treasury and its deposits acted as a kind of State bank for the population.

In his struggle with the Ptolemies, Antiochus III bid high for Jewish support. He was first beaten at Raphia; and afterwards the victor, Ptolemy Philopater, visited Jerusalem and wishing (as he thought) to make the polite gesture of rendering thanks to the local chief god, Yahweh, went to the Temple and entered the Holy of Holies, forbidden to all except the High Priest. There was a riot of pious Jews, which was brutally repressed. Hence Antiochus was able to pose as the defender of Jewish religious independence. He promised to free the Temple from the payment of 'pagan' taxes, to provide for the repair of the fabric and the expense of the services out of imperial funds, and to exempt scribes, priests and elders from any imposts. Hence, says Josephus, the Jews helped him to eject the Ptolemaic garrison from Jerusalem.

But there was no real possibility of Jews and Greeks getting on together. The conflict between them has rightly been called one of the great cultural epics of mankind. Of course they had much in common: their sense of history, for instance. Both were increasingly individualistic: the Greeks to compensate for the lost communalism of the city-state, the Jews for their lost nationhood. Both believed fiercely in freedom but they meant different things by it. To the Greeks freedom was the self-governing community, making its own laws by public deliberation and worshipping what gods it fancied. For the Jews, it meant freedom to follow their own law, divinely given and unalterable by any human agency, whether democratic senate or Hellenistic despot. Where the Greeks exalted man and depressed their deities, the Jews regarded God as infinitely greater than the entire cosmos, of which he was the sole creator. Man had rights by virtue of the fact that he was created in God's image; they therefore placed the highest value on human life and rejected a variety of Greek notions and practices in consequence – above all infanticide, which they regarded with horror. They were also scandalized by the Greek games, which involved nudity. Many of the acts regularly required for full participation in the Greek *polis* were forbidden by the Jewish law and so pious Jews could in no circumstances be Hellenized.

The Greeks might have respected Jewish susceptibilities had they known more about them. But they were woefully ignorant. It is odd that Greek inquisitiveness about the universe did not extend to the cultures of their subject peoples. In Alexandria they had the finest library in the world and the city teemed with scholars. But none, so far as we know, learnt Egyptian or how to read the hieroglyphics; Manetho's attempt to present pharaonic history and culture to the Greek world met with little response, as did a similar presentation of the Mesopotamian past by Berossus of Babylon. The first Jew to make his national history available to a Greek audience was Josephus, in Roman times. So the Greeks knew nothing except wild travellers' tales. Most of all, they did not even suspect the existence of Jewish literature. This is all the more remarkable in that the literary output of the Greeks in Hellenistic times was prodigious (though only a pitifully small proportion of it has survived). In Alexander and Antioch, writers were subsidized and honoured; they occupied a much higher social position than hitherto, and for the first time were widely known by name, and in large numbers – a list of over 1,100 Hellenistic authors has been compiled. Many of them, such as Antipater of Sidon and Philodemus of Gadara, lived close to the Jewish heartland. The Jews were the only indigenous people of the Hellenistic empire with a really mature literature, in size and range rivalling Greece's own, and who were still producing books in huge quantities. Yet no Greeks (so far as we know) troubled to learn Hebrew or Aramaic, and few read Jewish authors who were now writing in Greek. Even the energetic and normally well-informed Strabo wrote nonsense about the Jews in the sixteenth book of his *Geography*. He thought Moses was an Egyptian priest who rejected alike Egyptian animal-cults and the anthropomorphism of Greece. Moses forbade the worship of idols and redefined God as one and invisible, led a few enlightened spirits out of Egypt and established the new religion in Jerusalem. But in time Judaism fell into the hands of superstitious priests and tyrants who introduced the barbarity of circumcision and terrorized the Greeks, the true heirs of the original Canaanite population. Thus Strabo.

Strabo was writing, of course, in the light of what little he knew of the Maccabean revolt against the Seleucids and the history of the independent Jewish kingdom of the Hasmoneans. He was right, however, to detect the existence of more than one party among the Jews. Most of the oriental peoples who fell under the sway of the Greeks did not resist some degree of Hellenization, at least in the towns. Jewish attitudes varied. The Jews of the Diaspora had long since conformed in part to the local *ambiance*. In Babylonia their womenfolk had learnt to wail for Tammuz and make cakes to the Queen of Heaven. In Alexandria, Jews picked up the Greek language and many Greek forms of behaviour. In Asia Minor and Syria Jews adopted Graeco-Oriental cults. They equated Yahweh with Theos Hypistos, God the Highest, at any rate in Asia Minor and at Delos, as is confirmed by a synagogue inscription. There are references in both Jewish and Christian literature to 'synagogues of Satan' and to men 'who say they are Jews and are not.' Some Jews, like some Greeks, patronized the swarming literary underworld of magical writings which later took shape as the Cabbala; the Acts of the Apostles says that St Paul made a bonfire of such books at Ephesus.

Even in orthodox Judaea, Jewish culture had been eroded to the extent that Aramaic, the *lingua franca* of the Asian empires, had largely ousted Hebrew as the spoken tongue (though not, of course, as the language of religious literature). When the Greeks came, there were many educated Jews, including the Zadokite or Sadducee descendants of priestly families, who learned Greek and adopted Greek names. The Sadducees were, in fact, a collaborationist party. Even before the rise of the Sadducees, who were mainly rich aristocrats and landowners, there had been division between the House of Onias, the purists, and the conformist House of Tobias. The Book of Daniel reflects the problem of double allegiance. Both Ptolemies and Seleucids governed Judaea through the High Priest, whom they appointed, though he had to be drawn from certain families. Rich Sadducees not only paid for the office but, it was claimed by the orthodox, collaborated in a gradual paganization of the Temple. Thus the High Priest Jason (the Hellenization of Joshua) built a gymnasium in Jerusalem and levied taxes for the game of Heracles, a Greek adaptation of the semitic deity Baal Malkart, to be held at Tyre. As learned Jews knew, this was the wicked god of Jezebel. His

In the foreground the ruins of the city of Qumran and behind, the caves where, in 1947, the Dead Sea Scrolls were discovered in eight earthenware jars by a young bedouin.

successor, Menelaus, was 'not so much a High Priest as a wild beast.' Such men provoked the departure from Jerusalem of puritan sectaries like the Qumran monks, whose Dead Sea scrolls refer to evil priests and the polluting of the Temple.

Even in Jerusalem, then, the Jews were divided over Hellenization. Hitherto the Seleucids had left Judaea alone, apart from picking high priests well disposed towards the Greeks. In 175 BC, however, Antiochus IV Epiphanes siezed power and began a new policy. The Seleucid empire was falling apart, partly for lack of cash, partly because Greek religion was dying. The stress Antiochus put on his own divinity was probably a deliberate attempt to impose some kind of political cohesion on his crumbling empire, as were his desperate and brutal efforts to raise money everywhere. His determination to bring Judaea fully within the tax-system therefore coincided with an equal resolve to end Jewish religious separatism, both being symbolized by raiding the Temple treasury and Hellenizing its services. In 169 BC he destroyed the walls of Jerusalem and built an acropolis fortress on the western ridge to overawe the town. The Acra, as the Jews called it, accommodated not only the Seleucid garrison but the Hellenizing Jews. In 167 he abolished the worship of Yahweh and installed the cult of Zeus Olympus in the Temple, the god having, it was said, his own features – hence Daniel's phrase, 'abomination of desolation', the parodic opposite of Zeus's title 'lord of Heaven'. The new statue was insultingly dedicated by the sacrifice of a sow.

In the ensuing uproar, a party of Hasidim, 'those loyal to the Covenant', emerged as the focus of religious nationalism and populist opposition to Seleucid exactions. In general, the priests sided with the Seleucids; the nationalists were drawn from the scribes – Eleazir, the ninety-year-old martyr who was beaten to death by the Seleucids, was described as 'one of the principal scribes.' As the Jews were forced to make sacrifices on pagan altars as a test of loyalty to the State, and executed if they refused, the number of martyrs grew. It is indeed from this period that the concept of religious martyrdom developed, accounts of their sufferings and fortitude in the faith (for instance the tale of the seven brothers in the Second Book of Maccabees) becoming part of the sacred literature, as well as the propaganda of nationalism.

According to that brilliant essay in nationalist history, the First Book of the Maccabees, the revolt began at Modein, in the Judaean foothills six miles east of Lydda, when Mattathias Hasmon, the head of an old priestly family, from the Watch of Jehoiarib, murdered a conformist Jew and the official who was superintending his sacrifice. Mattathias's five sons, led by Judas the Maccabee (or 'hammer'), then conducted a guerrilla campaign against the Seleucid officials and garrisons. In two years 166–164 BC, they cleared Jerusalem and the surrounding area of Greeks, except for the Acra citadel. The Temple was cleansed and rededicated in December 164, an event still celebrated by the Jews as the Feast of Hanukkah, or Purification. The success of the revolt was made possible by the general decay of the Seleucid State, its divisions and endless family disputes, and not least by the enterprise of the Hasmonean family, who signed a treaty with the rising power of Rome. In 152 BC the Seleucids effectively abandoned their attempt to Hellenize Judaea by force, and recognized Jonathan, now head of the family, as High Priest – an office the Hasmoneans were to hold for 115 years.

This was as far as most orthodox Jews wanted to go. Indeed, it is important to realize that, throughout this period, the Jews were badly divided among themselves; the real nationalists were probably in a minority. The leading families, the Hakotz, Tobiads, Bilga and so on, favoured some degree of cooperation with the Greeks. So did a great many ordinary people, for the advantages of Greek city life were manifest. It is not without significance that the first man the Hasmoneans killed was a fellow Jew – and moderate Jews of all classes remained the chief victims of Jewish irridentist nationalism. The Sadducees were quite content that the Seleucids should continue to appoint the High Priest. The Hasidim were interested only in full religious liberty and when they got this they refused to follow the Hasmoneans further and became 'those who separated themselves', the Pharisees.

But the Hasmoneans pushed on to conquest and the recreation of the Davidic kingdom. Jonathan's successor, Simon, was effectively the first independent king of Judaea since the fall of Jerusalem. He stormed the Acra, massacring its Seleucid garrison and the Hellenizing Jews who lived within its walls. He took Gezer, Solomon's dowry city, and Jaffa – important, because it kept open the links to Rome by sea. The triumphant Jews entered the Acra 'with psalms and palm branches and with harps and cymbals and with viols and hymns and songs, for a great enemy had been destroyed out of Israel.' The notion of the kingdom of the Israelites, the Promised Land, was revived: 'We have neither taken other men's lands nor have we taken possession of that which belongeth to another but only of the inheritance of our fathers; howbeit was held in the possession of our enemies wrongfully for a certain time.' (1 Maccabees 15:33). An important text for Israel!

Simon's son, John Hyrcanus (134–104 BC) was the first of the Hasmoneans to issue coins, stamped 'Jehonhanan the High Priest and the Community of the Jews'; his son, Alexander Jannaeus (103–76), actually called himself king on his coins, inscribed 'Jonathan the King' in proto-Hebraic script. They recreated the structure of a military kingdom. At Jerusalem, Josephus says that when Simon took the Acra in 141, he not only tumbled over its walls but went on 'to level the very mountain itself upon which the citadel happened to stand, so that the Temple might be higher than it' – a work which took three years and also involved the total destruction of the quarter of the Hellenizing Jews, the 'sinful nation, transgressors of the law.' From this and other evidence, Kathleen Kenyon deduces that the Acra was on the west ridge of the city. This was now covered by an extended city wall which had to be pushed west not merely to protect the growing city but to deny the ground to the Hellenistic artillery, which had a range

The Tower of David in Jerusalem, a fine example of Hasmonean construction, was probably built by Jonathan.

of 350–400 yards (greater if fired from higher ground) and could thus have bombarded the Temple from the west ridge. However, not much Hasmonean construction remains at Jerusalem, or is visible, the so-called Tower of David, probably built by Jonathan, being an exception. Their big fortress, Baris, near the Temple, is concealed beneath Herod's later fortress of Antonia. Herod also rebuilt their castles and fortress-towns outside Jerusalem, with the exception of Alexandreion (Qarn Sartaba) and Bethsur (Khirbet et-Tubeiqeh), both of them built by Judas Maccabeus.

The Hasmoneans began with a people's army of mountain guerrillas; they ended with mercenaries. The story of their rise and fall is a study in *hubris*. Josephus speaks highly of John Hyrcanus, whom he calls 'the last ruler of the Jews to combine in his person the old ideal of sacral kingship, though not having the title of king.' But Hyrcanus was a reckless man: it was he, rather than Judas or Jonathan, who revived the fundamentalist notion that the whole of Palestine was the ancestral inheritance of the Jewish nation. It followed from this that there was no room for foreign cults or heterodox sects. John's armies raged through Samaria, and destroyed the Samaritan Temple on Mount Gerizim. He stormed and destroyed the city of Samaria itself, now a wholly Hellenized community; and he likewise pillaged and burnt the Greek city of Scythopolis.

These wars of fire and sword resembled the conquest wars of Joshua and were accompanied by the massacre of city populations whose only crime was that they were Greek-speaking. The province of Idumaea, with its two main centres, Adora and Marissa, was conquered and forcibly annexed to Judaea, its inhabitants being given the choice of death or conversion to Judaism. Marissa, which survives as Tell es-Sandahannah near old Lachish, is a good example of a Hellenistic-oriental provincial town, with a rectangular town plan not Jewish at all.

Alexander Jannaeus continued this policy of expansion. He defeated the Nabataeans of Petra, invaded the territory of the Decapolis, the league of ten Greek cities around the Jordan, and the province of Gaulanitis. Within the limits the Hasmoneans set for their kingdom, pockets of non-Jewish people were eliminated by forcible conversion, massacre or expulsion. In the Greek world these savage wars destroyed the initial Hellenistic respect for Judaism and led to the prevailing view in the Graeco-Roman world that the Jews were fanatics; and in Palestine and Syria itself, the Greek-speaking non-Jewish communities came to regard the Jews with implacable hatred. It was these terrific mutual hatreds which St Paul, the Jew from the Pharisee family who became the missionary to the Greeks, set himself to bridge. Alas, as he noted, 'the Greeks seek for wisdom and the Jews look for a sign': the conflict between religion as a philosophical enquiry and religion as revelation is not easily resolved.

The tragedy is that the Hasmonean kingdom, at any rate under John and Alexander, almost certainly did not have the support of most Jews. Of course the kings could rely on the Sadducees, the aristocratic party which tended to form itself around any power, native or foreign, which could guarantee order and the protection of property. But Sadducees followed the written Torah only: they would not admit the oral commentaries on the Law (later to find written form), which humanized and updated it. Hence the middle classes and the poor turned to the non-priestly sages, who claimed to be 'disciples of Aaron' rather than 'son of Aaron', among whom the Pharisees were the most important and numerous. As Josephus noted, 'the Sadducees draw their following only among the rich, and the people do not support them, while the Pharisees have popular allies.'

The Pharisees, belonging – unlike the Sadducees – to a developing religion, were not stuck with the notion of a physical kingdom of God on the lines of the kingdom of David, or the attempt to reconstruct it in Hasmonean form. Their inspiration was the great Book of Daniel, which seems to distinguish between earthly and heavenly kingdoms in a way which prefigures St Augustine's masterpiece of the fifth century AD, *The City of God*. The Book of Daniel is a retrospect of the various empires which held sway since the destruction of Jerusalem, and in particular of the Ptolemaic and Seleucid regimes. How unstable and impermanent they all are! Are not all earthly empires doomed to destruction in turn? They are mere historical episodes, but behind and above them is the Kingdom of God in Israel, immortal and unassailable. Indeed, the Kingdom of

God is necessarily in conflict with the kingdoms of the earth; a great struggle is impending between the forces of darkness and God, between God, who creates the world and makes history, and the armies of chaos.

Such notions were reflected in their most extreme and potentially violent form in the writings of millennarian sects like the monks of Qumran, who foresaw what they called 'the War of the Sons of Light against the Sons of Darkness', and who were at this time beginning to prepare for it by writing what can only be termed a military manual. But anticipation of an apocalypse was spreading among all Jews and was increasingly reflected in their literature. And if there were an apocalypse, what thereafter? The Sadducees believed in free will but they could not entertain the notion of eternal life because it was not in their texts. The Pharisees, by contrast, seized upon the Book of Daniel, especially 12:2: 'Many of those that sleep in the dust shall awake, some to everlasting life and some to everlasting contempt', to develop the idea of resurrection followed by personal immortality. The idea had been familiar to the Egyptians for nearly two millennia, but had been hitherto neglected by the Jews who were preoccupied by their obsession with the earthly interventions of the deity. Belief in immortality became the great doctrinal dividing line between Sadducees and Pharisees, and of course in time it became the central dogma of Christianity.

Whether the idea of eternal life came to Judaea from the Greek-speaking Diaspora in Egypt is impossible to demonstrate. It is also hard to prove penetration by any Greek ideas of 'the fence around the Torah.' But this undoubtedly took place in the Diaspora, and even Judaea, with its intellectual xenophobia, was not immune. Greek religion itself was dying in the Hellenistic phase of the civilization: the one Greek temple a Greek city planned to a purely Greek god, Apollo's Temple at Didyma, was still unfinished 400 years later. It may be that Antiochus Epiphanes's reckless assault on the Temple of Jerusalem reflected his view that drastic measures were needed to keep Greek religion going at all – rather like the Emperor Julian's desperate and unsuccessful attempt to resuscitate Roman paganism 500 years later. In fact the only successful Greek deity of the age was Fortune, who attached herself to select individuals as a *daemon*. She was the goddess of the Seleucid capital of Antioch and her magnificent and enormous statue by Eutychides was the city's artistic glory. But Hellenistic civilization in the Holy Land produced a more permanent memorial in the Phoenician Zeno, of Citium in Cyprus, who created the Stoic philosophy. Zeno, like most of the Jews themselves, believed both in free will and determinism. He resolved the conflict between them by his doctrine of duty. Happiness (the Jews, following Job, would have termed it wisdom) lay in conformity with divine will, which must be accepted even when apparently hateful. To reach this state of mind, man should systematically pursue virtue. In the end, virtue constituted its own reward, in the shape of happiness, because man gradually grew into harmony with God and conformity with his will became natural and delightful. The idea of deliberately cultivated moral growth, which underlay the idea of conscience, clearly influenced Pharisee thinking at this time, and of course it fitted in neatly with their developing doctrine of personal immortality, which the conscientious pursuit of virtue earned. And the notion of the individual conscience, whether Stoic or Pharisaic or both, was one of St Paul's greatest contributions in interpreting the Christian message.

In seeking to turn Judaism away from the search for earthly manifestations of divine favour and towards the idea of an eternal kingdom to be shared by the conscientious, the Pharisees made a far more lasting contribution to the survival of Jewish civilization than the Hasmonean pursuit of brutal religious nationalism. Moreover, they could hardly escape noticing that the new royal family of Judaea were rapidly corrupted by worldly success to the point where they neglected the religious certitudes which were their very *raison d'être*. Alexander Jannaeus, hereditary High Priest as well as Jewish imperialist ruler, made an elementary ritual mistake while officiating at the Feast of Tabernacles. The pious Jews pelted him with lemons. In revenge he turned his mercenaries on them. He crucified 800 Pharisees in Jerusalem and had their wives and children killed before their eyes as they hung on their crosses – while, in full view of the populace, he feasted in his harem. There is an obscure reference to this horrific episode in one of the Qumran scrolls: 'The lion of wrath ... when he hangs men up alive.'

Alexander died in 76 BC, by which time Roman power was just over the horizon in Syria. Rome would come to the aid of a subject people's struggle against a large state, and it would tolerate the existence of small and weak independent states. But an expansive-minded Jewish irredentist kingdom, forcibly converting its neighbours, was not acceptable for long. Rome waited for the opportunity to intervene. It came when the Hasmonean family split into warring fragments, as the Seleucids had done before them. Alexander's queen, Salome, reigned for a time after him, and sought to win popular support by befriending the Pharisees. But she died in 67 BC and her sons fell out. Ironically it was the chief minister of one of the claimants, Hyrcanus, who brought in Pompey, the Roman general. The Minister's name was Antipater, and he was descended from the Idumaeans who had been forcibly converted to Judaism under the policy of John Hyrcanus. Such men had never accepted the spirit of Jewish exclusivity and could not understand its fear of alien cultures. It was natural for Antipater to seek to avoid civil war and possible anarchy by invoking the aid of the great ordering force of the universe. His son, Herod the Great, was to lock the Holy Land into the imperial system of Roman civilization.

4
The Age of Herod and Jesus

ROMAN INTERVENTION BECAME DECISIVE in 63, when Pompey and his army occupied Jerusalem on behalf of one of the Hasmonean candidates, Hyrcanus. The real victor was Antipater, who led the post-Western party among the Jews and the non-Jews. One of his sons, Phasael, was made Governor of Jerusalem; another, Herod, governor of Galilee. These were active and clear-sighted men, determined to impose order and Roman civilization on a backward, divided and sectarian country. Herod's first act in Galilee was to smash a vicious band of semi-religious bandits, under one Hezekias, and execute its leaders. This was an offence under strict Jewish law, since capital punishment was reserved to the Sanhedrin, or council of religious elders. When his father, Antipater, was poisoned by another fanatic, Malichus, Herod got the authority of the Roman proconsul in Syria, Cassius, to put Malichus to death. In consequence, he barely survived an arraignment before the Sanhedrin, which he attended with an armed guard and he was nearly murdered when Antigonus, the other Hasmonean claimant, formed a conspiracy with the Jewish nationalists and the Parthians, to seize Jerusalem. Herod's brother Phasael was arrested and committed suicide in prison. Their Hasmonean candidate, Hyrcanus, was mutilated to make him ineligible as High Priest – Antigonus bit off his uncle's ears himself.

Herod fled to Rome, where the Senate made him king, *rex socius*, 'friend and ally of the Roman people', and he went back to Palestine with all the might of the empire behind him. He retook Jerusalem with 30,000 infantry and 6,000 cavalry. Antigonus was handed over to Mark Anthony, who executed him. When Anthony met disaster at Actium, Herod made his submission to Octavian in Rhodes and was confirmed as king. Thereafter, he never lost the favour of Rome, whoever ruled there, and his personal alliance with Octavian, later the Emperor Augustus, was one of the chief factors which brought peace to the eastern Mediterranean. In Palestine he reported directly to the Emperor, though he was always careful to consult the Roman legate in Syria. He carried on his anti-bandit

Herod, governor of Galilee and later King, on his throne, from a thirteenth century AD mosaic.

LEFT Pompey's occupation of Jerusalem in 63 BC marked the beginning of effective intervention in the lives of the peoples of Palestine by the Romans.

BELOW A representation of a meeting of the Sanhedrin, or Jewish Counsel of elders, taken from the Stackhouse Bible of 1733.

THE AGE OF HEROD AND JESUS

activities, thereby expanding his territory into the Hauran and the Jebel Druze; Augustus gave him Samaria, which he had pacified, as a personal fief. His territories were not much less than the Hasmonean kingdom at its height, and he held them, 31–4 BC, more securely.

Herod had a large standing army, he trained tribes as efficient native regiments and he had a personal bodyguard of 400 Celts which had once belonged to Cleopatra. His civil service was modelled on Egyptian lines. Herod was half-Arab, perhaps born in Nabatean Petra. He was oriental, part-savage, liable to spasms of appalling cruelty. But he was a first rate administrator. He took advantage of the growing population of Syria-Palestine, which had been increasing steadily throughout Hellenistic times, and the new wealth and prosperity of the Diaspora – promoted by the Augustan peace – to try to bring the Jews to the forefront of world culture. His expenditure was colossal, but he nevertheless always had a surplus in his treasuries. He did not hesitate to spend it during the great famine of 25 BC, when he melted down his gold and silver plate, and sold precious things from his palaces, to buy corn from Egypt.

It seems to have been Herod's policy, as it was that of many sensible Jews, to persuade Judaea to adopt the advantages of Roman culture without losing its essential Jewishness. In secular matters, he appointed Jews and non-Jews purely on their abilities. His wealth, like Solomon's, came chiefly from control of the trade routes, and his court was made up of many nationalities, speaking numerous tongues. It included many artists and scholars. One of his friends and advisers was the Aristotelian philosopher and historian Nicolas of Damascus, whose life of his master, which has since disappeared, was one of Josephus's sources.

Among Jews, Herod's chief supporters came from among the Diaspora, who were far more liberal and better educated than the Judaeans, and whose respect for Rome was akin to his own. Whenever possible, he chose Jews from the Babylonian or Hellenistic Diaspora for senior appointments, both religious and secular. He was determined to break the power of the old aristocratic and priestly families of Judaea, whom he saw as obscurantist, xenophobic, intolerant and barbarous – and who, of course, regarded him as a godless upstart, scarcely a Jew at all. One of his first acts, on assuming power, was to execute forty-five out of the seventy-one members of the Sanhedrin which had attempted to condemn him; Josephus says he eventually disposed of all but one of them. At the same time the institution was stripped of its secular jurisdiction and its decisions confined to points of doctrine. Herod also downgraded the office of the High Priest, which was made non-hereditary and subject to royal dismissal. Indeed, the High Priest was now merely an official in charge of the Temple services; even his ceremonial vestments were kept by Herod, who issued them only on special occasions.

For his choice of High Priest Herod naturally turned to the Diaspora. This brought him into conflict with his second wife, Mariamne, whom he had married for political reasons. She was a Hasmonean, the great-granddaughter of

Alexander Jannaeus. When Antigonus was executed, she expected the High Priesthood to go to her brother, Aristobolus III; instead, Herod gave it to a Babylonian, Hananel, and later had Aristobolus drowned in Jericho. Subsequently, he took High Priests from the Boethus family of Alexandria. According to Josephus, he had a jealous passion for Mariamne, which was returned with hatred. And eventually, under the impact of politics and the harem-conspiracies which were an inescapable feature of Oriental royal polygamy, he murdered Mariamne, his two sons by her, and her mother. Yet it is probably true, as Josephus remarks, that 'If ever a man was full of family affection, that man was Herod.' But affection, of course, for his own family: he founded cities and fortresses named after his father, mother and brother.

Herod was, by any standards, a great man. The New Testament and Christian tradition present him as a monster and the Massacre of the Innocents, though we have no secular authority for it, has a plausible ring from what we know of his paranoid suspicions and methods. Orthodox Jewish tradition also presents him as a tyrant. The pseudepigraphic *Assumption of Moses* says: 'He shall slay the old and the young, and shall not spare. Then the fear of him shall be bitter unto them in their land. And he shall execute judgments on them as the Egyptians executed upon them, during thirty and four years, and he shall punish them.' Herod was a kind of Peter the Great, dragging a conservative and obstinate people into the modern world. He realized that the suppression of piracy and banditry which Rome's power and new-found unity made possible was bringing in an economic golden age in which he wanted his country to participate. This meant knocking heads together, and in particular destroying the selfish oligarchy of families which controlled Jewish religion. He did this single-handed. But he also wanted to show the world that Jews were civilized and gifted people, capable of entering fully into the new expansive spirit of Mediterranean civilization.

Herod was a firm friend of Agrippa, Augustus's lieutenant and leading general; and it was this friendship which spread the protection of the Roman army over many threatened Jewish communities in Asia Minor and elsewhere. The Diaspora communities saw Herod as their firm friend and ally. He was also a generous patron. He provided funds for synagogues, baths, libraries and Jewish charities all over the Diaspora. He was the archetype of the big Jewish philanthropist and it was in his time that the Jews scattered through the Empire first became famous for the miniature Welfare States they set up among their own communities, in Rome, Alexandria, Antioch, Baghdad and elsewhere – a pattern which was eventually copied by the Christians and which was one reason why their faith spread so rapidly.

But Herod sought to counter any jealousy of the Jews by his generosity to other communities. He was a fanatical sportsman: wonderful on a horse, a reckless hunter, a superlative archer and javelin thrower. Athletes and wrestlers were always welcome in his palaces. Herod's vigorous patronage and powers of organization, and his endowments, rescued the Olympic Games from oblivion,

THE AGE OF HEROD AND JESUS

and ensured they were regularly and splendidly held. Thus his name was revered in many small Greek communities and islands. But he made gifts to the big cities of the Greek mainland and Diaspora too. He rebuilt the Temple of Apollo in Rhodes. He made big donations to Athens, Sparta, Lycia, Pergamum and Nicopolis. Nearer home, he gave theatres to Damascus and Sidon, built a wall in Byblos, gave Tyre and Beirut market-places and temples, gymnasia to Tripoli, Ptolemais and Damascus, an aqueduct to Laodicea, baths and fountains to Ascalon. And in Antioch he paved the muddy main street, two and a quarter miles long, with polished marble, and provided colonnades the whole length to shelter the citizens from the rain.

Herod's works in Palestine itself are still more impressive. He rebuilt Samaria, which had been wrecked by John Hyrcanus, and renamed it Sebaste – the Greek for Augustus. He gave it a temple of Augustus, the stairway of which survives;

At Caesarea, Herod created an artificial harbour in order to take advantage of the growth of commerce in that part of the Mediterranean. The Roman columns of granite were used to buttress the Crusader port town.

and so do two towers of the West Gateway, and a colonnaded street. There was another temple to the Emperor at Banias, of Egyptian granite. He built many fortresses and fortress-palaces, some of them on top of Hasmonean structures, like the Antonia (called after the emperor), built in Jerusalem on the site of the Hasmonean fort of Baris; others were Herodium, Masada, Machaerus, on the east side of the Dead Sea, and, near Jericho, Cypros, called after his mother. At Masada, he had a villa cut out of the rock with a spectacular view over the Dead Sea. Masada is the only one to have been fully excavated; but work has also been done on Herod's winter palace on the south banks of the Wadi Qilt, west of Jericho, where Herod had his wife's brother drowned, and where he himself probably died.

His most lasting memorial was the new maritime city of Caesarea, on the site of Strato's Tower. To take full advantage of the growth in commerce, Herod, according to Josephus, created an artificial harbour 'bigger than the Piraeus', by lowering 'into 20 fathoms of water blocks of stone mostly 50 feet long, 9 feet deep and 10 broad – sometimes even bigger.' These huge blocks formed the foundations of a breakwater 200 feet wide, which created a secure harbour. The town was of 200 acres and 'built entirely with limestone and adorned with a splendid palace.' It had a theatre, market-place and amphitheatre, with games every four years; and topping all was a colossal statue of Caesar which Josephus says was not inferior to the Olympian Zeus, one of the Seven Wonders of the

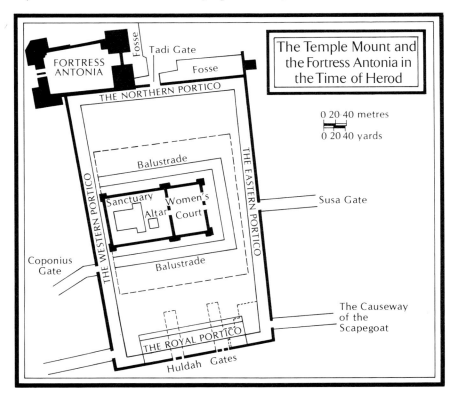

The lid of a decorated limestone ossuary, dated c. AD 100, represents an eight-arched arcade and probably imitates some well-known building, very likely Herod's Temple, completed in AD 64.

Ancient World. When Herod's kingdom had to be broken up at his death, the Romans picked the new city as the headquarters for their Procurator of Judaea.

In 22 BC, at a national assembly, Herod announced his life-work, the rebuilding of the Temple on a magnificent scale. He had already started work in Jerusalem by building a theatre and an amphitheatre, presumably outside the walls, so as not to anger the orthodox Jews too much – though they would have regarded the mere existence of such pagan institutions near the holy city as a provocation. Knowing in advance that his building plans would run into opposition, he first set up, on a corner of the Temple Mount, his formidable Antonia fortress, and later huge towers called Phasael (identified with the 'Tower of David'), Hippicus and Mariamne. Then he spent two years training regiments of Jewish stonemasons and other craftsmen. The actual building started in 20 BC, and the Temple structure itself was put up in a mere eighteen months. But the rest of the vast complex took forty-six years. Work was still going on in Jesus's day, and the Temple and its surrounds had not long been completed when the revolt broke out which led to its utter destruction.

Herod's Temple was built primarily for the millions of Diaspora Jews, for whom it was a great tourist and holiday as well as a religious attraction, and for whose sacrifices and accommodation he made ample provision. After the Roman destructions of AD 70 and AD 135, not a stone of the Temple itself survived, only the immense platform remained. But we have several descriptions of what it looked like and how it functioned. Herod kept to the original plan but he added the majesty which the post-Exilic builders had been unable to supply. The dimensions were greater than Solomon's – the portico, for instance, was 100 cubits high and 100 wide, whereas Solomon's was a mere 60. Beyond the three rooms of the Temple proper were three courts: a court which only priests could enter; a secular court for males of pure Jewish blood; and a court for Jewish women. Outside the gate to this last were tablets in Greek and Latin, warning gentiles that the penalty for going any further was death.

Since Herod was not of a priestly family and could not enter the inner court, he paid less attention to its fittings than to the externals. The Holy of Holies itself was bare, though there were sheets of gold on the walls. The stonework itself, says Josephus, was 'exceedingly white', and the decorative theme was dazzling white and gold throughout – even on the roof there were gold spikes to prevent birds alighting and discolouring the stonework. In a way, more impressive than the Temple itself was the prodigious platform on which it was built, twice the size of its predecessor. It was thirty-five acres in area and nearly a mile in circumference. Even more striking was its height: more than twice the height as seen today from the bottom of the valley, for all the lower courses of blocks are covered in the rubbish of many centuries. These huge blocks, some of them, according to Josephus, '45 cubits in length, 10 in height and 6 in breadth', were the work of imported craftsmen, and are to an exceptionally high standard. The top forty feet or so consisted of vaulted corridors, later known as 'Solomon's Stables', and above them, on the immense flat surface of the platform, were magnificent cloisters, with hundreds of Corinthian pillars twenty-seven feet high and so thick that three men with arms extended could barely encompass them. To look to the depths below from the cloisters of the platform, said Josephus, made you giddy.

At great feasts, many thousands of pious Jews, priests, Levites and others,

The recently uncovered stairway, outside the Temple in Jerusalem, which is probably the site of Jesus' final preaching.

worked in and around the Temple, which had various great gates, and an enormous bridge and staircase which connected it to the city below. The Temple itself was a place where vast numbers of sacrificial animals were despatched, blooded and butchered in a highly expeditious manner. Hence the platform was not solid. It was a gigantic cleansing system, containing thirty-four cisterns, which stored the winter rainfall and in summer water brought by aqueduct from the Pools of Solomon off the Bethlehem Road. The largest cistern, the 'Great Sea', held over two million gallons. A multitude of secret pipes and drains brought water up to the surface of the platform and carried away the torrents of blood in sluices. We have a valuable description of the Temple, written in Jewish Alexandria and called the *Letter of Aristeas*, by a Diaspora pilgrim. He says he saw 700 priests on the job, working in silence, handling the heavy carcases with professional skill and placing them on exactly the right spot on the altar. He was most impressed by the water-system, which was being operated all the time to keep the Temple clean: 'There are many openings for water at the base of the altar which are invisible to all except those who are engaged in the administration, so that all the blood of the sacrifices which is collected in great quantities is washes away in the twinkling of an eye.'

The priests, who conducted the sacrifices and the ceremonies, and the Levites, who provided the musicians, choristers, cleaners and maintenance men, were each divided into twenty-four watches; these *mishmarots* or shifts were often reinforced for feasts by priests from other Judaean towns and from the communities of the Diaspora. But the governing body was drawn purely from the upper, and usually hereditary, ranks of the Jerusalem priesthood: the High Priest, the *segan* or deputy, who was usually a pharisee, the *gizbarim* or treasurers, the *amarkalim* or trustees, and the *catholicos* or controller. Thirteen priests were required for each sacrifice. There was a regular sacrifice of two lambs at dawn each day and another two at sunset. Ordinary Jews were not allowed into the sanctuary during the service but the doors were kept open. Each service ended with the ritual drinking of wine and a performance by the Levite choir and orchestra of psalms, the reading of scripture, and hymns. On festivals they sang the hymn of praise or *hallel*, still used in synagogues today. So far as we know, the Herodian Temple musicians performed on the *kinnor*-lyre of ten strings, bronze cymbals, the *chalil* or double pipe, the twelve-string harp or *nevel*, and two kinds of trumpet: the *chotzetzerah* or silver-trumpet, blasts from which marked pauses in the ritual, and the *shofar* or ram's horn, still used on modern feastdays.

From the Egyptians the Jews had taken the notion of the altar-fire which never goes out and, in the form of sanctuary lamps, they passed it on to the Christians. They also adapted the Egyptian habit of incensing the most secret, dark part of the Temple. Temple incense was used in quantity, over 600 pounds a year, and it was very expensive; its recipe was a closely-guarded secret, the property of the priestly family of the Avtinas, whose womenfolk were not allowed to use scent to avoid any charges of corruption. From the evidence of Josephus and the

Babylonian Talmud, and from the directions in the Book of Leviticus, it looks as though the incense at this period was made by mixing stacte (gum balm) frankincense (gum resin from the terebinth tree), myrrh (gum resin from the camphor bush), onychas (ground-up shell), galbanum resin, cassia from the cinnamon plant, spikenard, saffron, Cyprus wine, Sodom salt, stuff called *maalah ashan*, which made rising smoke, and a kind of cyclamen called *kipat ha-yarden*.

The expenses of the Temple were met by a half-shekel tax on all male Jews over twenty, including the Diaspora and there were also endowments. But Herod, unlike Solomon, paid for the actual building of the Temple out of his own revenues. Such generosity did him no good with the orthodox Judaeans, and in his last months he quarrelled even with the Pharisees, who had generally been his allies. Over the main entrance to the Temple he set up a golden eagle. This did not worry the Diaspora Jews, but it outraged the orthodox of Jerusalem, Pharisees included. Some Torah students climbed up and hacked it to pieces. Herod was already on his deathbed, in his palace near Jericho; but he had his revenge nonetheless. The High Priest was deprived of office as accessory to the crime and the students themselves were dragged down to Jericho and tried in the Roman theatre there, being burnt to death immediately afterwards. After the slaughter, Herod was taken by litter to the hot springs at Callirrhoe, where he died in the spring of 4 BC.

Herod was attached to the sons he had by his first Nabatean wife, Doris, and his will divided up his kingdom among them: Archelaus got Judaea, Herod Antipas Galilee and Philip Ituraea. But Archelaus was a failure, and in AD 6 the Romans deposed him. Thereafter the province was ruled by Roman procurators residing in Caesarea, and ultimately responsible to the legate in Antioch, who was in charge of the whole of Syria. The only one of Herod's progeny to show conspicuous political ability was his grandson, Herod Agrippa. He was a Roman protégé, and in AD 37 he was awarded Judaea, which he successfully administered until his early death in AD 44, when direct rule was reimposed.

The Romans could not, or at any rate did not, solve the problem of how to administer Jewish Palestine with the consent of the inhabitants. The other oriental communities, and the Greek-speaking cities of the coast and the Decapolis, were reasonably content with the rule of Rome; indeed, they flourished mightily. But Rome could not reach a *modus vivendi* with the hard core of orthodox Jewry. One reason for this was that the Jews themselves were increasingly divided on what their religion was supposed to be about. It is true that many learned Jews believed that their main task was to study to understand the existing scriptures and set down commentaries as to God's precise meaning. The study of scripture, based on Deuteronomy 6:6–8, goes back to the time of Ezra at least, and proceeded industriously throughout the Roman period and beyond. Between 100 BC and AD 200 some of the basic forms of scriptural commentary took their classical shape. *Halaka* was the legal tradition of

THE AGE OF HEROD AND JESUS

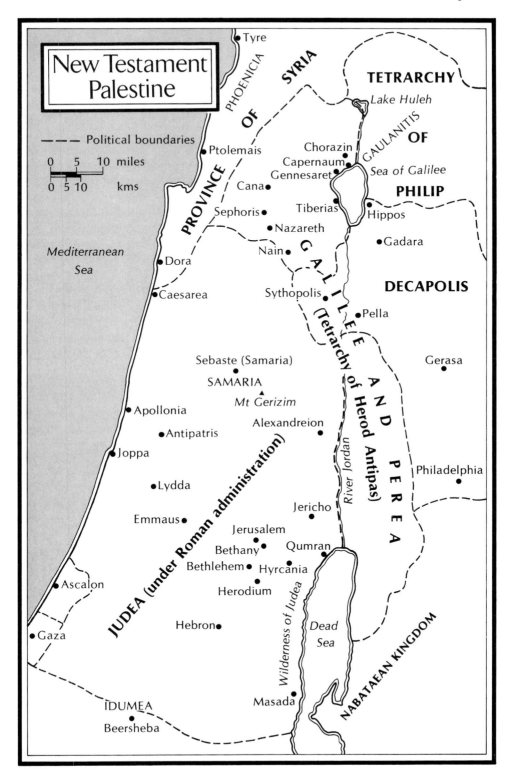

exegesis; *Haggadah* was the non-legal – narrative, homily, diet and medicine, and was popular rather than technical. General exposition of scriptural meaning was called *Midrash*, from the word 'to investigate'. There was also simple, literal interpretation known as *Peshat*. Hence the Midrashic *Halakah* was the expert corpus of scriptural exegesis of the law; Midrashic *Haggadah* was the corpus of homelitic exegesis. A systematic digest of the mass of legal tradition, the *Mishnah*, was completed by the Rabbi Judah, the Nasi or Prince, at the end of the second century AD; and this forms the heart of a much larger work, the Talmud (from the word 'to teach'), combining the Palestinian and Babylonian Talmud, and adding the *Gemara* or commentary on the *Mishnah* itself, which emerged in AD 400–500.

Had all Jews, or even all learned Jews, contented themselves with this pacific task of interpreting the Old Testament, and the parallel search for the ideal texts, which I have already mentioned, the Romans would have had no problem. But there were other Jewish traditions, including literary traditions of activism. What had happened to prophecy? It seems to have become a lost art in the quiet, contented years of Persian rule. Ben Sirach, who wrote Ecclesiasticus about 200 BC, said 'I will again pour out doctrine like prophecy and bequeath it to future generations.' But he was not accepted into the canon, and nor was Daniel, written in about 160 BC. Apparently men did try to prophesy, but were not heeded. About this time an addition was inserted into the text of Zechariah, 13:3ff:

Thereafter if a man continues to prophesy, his parents, his own father and mother will say to him: 'You shall live no longer, for you have spoken falsely in the name of the Lord.' His own father and mother will pierce him through because he has prophesied. On that day every prophet shall be ashamed of his vision when he prophesies, nor shall he wear a robe of coarse hair in order to deceive. He will say: 'I am no prophet, I am a tiller of the soil who has been schooled in lust since boyhood.'

Even when the Maccabee rising introduced a new episode in Judaic religious enthusiasm, no prophets arose; indeed, there is a reference in the First Book of Maccabees to 'the day when prophets ceased to appear', and Simon Maccabee was made leader and High Priest 'until a true prophet should come.'

The prophets had been statesmen as well as religious leaders and on the whole the political advice they had given had tended to be sensible and moderate. Unfortunately, prophetic literature was succeeded not by a gap but by the rapid growth of apocalyptic, that is by detailed visions of a great armageddon, involving a violent settling of accounts between the forces of good and evil. Of course this element had always been implicit in prophecy. But it inevitably grew strident under foreign rule. The Romans were the fifth imperialist power in turn to batten on the Jews, and commentary on the prophets tended to take an increasingly anti-imperialist line. Thus, in Habbakuk 1:7, the reference to the Babylonians, 'Terror and awe go with them; their justice and judgment are of their own making', is commented on as follows in one of the Dead Sea Scrolls from Qumran: 'This concerns the Kittim [Romans], who inspire all the nations

THE AGE OF HEROD AND JESUS

with fear ... All their evil plotting is done with intention, and they deal with all the nations in cunning and guile.' Superimposed on warlike prophetic commentary, apocalyptic literature, which was rejected by the canonists but was tremendously popular among the young enthusiasts, filled with the spirit of militant Judaism, suggested that history was speeding up and that the final battle would soon commence. As the Romans were responsible for maintaining order, for holding as it were the divine forces of history in check, they were the inevitable enemies, quite apart from their paganism and exactions; and any ill-judged move by a Roman procurator or governor might be taken as a sign that armageddon was about to commence. But collaborationist Jews, who denied the fiery signs and portents, were almost equally hateful.

Jewish religious sects, most of them millennarians, abounded in Syria-Palestine around the beginning of the Christian era. There were the exclusive Therapeuta, who came originally from Egypt, and who are mentioned by Philo. He also refers to the Essenes, as do Pliny and Josephus. These were quite a large sect, living a communal life in towns and villages, as well as caves. Another group of cave-dwellers were the Magharians, chiefly in Syria, and there were various Baptist groups, living in caves near the Jordan and performing lustration ceremonies in the river. The sect about which we know the most are the Qumran monks, because their monastery was excavated by G.L.Harding and Père de Vaux in 1951–6, and because we possess their library which was put for safety in tall jars concealed in nearby caves when they were menaced by the Romans in about AD 66. They lived in tents in summer, caves in winter; they had central buildings of a watchtower, dining-room, kitchen, bakery, pottery shop, scriptorium and meeting-room, and they had elaborate plumbing arrangements for their ritual lustrations. The sect was originally a protest against the defiling of the priesthood under the Seleucids and the Hasmoneans, but it illustrates the tendency of such sects to become increasingly militant in isolation. Their document, 'The War of the Children of Light against the Children of Darkness', was not just vaguely apocalyptic but was a detailed training manual for a battle they believed to be imminent, and which indeed came in 66–70, when their settlement was destroyed.

Since Judaism is not only a historical but a historicist religion, it tends to radiate a consciousness of impending events at all times, and especially in what appear to be periods of crisis and suffering. In particular, it has the notion of a saviour figure or Messiah. The idea is surrounded with ambiguities and it has led to disputes almost since its conception – plus, in classical, medieval and modern times, an ever-growing and acrimonious library of elucidation. It is never clear what the Messiah was, or what he was supposed to do. He might be a saviour-king, like Saul, David or Zedekiah; or even a friendly foreigner, like Cyrus. He was supposed to come from the line of David, though the passages which mention the Messiah regard the Davidic monarchy as a concluded episode. One of the psalms describes him as the son of the God, though Hebrew kingship had never been divine. The Messiah was to live among the people, die and be exalted, and so bear away their sins, though Hebrew kings had never been thought of as the embodiment of their people. But the Messiah is not always referred to as a king, but as the Son of Man, the Servant of the Lord, the Seed of the Woman and the Suffering Servant. His function was to sign a new Covenant and to bring salvation to all those in need. But he might be interpreted as the symbolic name for the 'faithful minority', the 'ideal Israel', the 'true remnant', or other collective group.

The notion of the Messiah and the apocalypse – together with the developing idea of eschatology, the Four Last Things, in more orthodox Jewish theology, inevitably became confused in the minds of ordinary Jews. When Jews thought of the Messiah, they thought of kingship, political change, revolution, the end of empire, the coming of some kind of kingdom approved by God. When Herod the Great learned that the Christ was born, he reacted violently, as to a threat to his throne and dynasty. A Jew who heard a man make messianic claims assumed they included some kind of political if not military programme. The Roman authorities, the Sanhedrin, the Sadducees, the Pharisees, would in varying degrees regard with intense suspicion a being who came to overthrow and replace the existing order with which all of them were to some extent associated. And the poor people of Jerusalem and Judaea, and of Galilee and Samaria too, would naturally suppose that a Messiah who preached fundamental change would be talking not in spiritual and metaphysical terms, but of actual government and real taxes and everyday justice.

It is evident that Jesus of Nazareth fitted none of these Messianic stereotypes. His mission was adumbrated in chapter 53 of Isaiah: he was to be the 'tender plant', the 'despised and rejected of men', the 'man of sorrows', who would be 'wounded for our iniquities, bruised for our transgression', 'oppressed and afflicted and yet he opened not his mouth.' Such a Messiah, the 'suffering servant', would be 'taken from prison and from judgment', 'brought as a lamb to the slaughter', should make his grave with the wicked and be 'numbered with the transgressors.' This meek Messiah was not a danger to any existing order, or any particular throne, or ruling class or bureaucracy – at any rate in a direct or immediate sense. He was not a mob leader, a democrat, or a guerrilla. He was talking, it is true, of freedom. But it was not the freedom of Republican Rome, the freedom, within a firm framework of orderly government, to move, trade and worship where you will; nor was it the freedom of the Jewish priesthood to carry out the commands of the Torah free from external interference; it was, rather, the internal freedom of the conscience at ease with itself, a form of freedom later preached so eloquently by St Paul. And this new freedom, which could not be measured in terms of frontiers and forms of government, and would be won by a degraded sacrifice of the Messiah himself, was offered not merely to Jews but to all mankind in accordance with the prophecy of Isaiah: 'In thee shall all families of the earth be blessed.'

THE AGE OF HEROD AND JESUS

This inscription in the theatre at Caesarea is the only concrete historical evidence associated with Pontius Pilate.

Seen in the apocalyptic and political context of his time, the movement led by Jesus was thus initially an anticlimax. When it came to the point he repudiated popular Messianism. The authorities sighed with relief and dispatched him without hesitation or compunction. The mob was disappointed. It is possible to see the cry, 'Crucify him!' as prompted by disgust. So the Messiah had not come to liberate the Jews, but to preach self-sacrifice and resignation to all! That was not the message the Jews in the streets had been expecting, or wanted.

Hence it is not entirely surprising that Jesus of Nazareth left so little trace on the secular history of his day. Even his reluctant executioner, Pontius Pilate, the Procurator of Judaea, is known in Palestine only from one inscription in the theatre at Caesarea. The first Christian texts, St Paul's early epistles, date from the 50s, about twenty years or so after Jesus's death; the gospels were probably written after, rather than before, the fall of Jerusalem in AD 70. Jesus is not mentioned in Roman historical texts and the earliest mention of Christians comes in a passage dealing with the Neronian persecution. Josephus undoubtedly wrote about Jesus, but presumably in an unfavourable sense for all copies of his manuscript went through Christian control at some stage, Josephus's actual words were censored and an adulatory passage was inserted. Early Jewish sources are unenlightening. It is possible to identify sites connected with Jesus with some degree of confidence, as we shall see, but for practical purposes our knowledge of his life and works is confined to the New Testament.

One reason why Jesus was not much remembered at the time, except by his immediate followers, was precisely because he turned away from Messianic or apocalyptic politics. Other, more active and violent figures crowded him off the stage. Examining the doleful historical record of Judaea in the first century AD, one comes to the conclusion that it was becoming ungovernable, whether by orthodox Jews, of Hasmoneans, or Hellenized Jews like Herod and his family, or by direct Roman rule. With the best intentions, Herod had turned Jerusalem into a huge, overcrowded pilgrimage city, packed with religious fanatics of all kinds. The Jewish law forebade so many things, and breaches of it caused such outrage, and people of such different races and customs congregated in the city, that a riot could flare up at any moment; and in an age of apocalypse a riot could become revolution overnight.

There was a revolt of some Jews immediately after the death of Herod the Great. The brief reign of his son, Archelaus, was a time of continuous rioting and unrest, which provoked Roman intervention and direct rule; and this, in AD 6 set off the rising of Judas of Gamala, a militant Messiah. Risings occurred when the Romans marched troops bearing their eagles into the city, or moved about near the Temple area, or when, for instance, Pilate tried to use Temple funds to improve the city's water supply. In the time of the procurator Cuspius Fadus, sometime after AD 44, there was a rising by one Theudas, who marched his mob down the Jordan valley. Again, in the time of the procurator Felix, 52–9, an Egyptian Jew mustered 4,000 rioters and waited on the Mount of Olives for the walls of the city to fall down miraculously.

The Jews were divided into innumerable sects, always mutating and waxing and waning in strength, but in these years power shifted steadily towards the religious extremists and the men of violence. Judas of Gamala had been the first zealot leader; by the late 60s, the zealots had come to dominate Jewish political activism. Indeed it was hard for anyone, and impossible for the Romans, to distinguish between the religious terrorists and ordinary brigands, since the two types often worked together. One particular sect, called *sicarii* after the *sica*, or short dagger they used, specialized in murdering Jewish moderates, especially during religious processions and ceremonies, thus calculated to cause the most trouble. The way of the peacemaker was hard. Agrippa II, tetrarch in AD 66, with the right to appoint the High Priest, did his best to mediate between the Romans and the religious Jews, but as Herod's great-grandson, he was suspect; he was accused of incest with his sister Berenice and, more seriously, of building an extension to his palace which overlooked the Temple. When the Temple priests erected a wall to protect their privacy, Agrippa got the procurator to knock it down; the Jews appealed to Nero, who sided with them and so Agrippa sacked the High Priest. The last straw was when the procurator, Florus, confiscated money in the Temple treasury to cover arrears in the tribute of Judaea. In the atmosphere of apocalyptic Jerusalem, such relatively trivial matters could ignite a monstrous conflagration.

THE AGE OF HEROD AND JESUS

The graphic historical accounts of Josephus, although often subjective, do much to enlighten historians about events, particularly where he was personally involved.

The great Jewish War was one of the most horrifying episodes in history. We have a very full and graphic account of it by Josephus. He had been a priest in Jerusalem and during the early stages of the revolt was a Jewish general in the fortress of Jotapata in Galilee. After its surrender he went over to the Romans, as did many moderate Jews, and acted as interpreter for them during the siege of Jerusalem. He later settled in Rome on a court pension and was encouraged to write official history. He was personally involved and his version is suspect; but he was also an eyewitness and his tale is told in shocking detail and with sombre magnificence.

The start of the revolt was the deliberate cessation of Temple services in the name of the Emperor, and in one sense the war was an anti-colonial explosion. But it was also a protest against anti-Jewish pogroms in Greek-speaking cities, especially in Caesarea; and it was, finally, an uprising against the Jewish moderates and men of property, who argued that the only hope for Jewish survival and religious independence lay in an accommodation with the Roman authorities and their Greek-speaking neighbours. In 66, the zealots, under Menachem, took control of the main city. They had already captured the Masada fortress and slain its garrison. When they stormed the Antonia and broke into the upper city, they burned the palaces of Agrippa and his sister Berenice and the official residence of Ananias, the High Priest. He himself was murdered in the streets. The Roman garrison surrendered on terms, but was massacred.

At this stage there were still many moderates in the city, not only willing but anxious to restore Roman authority. They would undoubtedly have cooperated with the legate in Syria, Cestius Gallus, had he acted decisively. He moved up towards the city with two legions, and he had 7,000 of Agrippa's troops in addition. He might have taken the city by a direct and immediate assault, but he hesitated, retreated and was lost. This wholly unexpected victory over the principal Roman army in the Near East was hailed by the extremists as an act of Yahweh, and fanatical bands from all over the country swarmed into the city. Once inside, they set about the moderates. Josephus says that 8,500 moderates were murdered by Idumaean extremists, and that a further 12,000 were dispatched by the zealots, many after summary trials. He gives a chilling account of the reign of terror within the walls.

Vespasian, in charge of the Roman security forces, took his time about laying siege to Jerusalem, rightly believing that the Jews were divided and would spend most of their efforts in slaughtering each other. He pacified the rest of Judaea first. In AD 69 he was elected Emperor, and left his stepson Titus in charge, with four legions and strong forces of auxiliaries. Titus put up walls of circumvallation around the city and prepared to starve it; right up to the end, he hoped for a negotiated surrender. The siege itself lasted from April to September 70, and at last Titus had to take every inch of the city by force: first the outer north wall, then the second wall, then the inner city wall which coincided with the wall of the Temple precinct, then the Antonia fortress, then the Temple itself – which was destroyed by fire despite Titus's attempts to put it out, on 29 August 70. Finally the rebels were driven into the Upper Citadel and hunted down.

Josephus says that in all a million people died in the war. Some 97,000 were captured at Jerusalem. Those under seventeen were sold as slaves; most of the able-bodied survivors were sent to the Egyptian mines, or to die in various arenas. There was a gruesome victory celebration in the amphitheatre at Caesarea; to mark his brother's birthday, Titus had 2,500 Jewish prisoners burnt to death, slaughtered in combats with wild beasts, or between each other. One of the leaders, John of Gischala, was imprisoned for life; another, Simon Ben Gioras, was taken to Rome for a triumph, and then hanged. The triumphal arch of Titus in Rome, with its captured candelabrum, still stands, and Titus kept in his palace the curtain from in front of the Holy of Holies and the copy of the scriptures he found there – but these, of course, have long since vanished.

There was an aftermath. Some of the zealots, led by Eleazar Ben Yair, escaped from the doomed city and shut themselves in the rock fortress of Masada, from which they conducted a guerrilla campaign, until it was invested and circumvallated by Flavius Silva and the 10th Legion in AD 72. The summit of this natural plateau-fortress is 600 by 200 yards, and it was first fortified by the High Priest Jonathan. In the years 37–30 BC, Herod the Great gave it a casemate-type wall and built a palace there. The wall, says Josephus, was eighteen feet high and twelve feet wide, and there were thirty-seven towers, each seventy-five feet high.

THE AGE OF HEROD AND JESUS

Herod also carved out huge water-cisterns in the live rock, and there were, according to Josephus, immense stores of food and arms. Masada was the last stand of the zealots and *sicarii*. Since the site was excavated the story of the siege has become a popular Israeli folk-tale, but the truth is that it illustrates pointless fanaticism as much as heroism.

To get his engines to a height from which he could attack the fortress, Silva seized the White Cliff, 450 feet below the level of Masada, and on it built a 300-foot high platform of earth, a stone pier 75 feet high, and then an iron-plated tower, 90 feet high. From this he cleared the battlements, and smashed a hole in Herod's casemate with a giant ram. The defenders then built a second wall within the first, compacted of earth to take the ram-blows. But the Romans succeeded in burning it and after this final setback Eleazar decided that the defenders, who numbered about 1,000, including women and children, should destroy themselves rather than become human booty for the Romans. Josephus gives

The depiction of the capture of the candelabrum
on the triumphal arch of Titus in Rome.

what purports to be the full text of two speeches he delivered the night before Masada fell, in which he eventually persuaded the garrison to kill their families as well as themselves. Josephus says that two women and three children hid in the conduits and so were found alive when the Romans entered the fortress the next morning; and that one of the women, 'in intelligence and education superior to most', was able to tell what had happened and summarize the speeches. The second was a passionate plea for freedom, and a remarkable early example of anti-colonialist rhetoric (or rhodomontade). But it seems less impressive if we bear in mind that, while Eleazar and the hard core were certain of death when the Romans broke in, most of the women and children would have been spared, so he whipped up his men into murdering their wives and offspring. The crime was

An aerial view of the zealots' stronghold at Masada, the rock fortress where some 960 Jews committed mass suicide in AD 72, sooner than be taken by the Romans led by Flavius Silva.

THE AGE OF HEROD AND JESUS

peculiarly un-Jewish. Excavation of the site revealed some touching indications of the last hours, and also the first surviving synagogue, orientated towards Jerusalem; scraps of Genesis, Leviticus, the Psalms and some of the Apocrypha, and the earliest fragments of Ecclesiasticus we possess; even coins, for the Jewish nationalists minted their own currency during the revolt.

The fall of Jerusalem in AD 70 had tremendous consequences for Christianity, for it led to the impoverishment and dispersal of the Jerusalem Christian Church and the transfer of leadership from its Judaizing elements to the Greek-speaking Christians of the Diaspora and the gentiles. The consequences for the Jews might not have been catastrophic. There is no evidence that Vespasian wanted to destroy the Jewish nation and its religion. He had coins minted with his features

Herod's palace at Masada which he built from 37–30 BC.

A bronze Roman coin, minted at Caesarea, shows the founder of the city ploughing.

A bronze coin of the leader of the Jewish revolt, Simon Bar-Kochba.

on one side, and on the other a sorrowing woman under a palm tree, with the legend *Judaea capta*. One of the Pharisaic rabbis, Johanan Ben Zakkai, had smuggled himself out of besieged Jerusalem in a coffin and went to the Roman camp; he got permission to constitute a synod of rabbis at Jamnia (Yebneel), which replaced the old Sanhedrin and constituted Jewish legitimacy. With the Temple in ruins, the synagogue began to become the focus of devotion, and the rabbi of authority, but Jerusalem was still to some extent a Jewish city and full of activists. Some at least of the rabbis supported the physical and earthly manifestation of Yahweh's omnipotence. In 132 there was a further revolt. We know little about it, but it seems to have been provoked by Jewish fears that the Emperor Hadrian intended to establish a Roman city grouped around a Temple of Jupiter, on the ruined Temple mount and by another rumour that circumcision was to be made illegal.

At the first sign of trouble the Roman 10th Legion, which had been camped on the south-west hill, withdrew from the city, which had no walls and was barely defensible. The rebels reoccupied it, and endeavoured to rebuild some of the walls. Temple services were resumed after a fashion. The rebel leader was Simon bar Kosiba, known as Bar-Kochba, 'son of a star', after the saying in the Book of Numbers: 'There shall come a star out of Jacob, and a sceptre shall arise from Israel.' Coins were minted, with the legends 'Shimeon, Prince over Israel' and 'For the Deliverance of Jerusalem' in old Hebrew script. This time the horrific terrorist element was absent and the Jews were reasonably united, though most rabbis refused to accept Simon's Messianic claim. An exception was the famous Rabbi Aqiba, who joined him, though his colleague, Rabbi Johanan Ben Torta, is traditionally supposed to have said: 'O Aqiba! grass will sprout between your jaws sooner than the Son of David will appear.'

Simon's men carried out raids against Roman settlements of veterans, and massacred Christians, according to Justin Martyr, of Neapolis (Nablus). But when the Romans reorganized under Julius Severus, sent from Britain to take command, the Jews were forced out of Jerusalem and Simon was finally bottled

THE AGE OF HEROD AND JESUS

up and killed in Bethar or Battir, a few miles to the south-west (probably Khirbet al-Yahud). In recent years papyrus fragments have emerged in caves within a ten-mile radius of the Engedi oasis; the caves were evidently the last refuges of the rebels and the fragments the remains of their archives. They are mostly legal documents, in the name of 'Shimeon bar Kosiba, Prince of Israel' and testify to an intense desire for legality and still more, for the details of religious rituals, even in the last desperate phase of the rising.

Simon was killed in 135; Aqiba was tortured to death. The historian Dio Cassius claims that 580,000 Jews were killed and in the late fourth century AD, Jerome reported a tradition from Bethlehem that there were so many Jewish slaves for sale that the price fell below that of a horse. Hadrian, the great engineer-emperor, literally replaced Jewish Jerusalem with a city named after himself, Aelia Capitolina: that is, he buried the hollows of the old city in rubble to level the site for his regular layout, and outside the limits he set he removed the debris in order to excavate the rock below it to provide massive stones for the public buildings he set up. The new city was the first to be broadly on the plan of the present 'Old City'. The main road from the north entered through the Damascus gate, and then became a colonnaded street bisecting the interior, as it is shown in the sixth-century plan of Jerusalem on the mosaic floor of the church of Madaba, near Amman. Inside the Damascus Gate was a semi-circular piazza, with a column, which still gives the gate its Arab name, *Bab el-Amud*, Gate of the Column. The main east gate was the one later called St Stephen's Gate, to the north of the Temple platform; it was spanned by a triumphal arch, the so-called Ecce Homo Arch, the remains of which are now within the Convent of the Sisters of Sion. Greek-speaking settlers were moved into the city and Jews were forbidden to set foot in it on pain of death, though apparently they contrived to bribe their way to what now became the Wailing Wall on the anniversary of the city's destruction.

The disintegration of the Jews as a militant community with a nation-building potential inevitably placed the leadership in the hands of the rabbinical element, who had decided even before the fall of Jerusalem in AD 70 that resistance to the Romans carried no divine sanction. After 135, the rabbinical authorities transferred themselves from Jamnia to Usha, a town in western Galilee, where the President of the Rabbinical school was given the title of Nasi or prince, and recognized as a patriarch by the Roman authorities, with responsibility for the spiritual welfare of all the Jews of the province.

Henceforth Jewish life was dominated by the synagogue and the Torah, with the rabbi as guide and spokesman. The remains have been found, and in some cases properly excavated, of a surprising number of early synagogues in Roman Palestine and Syria; and, what is still more remarkable, the decorative element is prominent and in some cases contains images. Evidently the Talmudic censure of the image was not always strictly applied, especially outside Judaea, and more particularly after the Jewish communities reconciled themselves to Roman rule.

ABOVE LEFT St Stephen's Gate, the main east gate of Jerusalem. This is often called 'The Lion Gate' because of the four lions represented on either side of the archway.
ABOVE RIGHT The remains of the Ecce Homo Arch, now within the Convent of the Sisters of Zion, under which Jesus is supposed to have walked on the way to his death.

Not all Jewish intellectuals were opposed to religious art. The great Jewish philosopher Philo was, in fact, sympathetic towards artists: 'They have often grown old in poverty and disesteem, and tragedies have accompanied them to the grave, while their works are glorified by the addition of purple and gold and silver and other rich embellishments which money supplies.' There is literary evidence, too, that in the imperial period images were regarded as permissible among some Jews.

Of the Jewish centres, Tiberias, settled by them in the time of Vespasian, was one of the most important. It had been founded fifty years earlier by Herod Antipas, on account of its sulphur springs. There the *Mishnah* was completed about AD 200, the Jerusalem Talmud in the fourth century, and the so-called Tiberian or Western system of vocalizing the Bible (that is, supplying the pointing for the missing vowels, which involved reconstructing the ancient pronunciation) in the sixth and seventh centuries. The fourth-century synagogue has a mosaic floor with both human and animal figures and, like some other synagogue mosaics of the Roman period, the signs of the zodiac. At Tiberias, too, Jewish history is written in the tombs: the martyred Rabbi Aqiba is buried on the hill near the town and there are other important tombs of the Roman period, those of Ben Zakki, Rav Ammi and Rav Assis, and, two miles down the Lake of Tiberias,

THE AGE OF HEROD AND JESUS

he miracle-working tomb of the second-century Rabbi Meir, with chapels for both the Sephardi (Spanish) and Ashkenazi (central European) Jewish cults. Above all, there is the early thirteenth-century tomb of Saladin's famous doctor, the Rabbi Moses Ben Maimon, known as Maimonides or 'Rambam', Torah-scholar and philosopher.

Not far from Tiberias is Capernaum, where Jesus healed the Centurion's servant. The Centurion was described to Jesus as 'worthy ... for he loveth our nation and hath built us a synagogue.' This first-century building has disappeared, but its successor, built in the second and third centuries and excavated between 1905 and 1926, has a stone carving of the regimental crest of the 10th Legion and a Roman military award, evidently in memory of this officer. The building had a courtyard and a rectangular hall with aisles formed by Corinthian columns, with the galleries of the women above; the carvings include the shofar and the menora candlestick, the shield of David, the manna-pot and the palm-tree.

Along the Lebanese frontier of modern Israel there are remains of synagogues

Tiberias, with the second-century Tomb of Rabbi Meir in the foreground and the city of Safed on the skyline.

LEFT An example of the unique collection of wall paintings from the synagogue of Dura Europus, AD 245.

BELOW The late second- or early third-century AD synagogue at Capernaum is elaborately decorated with carved stone ornament, and stands on the site of an earlier synagogue where Jesus taught and healed.

from Roman times at Sasa, Gush-Halav and Kefar-Biram, the last of which has the traditional burial-place of Queen Esther. Just off the Haifa–Nazareth road is Beth She'arim, another early centre of Judaic scholarship, and the seat of the Sanhedrin in the second century AD. The place was destroyed by the Romans in AD 352, after yet another Jewish revolt, but the extensive ruins and the catacombs of the synagogue have been excavated. Rabbi Yehuda Hanassi, the principal compiler of the *Mishnah*, was buried there about AD 220, in the rock-cut cemetery. His tomb has not yet been identified, but hundreds of inscriptions in Hebrew, Aramaic, Greek and Palmyrene have been found and translated, and what is most striking about this important early cemetery is the sheer wealth of figurative art, especially carvings and painting of Biblical scenes. Again, at Beth Alpha in the valley of Esdraelon figurative work was discovered in 1928 on the mosaic floor of a sixth-century synagogue, one panel of which shows the sacrifice of Isaac.

The greatest depository of Jewish art of the Roman period, however, was found hundreds of miles to the east in the Roman caravan city of Dura Europos on the right bank of the Euphrates. Dura was founded by the Seleucids around 300 BC and occupied by Rome in AD 165. The Jewish colony there was formed by descendants of Jews deported by the Assyrians in 722 BC, and others hunted from Palestine by the armies of Vespasian, Trajan and Hadrian. They got on reasonably well with the Parthians and Sassanids and seem to have been tolerated when the Romans took over the city. In 1932, excavation uncovered the remains of a synagogue, dated AD 245, with inscriptions in Aramaic, Greek and Pahlevi, the Parthian script. The architecture is Hellenistic, after the Macedonian pattern, but the most remarkable find was a series of thirty painted panels (now removed and displayed in the National Museum in Damascus), which systematically illustrate the Messianic idea of the return, the restoration and salvation. The imagery probably reflects that of the illustrated Bibles which are believed to have circulated in the second and third centuries AD, and the paintings are a digest of some of the great events of Jewish history – the patriarchs, Moses and the Exodus, the loss of the Ark and its return, the reign of David, Esther and Ahasuerus, and so forth. This unique survival, taken in conjunction with other evidence of Jewish figurative art in these centuries, indicates that early Christian ecclesiastical art, like so much else of the ecclesiology and liturgy of the Church, was produced from a Jewish foundation – though the Jews themselves later abandoned the tradition, which was never strictly orthodox.

At what precise point Jews and Christian diverged it is not easy to say. About AD 100, the rabbinical leaders at Jamnia drew up a formal indictment of Christianity and sent letters to all the synagogues of the Diaspora, enclosing a new malediction of Christians devised for insertion into the three daily benedictions. Not long afterwards, in turn, Jewish-Christianity was proscribed as heretical among orthodox Christians and after the end of the AD 135 Jewish rising, Christian groups claimed official treatment as non-Jews.

The Jewish rift with the Samaritans was never healed either. They continued to worship Yahweh on Mount Gerizim even after John Hyrcanus smashed up their sanctuary there; the inept and brutal Pilate was recalled by Rome because he ordered his troops to massacre an unarmed party of Samaritans who were on their way there. The Samaritans, in turn, killed a group of Jewish pilgrims travelling through Samaria to Jerusalem in AD 52. They rose with the Jews in 66–70, and 11,000 of them were slaughtered in consequence; but in 132–5 they sided with the Romans, who rebuilt their Mount Gerizim temple in gratitude. Samaritan history is largely unwritten, but there are still a few hundred of them in the region of Nablus. Their priests have long hair and beards and their Torah is written in the Samaritan script. Every passover the entire community goes up Mount Gerizim and kills a sheep, while their High Priest stands on a rock and recites the twelfth chapter of Exodus; the carcase is roasted and everything is eaten except the bones, horns and hooves.

Other small independent peoples enjoyed mixed fortunes under Roman imperialism. Some sixty miles above the head of the Gulf of Aqaba was the city of Petra, in a sandstone fault in the limestone plateau, 2,800 feet above sea-level. This was an ancient city of the Edomites, until in the sixth century they moved to south Palestine, where they became known as Idumaeans. Arab nomads, called Nabateans, moved in, and in late Seleucid times they built up a formidable kingdom around Petra. Of course they straddled the north–south inland caravan route; and, though it is not clear whether they were a properly organized commercial state or merely disciplined brigands preying on caravans, they were certainly prosperous in the early Roman era. Pompey accepted them as a vassal-state, and at one time their authority stretched to Gaza; indeed in St Paul's day the Governor of Damascus was a Nabatean. Trajan ended their independence in AD 106, when he created the province of Arabia Petraea beyond the Jordan and thereafter Petra's fate was linked to the fortunes of Rome–Byzantium. At some point the southern desert route became untenable, the Nabateans disappeared and their city was overtaken by the desert – not to be rediscovered until Burchhardt found it in 1812. It has been only partially excavated by the British School of Archaeology, and even today it is difficult to get to. The road ends at El-Ji, and entrance to the 'rose red city' is through a rocky rift called the Sik, only ten feet wide in places. The site straddles the best part of a millennium: the original High Place sacred to the Nabatean god Dushara, the Temple, built just before the time of Christ, rock-cut tombs of great splendour, mostly Nabatean, but some from the Roman period, and a Roman street, theatre, baths and other buildings. It is weird, magnificent, and it has been dead for eighteen centuries.

The rise and fall of Petra illustrates a truth which applies to much of this area. The land is marginal. Prosperity depends on the patterns of trade and these in turn on the security which the paramount power provides. The period from the coming of the Greeks until the expulsion of the Byzantines about 900 years later –

The magnificent Treasury at Petra, the 'rose red city' in Jordan. Carved out of sandstone, the city was an ancient city of the Edomites.

the centuries of western control – was the most notable in the whole history of the Holy Land. The Jews fared badly as a nation and were diminished and dispersed and other small peoples vanished entirely. But for most of the inhabitants classical order brought modest progress. Population increased, towns flourished and expanded. Palestine and Syria and the desert which hems them in are littered with the sites of former towns, villages and farms which have not been inhabited since the Persian and Arab invasions of the seventh century. Not even the twentieth century, with its immense resources of technology, capital and manpower, has yet fully restored the damage inflicted at that time and after.

Even so, what remains still fills the traveller with wonder. At Baalbek, off the high mountain route which leads from Beirut on the sea to Damascus on the edge of the desert, the Romans built the holy sun-city of Heliopolis. The ancient city, which covers 200 acres, has never been scientifically excavated; we know virtually nothing about its history; but its buildings speak for themselves. The temple of Jupiter Heliopolitanus, with its forecourts and surrounds, formed an acropolis of itself. This great temple measures 290 by 160 feet, and originally consisted of 54 columns, each 67 feet high and 7 feet in diameter. The six which are still upright are probably the largest columns in the world, and in the nearby quarry there is an unused stone 60 feet long which weighs 1,500 tons. The 'small temple', known as the Temple of Bacchus (though actually dedicated to Venus the Semitic Atargatis) still has its ceiling slabs 30 feet square, and is actually larger than the Parthenon. The inert mass of these colossal ruins tells an undeniable tale of formidable powers of organization, huge resources over a prolonged period and enviable self-confidence.

Across the Jordan, in even less hospitable territory, were the Greek-Roman cities of the Decapolis, the urban standard-bearers of classical civilization right on the rim of the desert. Jerash, the ancient Gerasa, founded by an Alexandrine general about 320 BC, is still magnificent. The walled town of 235 acres is dominated by a great temple of Artemis. The Romans ended its autonomous status early in the second century AD, as part of Trajan's policy of expansion and consolidation in the east. That is when it acquired its North Gate, built in 115, and, shortly after, a fine triumphal arch outside the walls on the road to Amman, commemorating the Emperor Hadrian's visit in 129. The Ionic columns of the oval-shaped market place still stand in part; there is a big theatre, and a hippodrome, which the seventh-century Persian ravagers turned temporarily into a polo ground. This great city flourished for 900 years and then lay deserted in ruins for another 1,200 – it was only reoccupied in the late-nineteenth century. It is another silent witness to the power of the classical world to generate wealth.

As we penetrate further into the wild country, our respect for the sheer hardihood of Roman culture is strengthened. The Hauran massif, perhaps the country of Job, and certainly known to the Patriarchs, is now called the Jebel Druze. In fact the Druzes have been there only since the second half of the nineteenth century, when they were driven out of the southern Lebanon by

138   The remaining six enormous columns of the temple of Jupiter Heliopolitanus at Baalbeck are 67 feet high and 7 feet in diameter and are probably the largest in the world.

The ruins of Jerash, one of the Greek-Roman cities of the Decapolis.

Christian reprisals. Before that it was deserted for over 1,000 years; basalt country with no water, its topsoil carried away by the prevailing winds. But long before that, Roman power and organization had made it possible for large numbers of people to live there in apparent prosperity and even grandeur. The ruins of Kanawat, Bosra and Chahba, the three best sites, bear witness. Such hill cities on the verge of the wilderness cannot survive unless there is an active central power to prevent incessant Bedouin raiding. In 105 the Emperor Trajan decided to provide that power, and ordered the massif to be occupied and settled. Nova Trajan (Bosra) became the capital. Since there are no springs or rivers, the Romans also built a system of reservoirs and aqueducts – one of the reservoirs still exists at the modern centre of Soueida (Dionysias). The area attained a kind of fame in the mid-third century, when the son of a Hauran chief, Philip the Arab, became Roman emperor for five years. He was born in Chahba, which he transformed into Philipopolis, and much of what he built there remains. He raised Bosra to metropolitan rank, and there too the theatre and baths survive. At Kanouat there is an elegant temple. Once Roman-Byzantine power retreated, the Bedouin moved back into range, the water-system collapsed, and the combination of banditry and aridity brought silent oblivion.

These dead cities of the classical age stretch far into the Syrian and Mesopotamian deserts. They cover the original route along which Abraham moved from Ur of the Chaldees into Canaan. In antiquity, desert caravan traffic, provided it could be made secure, was enormously profitable. The profits from this trade helped to build Solomon's Temple, and his walled cities, and in time the Second Temple of Herod the Great. Under Roman control it flourished even more: they sank wells, in their methodical way, every twenty-four miles along the

THE AGE OF HEROD AND JESUS

route, digging ever deeper until they struck water. They had forts, and a regular convoy system, so that, says Strabo, there were caravans of 2–3,000 camels, like armies. At the centre of the system of routes was Palmyra, the grandest and most mysterious of all the dead desert cities.

Palmyra is approached from Damascus along the route of the Valley of the Tombs, built by the merchant-princes who flourished in the city, and adorned with fine sculptures of the first and second centuries AD. The town is, or was, enormous, and its main colonnaded street, the chief feature of imperial towns in the east, still has about 150 of its 375 columns standing. It became economically important in the first century BC as a caravanserai on the new route between the Mediterranean coast and the Persian Gulf. Its profits increased rapidly again after the Romans occupied Dura Europos and improved security over the whole area. Then followed over three generations the rise of the Odenath family, chief of the Palmyran merchant class, culminating in the reign of Queen Zenobia – said to be the Hellenization of the Arab word for 'merchant's daughter' – who challenged Roman power and eventually succumbed to it.

Zenobia's father, Odenathus, had trained a remarkable force of mounted archers who policed the desert routes and were famous as the best cavalry regiments in western Asia. Behind this screen his daughter created one of the most unusual courts of antiquity. The underlying structure was Arab and tribal, the essential culture more Iranian than Greek, but many of the forms and titles, and the coinage, were based on Roman norms. Our chief source says that she spoke Latin and Egyptian as well as the local tongues. She received visitors in the Persian manner, but went to public assemblies like a Roman emperor, with helmet and purple robe, sometimes with bare arms. She had dark skin, black eyes, and very white teeth. She rode and hunted like a man, and sometimes she drank with her generals, though she had her maidens and eunuchs too. It was said that the gold and silver plate she used at banquets had once belonged to Cleopatra. She dabbled in Neoplatonism. 'She had drawn up for her own use,' wrote Gibbon, 'an epitome of oriental history, and familiarly compared the beauties of Homer and Plato under the tuition of the sublime Longinus.' In fact it is now doubted whether Cassius Longinus was the author of *On the Sublime*, which is rated a first-century work today. But he was a critic, a rhetorician, according to Eunapius 'a living library and a walking museum', and he was certainly at her court. So was Paul of Samosata, the famous heretic Bishop of Antioch, and other intellectual waifs and strays from the Empire. Alas, Zenobia the brave was tamed by the Emperor Aurelian, taken in golden chains to Rome, and married off to a senator, dying in Tivoli. Longinus was executed. Palmyra's prosperity gently declined, then vanished, and the silent city was washed by sands until English Turkey merchants rediscovered it in the late seventeenth century. The columns of Palmrya, like the walls of Dura Europos, remain to remind us how far the tentacles of Rome stretched when it was the dominant civilization of the Holy Land.

OVERLEAF The colonnaded forum at the powerful caravan city of Palmyra, with the later Turkish castle on the tell in the background.

144 The present Church of the Holy Sepulchre in Jerusalem stands on the site of both the early Church and Golgotha.

5
Byzantine Christianity

AFTER THE SUPPRESSION of the Bar-Kochba revolt, and the creation of Aelia Capitolina, the Roman authorities erased the very name Judaea from their map of empire and reorganized the Holy Land as the province of Syria-Palestine. The 500 years that followed, until the Moslem Conquest, was the longest period of peace and security which the peoples of the area have ever enjoyed. Of course prosperity was not continuous. Syria-Palestine, like most other parts of the Roman Empire, decayed economically in the mid-third century, and towards the end of it the Emperor Diocletian began that great reconstruction of the Empire which was to lead to the transfer of the capital to Constantinople and usher in the Byzantine Age. For himself he created the Eastern Prefecture, which had three dioceses: Asia, Pontus, and one formed from the provinces of Mesopotamia, Syria, Palestine, Arabia and Egypt. Palestine now consisted of Galilee, Samaria, the former Judaea, and most of the Decapolis across the Jordan. In the fourth century, Palestine was further divided into Palestine Prima (Judaea, Samaria and northern Idumaea) and Palestine Secunda (Galilee and Decapolis). There was another province of Arabia Libanensis, with its capital at Bosra, and one grouped round Petra, called Arabia Petraea or Palestina Tertia. These administrative divisions lasted until the Moslems destroyed Byzantine rule.

This rule was never exactly popular. The Byzantine Greeks were tremendous bureaucrats and ferocious legislators. From the time of Diocletian onwards, the Emperors assumed the powers and style of Oriental autocrats, maintaining a large mercenary army and a swarming secret police. In a way, it was a return to Seleucid times, but there was less freedom. The structure of government was corporatist. Everyone belonged to the corporation of his trade or calling in life, which he could not change at will and which was hereditary. The peasants were 'bound to the soil.' In the directly administered areas taxation was very heavy: there was a land-tax, a poll-tax, a tax on crops and, in addition, forced labour. The Emperor could levy exceptional taxes too, and he could billet troops at will.

Yet, in the Greek fashion, a very large number of towns and cities had the status of a *colonia*, that is they administered their own justice and were exempt from some taxes. Such, for instance, were Ascalon, Diospolis (Lydda), Aelia Capitolina, Elutheropolis (Beit Jibrin), Caesarea, Sebaste, Neapolis. Most regularly constituted and well conducted towns had privileges and self-government of some sort. Hence urban life flourished, even in the most unlikely areas, such as Sinai. The use of terrace cultivation and the building of water-cisterns, aqueducts and underground pipelines, ensured the maximum use of the surface area for agriculture. There were settlements everywhere and the population was undoubtedly larger in the first half of the first millennium AD than it has ever been since.

The language of government and culture was Greek but the people of the area were of very mixed descent and this is reflected in their religious disharmony which continued even after Christianity became the orthodox faith in the fourth century, taking the form of sectarian disputes. Indeed, one can truly say that Byzantine rule in the Near East was eventually destroyed by religious controversy.

The beginnings of Christianity are very obscure. The Epistle of James refers to the Christians of Jerusalem worshipping in an 'assembly' or synagogue. This may have been a house, located by tradition south of the summit of the south-west hill of Jerusalem, which Theodosius called 'the mother of all the churches.' During the 66–70 revolt, the Christians withdrew completely from Jerusalem and lived in the Greek-speaking city of Pella across the Jordan. This marked the real break with Judaism, and in time the Jewish-Christians were categorized as heretics – Ebionites or 'poor men' – a sect which migrated to Egypt and disappeared there. The Christians returned to the site of Jerusalem when the new city of Aelia Capitolina was built, which indicated that the authorities accepted their claim that they were no longer Jews but gentiles; but they were barely tolerated.

The first Christian settlement of any size in the Near East was in Caesarea, the Roman administrative centre where St Paul had been imprisoned and where St Peter baptized the centurion Cornelius. It had bishops at a very early date and it is likely that some of the earliest Christian writing were composed there. When Origen (*c.* 185–254), the first Christian scholar to achieve any international eminence, quarrelled with his superior in Alexandria, where he was head of the Catechetical School, he eventually found a refuge in Caesarea. There he wrote the *Hexapla*, which was an attempt to get at the final truth of the Old Testament by presenting all the known versions – the Hebrew consonantal text in Hebrew letters, the transliteration of this into Greek characters, the Greek version of Aquila (*c.* 132), the Greek version of Symmachus (late second century), the Septuagint, and the revised Septuagint made by the Jewish-Christian Theodotion – in six parallel columns. Origen created Christian Biblical scholarship as a serious discipline and, perhaps more important, he began to integrate Christian thought with classical philosophy, thus preparing the way for the

LEFT Origen, the first internationally known Christian scholar, did much of his writing at Caesarea.   RIGHT Eusebius became Bishop of Caesarea in AD 315. His *Ecclesiastical History* contains nearly everything we know about the early Church.

Christianization of the Empire at an intellectual level, and laying the foundations of the Christian philosophical tradition of Augustine and Thomas Aquinas. During his lifetime he was persecuted by various Christian prelates and pagan authorities, under the Emperor Decian; after his death he was treated as a heretic, which means that little of his work survives. His importance is that he educated the early Church and taught it to value scholarship and documentation. At Caesarea he founded a school and compiled the nucleus of a library, both of which were carried on and enlarged by his disciple, St Pamphilus, who was martyred in the city during the last persecutions before the conversion of Constantine. It was in the library at Caesarea that Eusebius, who became Bishop of the city in 315, and who acted as Constantine's ecclesiastical adviser, found the materials for his marvellous *Ecclesiastical History*, which contains virtually all we know about the early Church, from the age of the Apostles to his own day.

Caesarea was thus the administrative centre of the Church in the Holy Land for more than two centuries. It became an archbishopric but never a patriarchate. That honour was finally bestowed, as was natural, on Jerusalem at the Council of Ephesus in AD 431. But it was an honour not a sign of power – power in the Church had long since gravitated to Rome, Byzantium and Alexandria. The Christians were not a race or a nation, as the Jews were; hence they never regarded Jerusalem as their immutable capital. But they did inherit from Judaism the reverence for holy places and the cult of relics. From very early times serious and continuing attempts were made to identify and authenticate all the places mentioned in the New Testament and associated with Jesus Christ, and attention naturally concentrated in the first place on Jerusalem.

The earliest systematic gazetteer of Christian Palestine was the *Onomasticon*, compiled by the industrious Eusebius as a companion to his Church history. This, like his other works, was based on documents and information which were available in his day but which are lost to us. Of course the two most important sites were Golgotha and the Holy Sepulchre. Present-day visitors to the Old City of Jerusalem are often puzzled to discover that the Church of the Holy Sepulchre, which covers both these sites, is well inside the line of the city wall – is, in fact, not far from the centre of the city. The explanation, of course, is that the Old City walls are in fact those built by Suleiman the Magnificent in the sixteenth century, and encompass a much larger area, especially to the north and west, than the walls of the Herodian city in which Jesus was crucified. According to the New Testament accounts, Golgotha and the Holy Sepulchre were outside the north wall of the city, which is described in Josephus's *Jewish War*. But there are three possible alternatives as to the line of this wall, only two of which leave the Holy Sepulchre Church outside. However, Kathleen Kenyon's work on the archaeology of Jerusalem has established to the satisfaction of the majority of scholars and travellers the most likely line of route, and this leaves the church and the sites it covers well outside the walls as they stood in Jesus's day.

Of course, Miss Kenyon has merely shown that there is no insuperable difficulty in accepting the location of these sites. It is most unlikely that anyone can now prove that Golgotha and the Sepulchre lie under the Basilica. But there is no good reason why we should reject the tradition which places them there. The earliest Christians may have venerated these sites even before the disaster of AD 66–70, that is before most of the Gospels reached written form. They were certainly so venerated before the revolt of 132, because when Hadrian built the city of Aelia Capitolina, a temple of Venus, distinguished by its ritual sexuality and therefore of peculiar abhorrence to Christians (as well as Jews) was deliberately constructed over this place of Christian veneration. As the Christian community had been living at Pella between 70 and 132, they had presumably visited the site only as pilgrims, which suggests that the cult went back to before the beginning of the 66–70 war. Men and women were still alive then who had known exactly where Christ had been crucified and buried.

Certainly, by the mid-second century Christians were coming to Jerusalem to see the Holy Places. The first pilgrim we know of was Melito, Bishop of Sardis in Asia Minor, who must have been there shortly after the insulting pagan temple was set up. In the third century Origen – not a credulous man but an exacting scholar – toured the sacred sites, and several decades before, in 212, Alexander, a bishop from Cappadocia, came to Jerusalem 'to pray and gain information about the Holy Places.' He, too, was a scholar. He stayed at Jerusalem, became a bishop, founded a library and wrote a five-volume synchronization of general and Biblical history, now lost. Eusebius used both this work and the Jerusalem

BYZANTINE CHRISTIANITY

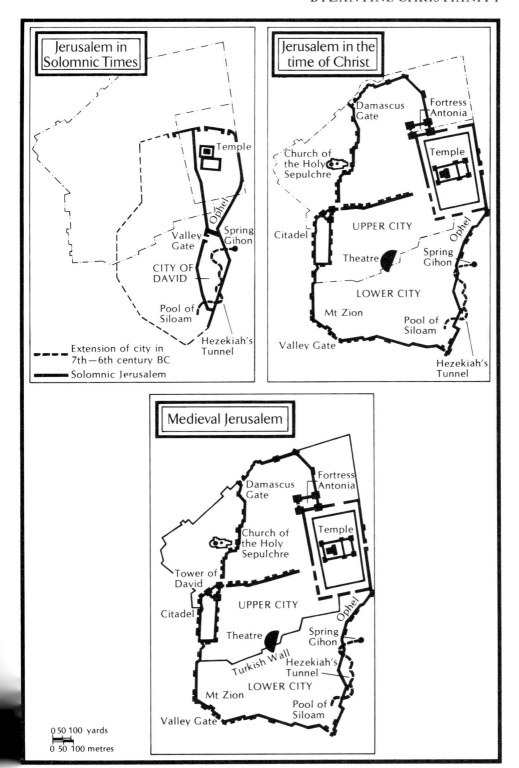

A statue of Constantine who legalized the practice of Christianity within the Empire.

library as source material. The fact that a number of scholars accepted the traditional location of the sites long before they were endorsed by official patronage must strengthen our belief in their validity.

When Constantine legalized the practice of Christianity within the Empire, and thus made it possible for Christian communities to build churches openly and in places of their choosing, the temple of Venus still stood over the supposed Holy Places, which were now, of course, well within the walls of the rebuilt city. In 325, at the Council of Nicea, a Bishop Makarius got permission to go to Jerusalem, remove the temple, and investigate what lay beneath. He did so, with Eusebius as an eye-witness, and seems to have found the tomb without difficulty. At about the same time, Constantine's mother, the Empress Helena, also visited the city and discovered the true cross. This second episode is much less well authenticated than the rediscovery of the Sepulchre. What is important about Helena's interest is that it led to the erection of the first church on the site.

The history of the Church of the Holy Sepulchre in Jerusalem is enormously complicated and very confusing. The church as we see it today is essentially a medieval building, but it follows roughly the outline of a vast Byzantine basilica which was built in the decade following the destruction of the pagan temple, and was dedicated in Constantine's presence in September AD 335. This church, then referred to as the Anastasis or Resurrection, was designed and built by two

BYZANTINE CHRISTIANITY

architects from Constantinople, and was situated off the main street of Hadrian's city. It was really two churches, or two shrines: a colonnaded atrium or entrance court led from the street into the basilica proper, which was rectangular, with four colonnaded aisles, and a high altar in the apse. This was called the Martyrium. Then there was another colonnaded court, with the site of Golgotha in one corner, and this court led, in turn, to a domed rotunda over the site of the tomb. The rock in which the tomb was cut had been levelled and tidied up, and it was located directly underneath the cupola of the rotunda. This church was greatly modified under Justinian and it was gutted in the early eleventh century during the persecution of the fanatical al-Hakim, when the tomb and the surrounding rock were smashed. What we now see, therefore, is the mere floor of the tomb, encased in marble, with the altar at its side. In Byzantine times pilgrims could also see the supposed stone which was rolled away by the angel.

The Anastasis-Martyrium church was not the only place of holy marvels. There was the Lithostratos (still shown within the convent of the Sisters of Sion), the pavement where Jesus was supposedly judged by Pilate within the already vanished Antonia fortress. There was also the blood-stained column against which Jesus was scourged, which was kept near the supposed site of the room where the Last Supper was held, on Sion Hill. The Christians also believed they had identified the place where the Holy Ghost descended at Pentecost, and the grotto where Jesus had instructed his disciples before the Passion.

Over this grotto, on the Mount of Olives, St Helena built a little church, the Eleona, one of many associated with her. Indeed, Constantine's edict, and the opening up of Jerusalem as a legal centre of pilgrimage and ecclesiology, led to the rapid expansion of the city to the south, an era of prosperity, and the erection of a multitude of churches. The first wave of sacred building came in Helena's day, and certainly included the Church of the Ascension on the Mount of Olives; these works were going ahead when the first visitor to leave a detailed record, the so-called Pilgrim of Bordeaux, came to the city in 333. By the early fourth century, as we know from one of St Jerome's letters, pilgrims were coming from Egypt, Pontus, India, Mesopotamia, Gaul, Syria, Persia, Armenia, Britain and Ethiopia. From the records left by these visitors, at least thirty-five distinct churches within the Byzantine walls can be identified.

Some of these were the work of the estranged wife of the Emperor Theodosius II, Eudoxia, who lived in Jerusalem, 456–60, as Governor of Palestine and spent its revenues on the city. She built the city walls which were restored by Suleiman the Magnificent in the mid-sixteenth century, the first patriarchal palace near the Holy Sepulchre Church, and many churches, including ones on the site of Pentecost, the site of the House of Caiphas, and above the Pool of Siloam – remains of this latter have been found underground. Outside the present Damascus Gate, on the site of St Stephen's martyrdom, she built a shrine which later burgeoned into St Stephen's Basilica. The Emperor Justinian added the great Church of St Mary the New, near the Temple platform. Nothing now

LEFT The Church of the Ascension on the Mount of Olives was built about AD 333.

RIGHT Constantine's mother, St Helena, whose interest led to the building of the first church on the site of the Holy Sepulchre. She also built a church on the Mount of Olives, the Eleona. Detail of painting by Paolo Veronese.

remains of it except some pillars which were incorporated in the Aqsa Mosque.

These buildings figure in the Madaba Map, which shows Byzantine Jerusalem as a walled city with six gates and twenty-one towers. It includes St Mary the New, which evidently had double doors of bronze, and the great basilica on the Pentecost site, 'the mother of all the churches.' Other buildings which, though unidentified, may be seen on it are the Church of St Anne, in the Temple area – later rebuilt by the Crusaders – and the sixth century Church of St John the Baptist. But the map does not include anything like all the basilicas and shrines which then existed. Theodosius, a pilgrim who visited the city at roughly the same time as the mosaic map was designed, mentions a Church of St Peter (probably the one built by Eudoxia over Caiphas's house), a Church of the Holy Wisdom on the site of the Praetorium, and Eudoxia's other churches at the Pool of Siloam and St Stephen's martyrdom. He also listed a Church of St Mary in the valley of Jehoshaphat, outside the city, and a church commemorating the Last Supper. He said that on the Mount of Olives alone there were twenty-four churches and shrines.

Perhaps the most famous of the Western visitors to the Holy Land in Byzantine times was St Jerome, and from his voluminous writings we can gather a good deal about its civilization and topography. He set out for the east, aged forty, in 372, to learn Greek, for by his day it was already rare for Westerners, even scholars, to know more than Latin (Augustine never learnt Greek properly). He spent a year in Antioch, then still an enormous and prosperous city, staying with the famous

BYZANTINE CHRISTIANITY

BELOW The valley of Jehoshaphat with the ancient Jewish burial ground containing the tombs of the Prophets rising up the Mount of Olives.

preacher Evagrius, and perfecting his Greek. It had a magnificent library, maintained since Seleucid time, full of classical masterpieces since lost to us. He spent more time in the library reading pagan authors, especially the Latins he venerated, than on his spiritual exercises, and ten years later, in a letter, he described his famous dream, in which he was dragged before a heavenly tribunal and asked what he was. 'A Christian', he answered. The Judge replied: 'You are lying. You are a disciple of Cicero, not of Christ. For your heart is where your treasure is.' He was then ordered to be flogged, and he awoke – so he told his correspondent – with his shoulders black and blue. As a result he spent the years 374–5 in the Syrian desert as a hermit in the neighbourhood of Chalcis (Qinnesrun), whose ruins can still be seen, half covered in mud huts, between Antioch and Aleppo.

Jerome was back in Rome in the years 382–5, where he became secretary to the famous builder and ritualist Pope Damasus, and might even have been Pope himself, had not his close association with various holy ladies led to scurrilous accusations which drove him eastwards again. On this occasion, in company with the wealthy virgin Paula, Jerome went on a regular tour of the places which feature in the Old and New Testament. Many of these had already been identified – the Bordeaux Pilgrim had followed the route of the Exodus and seen the stone on which Moses had broken the Tablets of the Law. Jerome and Paula toured the Holy Land by donkey, visiting the Negev and the wilderness around the Dead Sea. We can follow the trip in the saint's letters and in other references in his

The ruins of the citadel of ancient Antioch, where St Jerome stayed to perfect his Greek.

books. On the supposed site of Sodom, Paula burst into tears, recalling how Lot had been made drunk and seduced by his own daughters, and she warned the virgins who attended her to keep clear of wine. Jerome believed passionately in the holy sites he visited, and he defended the idea of the antiquarian pilgrimage

Just as Greek history becomes more intelligible to those who have seen Athens, and the third book of Virgil to those who have sailed from Troas by Leucate and Acroceraunia to Sicily and so on to the mouth of the Tiber, so that man will get a clearer grasp of Holy Scripture who has gazed at Judaea with his own eyes and has got to know the memorials of its cities and the names, whether they remain the same or have changed, of the various localities.

Of Jerusalem itself he wrote: 'I entered Jerusalem. I saw a host of marvels, and with the judgment of my eyes I verified things of which I had previously learned by report.'

BYZANTINE CHRISTIANITY

Jerome and Paula eventually settled in Bethlehem, which had been a Christian shrine from the very earliest times, and where Hadrian, as at Jerusalem, had built a temple of Venus and Adonis over the grotto or cave where the Christians worshipped. But Christian pilgrims continued to go there even while paganism was officially supported. Origen certainly visited the cave about the middle of the second century and says that a crib was on show. The pagan superstructure was presumably removed at the same time as the temple over the Holy Sepulchre in Jerusalem. Jerome wrote: 'Bethlehem, which is now ours ... used to be overshadowed by a grove of Tammus, that is Adonis, and in the grotto where the Christ-child once cried, people mourned for Venus's lover-boy.' He reported regretfully that, by the time he got to Jerusalem, the original earthenware trough or manger in which Jesus had been laid had disappeared and had been replaced by a magnificent silver replica. In his day, the first Church of the Holy Nativity, built over the grotto, was still standing. It was attributed to St Helena, but more probably was built by Constantine. It was destroyed, very likely by an earthquake, in the early sixth century, and rebuilt on a grander scale by Justinian in the 530s; but it may be that the red limestone columns of the nave, which come from the locality, were rescued from the wreck of the earlier church. At any rate,

The original Church of the Holy Nativity, Bethlehem, the building of which was attributed to St Helena, was probably built by Constantine. The present church was built by Justinian around AD 530.

Justinian's building, much altered, patched, adorned and added to, is the basis of the church we see today, with the circular steps leading to the cavern of the Manger, beneath the area immediately in front of the high altar.

Paula spent her fortune building two monasteries, one for men and one for women, in Bethlehem. There Jerome spent the rest of his life. He ran a school for children, translated the Bible into Latin, in a version known as the Vulgate, wrote innumerable Biblical commentaries and some of the earliest lives of the saints, and engaged in ferocious doctrinal controversies with fellow-churchmen, notably Pelagius. Paula ran her monastery, dividing her ladies into three groups, on class lines, for purposes of accommodation, but they all dressed alike, and her remedy for sexual desire was identical for all three classes – near-starvation. She died in 404, aged fifty-six, and Jerome buried her underneath the Church of the Holy Nativity close to Jesus's cave. There is still an altar which marks her cenotaph, the approach to it being by a staircase from the Franciscan friary to the north of the church. After her death, Jerome continued to preach in the church. We have nearly 100 of his sermons, which are wonderfully intelligent and learned, often foolish, and must surely have been well above the heads of his mixed congregation of pilgrims and locals. He died in 420, probably in his ninetieth year, and was buried close to Paula. In about the year 570 a visitor from Italy noted: 'At the very entrance to the cave [of the Nativity] the priest Jerome had an inscription carved on the natural rock and made for himself the tomb in which he was laid.'

It was about this time that the long-dead Jerome had foisted on him the story of the lion with the thorn in his pad, a story probably borrowed from the legend of St Gerasimus of Palestine. To the early hagiographers, Jerome's tame lion was linked to the period when he was in the Syrian desert as a would-be hermit, and it was associated in the minds of his devotees with a famous passage in which he described the temptations of the flesh:

How often when I was installed in the desert . . . I would imagine myself taking part in the gay life of Rome . . . Although my only companions were scorpions and wild beasts, time and again I was mingling with the dances of girls. My face was pallid with fasting and my body chill, but my mind was throbbing with desire. My flesh was as good as dead, but the flames of lust raged in it.

The imagery of the lion and the naked dancing ladies combined to make St Jerome the hero of the iconography of the Dark and Middle Ages, and right down to the Renaissance he was the most popular of all saints among the artists of Latin Christianity.

Jerome, in fact, was not a particularly austere monk. He had his library with him in the desert, and assistants and copyists. He nonetheless grumbled that there was nobody with him to speak Latin: 'Either I must learn the barbarous local gibberish or keep my mouth shut.' But by his day monasticism was already an established institution and had taken a variety of forms, though his biography of St Anthony helped to popularize it, especially in the West. The monastic idea

BYZANTINE CHRISTIANITY

St Jerome in the desert, with the legendary lion at his feet. Painting by Bono da Ferrara.

came from Egypt, where it was undoubtedly adopted from a pre-Christian pattern of desert living for farmer-priests whose job was to look after the tombs of long-dead pharaohs and grandees. It was the Egyptian St Anthony who first adapted the notion to Christianity, and organized houses of monks in the Nile deserts towards the end of the third century. His successor, St Pachomius (c. 290–346), invented the idea of cenobitic monks, that is of men taking vows and living according to a rule in a community. This discipline was quickly brought to Palestine by Hilarion, who established himself in Gaza, and thereafter it spread rapidly over the Near East.

Of course Palestine had its own quite separate tradition of desert-dwellers, mystics and prophets, from Elijah to St John the Baptist. It became, as it were, an experimental theatre for modes of austerity and asceticism. The great St Basil (330–379) from Cappadocia, studied the monks of Palestine and Syria, before formulating his rule, which became the basis for the cenobitic monasticism of the Greek Orthodox Church. Another group of houses, which followed the discipline of Pachomius, were called *lauras*, and had their constitution drawn up by St Sabas (439–531), who founded the monastery of Mar Saba by the Dead Sea, which still exists. Mar Saba was the training ground of Cyril of Scythopolis, who took its rule to Rome; and another early writer on monasticism, John Moscus, came from Palestine, and served as a monk in Jerusalem.

Palestinian and Syrian monks formed a substantial and certainly an active and vocal part of the Christian community, and they performed an important role in the hectic religious politics of the Byzantine Empire. Monks led the fourth-

The refectory of the Greek Orthodox Monastery of Mar Saba in the Judaean Desert, founded by St Sabas.

century assaults which demolished most of the great pagan shrines and temples of antiquity. They also attacked the Jews, who had unwisely cooperated with the Emperor Julian in his attempt to undermine Christianity in the mid-fourth century. Shortly after Jerome's death, the Syrian monk Barsauma led a band of militant cenobites who burned synagogues and villages, and terrorized the Jewish districts. Monks intervened violently in the internal controversies of the Christian Church. At the Council of Ephesus in 449, Barsauma's shock-monks murdered Flavian, the Patriarch of Constantinople. They terrorized Juvenal, Bishop of Jerusalem, for his part in the Council of Chalcedon, 451, which reversed the decisions of Ephesus. The Bishop finally hit back, with the support of the Emperor and the Pope, and wiped out the monks, using a mixed force of imperial troops and Samaritans who had also suffered from monkish fanaticism.

But not all monks were ruffians. Some led lives of almost unbelievable privation. Jerome says that he saw one who had lived, walled up, for thirty years, eating only barley bread. He met another who lived in an abandoned cistern and ate a mere five dates a day. These monks were normally illiterate, but some were learned and attracted visitors, and a few commanded the respect of the civil as well as the ecclesiastical authorities for their wisdom and saintliness. Of these by far the most famous was St Simon Stylites of Antioch. He was born in 390 and achieved notoriety as a holy man long before he left for the desert. In 423 he went to Kalat Seman to escape the pestering crowds, and to achieve yet further separation from the world of the distracting senses, he mounted permanently on top of a pillar. The idea of pillar-life was to draw physically closer to God and to maintain uninterrupted communion with Him. Simon gradually extended his pillar until it was sixty feet high, with a small platform on top. He was chained to the platform by an iron collar and all food was sent up to him in a basket – including the Eucharist once a week. He preached twice daily to the enormous crowds that streamed out from Aleppo, thirty miles away, and prostrated himself hundreds of times in between sermons. According to his biographer, Theodoret of Tyre, even the Emperor took his political advice, and when he died in 459, his body was taken to Antioch to be buried with great pomp.

Towards the end of his life, to accommodate the masses who came to hear him, the authorities began to build a vast basilica at Kalat Seman, the ruins of which still stand. They illustrate a revolution in church architecture. The very earliest Christian churches had been modelled on synagogues, or in gentile territory on the classical pagan *cella*, which was comparatively small, even if the temple which surrounded it was enormous. To accommodate the new crowds, the design was recast on the model of the king's hall of justice. Kalat Seman, built in perhaps the 450s, was not the earliest basilica by far, but it is one of the earliest to

The Octagon and the remains of the pillar, at one time sixty feet high, on the top of which Simon Stylites lived for thirty-six years.

survive in more or less its original form, and it is, in fact, one of the biggest Christian churches to be built anywhere, until the eleventh-century cathedrals of the Latin West. It has no less than four pillars, with fine acanthus-leaf stone carving, but of Simon's 60-foot pillar only the bottom bit is left. The church is merely the centrepiece of a large block of ecclesiastical buildings, including a former monastery, all set on a high rocky cliff, with fine views in all directions, and eminently defensible. It was, indeed, turned into a fortress as Byzantium declined, and finally stormed by the Moslems in 985.

Kalat Seman is merely one outstanding example of the ecclesiastical architecture built in the Byzantine centuries, which was once the glory of the Near East. Syria, in particular, is dotted with the ruins of Christian churches, monasteries, towns and villages, some of which have barely been touched since they were wrecked or abandoned in the early Middle Ages. A few of them may go back to the fourth century, though most date from the fifth and sixth, when Christianity first became the religion of the majority. Further south, in Palestine, the relics of Byzantine Christianity are less easily found, partly because they have been built over many times by successive phases of Christian devotion. Thus at Nazareth, which ought to be one of the glories of primitive Christianity, there is not much to show. On the supposed site of the Annunciation, there are two columns called the 'Column of Gabriel' and the 'Column of Mary', and visitors are shown the site of St Joseph's carpenter's shop and the cave where the Holy Family lived. But the fact is that Nazareth, unlike Bethlehem which became a Christian centre very early, remained a Jewish town well into Byzantine times. In 1960 the remains were found of a very early, pre-Constantine Christian place of worship, but it is of the synagogue-type. There is some seventh-century Byzantine work, and in the monastery museum fine carved capitals, twelfth-century Crusader work from the old Church of the Annunciation. But the present double church and all the other churches in and around the town, are modern.

Again, at Mount Tabor, 1,700 feet above the Plain of Esdraelon, which was accepted as the site of the Transfiguration as early as the fourth century, at least three churches had been built by the year 700; but the present Franciscan and Greek Orthodox churches are quite modern. At Tabgha, near the Mount of Beatitudes, the Basilica again is modern: but in 1932 a discovery was made of some fifth-century mosaics from the original Byzantine church on the site, showing the flora and fauna of the Lake of Tiberias below. On the savage Mount of Temptation, overlooking the Judaean wilderness, there were once buildings which housed over 6,000 monks and hermits, but the present Greek monastery on the side of the mountain is only 100 years old. Its grisly relics of Byzantine times are the fourteen skulls kept in the chapel, said to be of monks who were massacred by the Persian invaders in the early seventh century. A more visibly impressive reminder of Byzantine monasticism, at least from the outside, is the Monastery of St George, also in the Judaean wilderness, on the slopes of the ravine of the Wadi Kelt (Brook Cherith), near the site of Herod the Great's winter-palace. Travelling

RIGHT The magnificent Dome of the Rock, Jerusalem.
OVERLEAF The ceiling of the Dome of the Rock, richly decorated in gold and mosaics and encircled by sixteen stained glass windows.

An aerial view of the Mount of Beautitudes with its modern Basilica. In the middle distance is the Plain of Gennasaret and market garden of Galilee.

through the countries of the Holy Land, one is appalled at the ravages of time and warfare which have swept away so much of what the Greek-speaking followers of Christ once built.

Yet it must be said that, to a considerable extent, Byzantine Christianity was the architect of its own destruction. It tried to define and particularize too much, and therefore became increasingly divided and acrimonious. Like the state of which it was an integral part, it was too authoritarian, seeking exact credal definitions at packed church councils, later enforced by imperial police and troops. In the fifth and sixth centuries, it was increasingly disturbed by its efforts to suppress the monophysites, that is those Christians who held, in their desire to reconcile the concept of the Son of God with monotheism, that the divine and human elements in God made one nature, *monophusis*. The controversy created those Eastern Churches which are independent of both Rome and Greek Orthodoxy. Thus the Coptic church of Egypt was monophysite, as was its Abyssinian offshoot, and so were most of the eastern provinces. In Persia, too, the Church was independent of Constantinople, and eventually became Nestorian, acknowledging the teaching of the ex-Patriarch of Constantinople, who distinguished between the divine and human natures in Christ, and held therefore that Mary should be called 'Mother of Christ' but not 'Mother of God.'

The great Emperor Justinian, 527–65, might have accepted these arguable differences of view and worked for a policy of toleration and reconciliation, but he was a fanatical supporter of the anti-monophysite definition of the Council of

LEFT The silver-domed El Aqsa Mosque was largely reconstructed in the eleventh century.

Chalcedon, and used the whole power of his State in an attempt to enforce it. Throughout the Near East, he backed bishops termed Melkites by the monophysites (from the Semitic *melk*, meaning king or royal), who upheld the views of Chalcedon and Constantinople. But the effect of his persecution was to produce another monophysite church in Armenia, for the Armenians, uncomfortably placed between the antagonistic empires of Byzantium and Persia, chose the latter, and turned monophysite to convince the Persian authorities that their links with Constantinople had been finally severed. Even Justinian's Empress, Theodora, was a secret monophysite. It was her patronage which made it possible for Jacob Baradi, a monophysite bishop, to travel throughout the east, in the years 543–78, consecrating like-minded bishops and organizing the monophysite Church against the Melkites. It included vast numbers of Arabs on the fringe of the Empire and the desert, from the Romanized Nabataean Arabs of Petra to the tribes of Arabia, many of them hitherto pagan. From this persecuted desert Jacobite Church it was only a comparatively short theological step to the pristine monotheism of Islam.

In Justinian's day, the power of Byzantium and its Church nevertheless seemed formidable. It stretched as far as Rome ever had. At Halebiyah, beside the brown waters of the Euphrates, an enormous fortified camp was built, in the mid-sixth century, by John of Byzantium and Isidore of Milo, nephew of the architect who created Santa Sophia in Constantinople. The walls of grey gypsum survive, and within them are a governor's palace and two ruined Melkite churches. Some fifty miles to the west is Resafa, another remarkable desert site, rectangular this time, a famous pilgrim city as well as a fortress, for it was the scene of the martyrdom of St Sergius, the patron saint of Syria. Justinian built massive walls, which still stand, and cisterns and reservoirs which could supply the town and garrison with water for two years of siege. There are marvellous carvings on the Church of the Martyry and the triple North Gate, and the forty-acre site was packed with churches, State buildings and all the ostensible solidity of Byzantine-Christian power.

In fact this power collapsed in less than a generation. The Persians struck in 614, the first foreign invasion of the Holy Land for 600 years. Following them came the Arab Moslems, who in 636 beat the main Byzantine army at the Battle of the Yarmuk, the tributary of the Jordan on the far side of the Lake of Tiberias. Byzantine Orthodoxy had created a host of enemies for itself – Jews and Samaritans as well as Jacobites, Armenians and Monophysites of all kinds. Jerusalem was one of the few Eastern patriarchites which accepted the Chalcedonian definition, but even Christians of Judaea and Galilee, who hated Byzantine taxes and bureaucracy, were lukewarm in their fidelity to the Orthodox government. Most of the Melkite officials fled. Almost everywhere, the all-conquering Arabs were welcomed or at worst received with indifference. Thus the Western millennium in the Holy Land came to an end and Oriental civilization returned in a new guise.

The Emperor Justinian and the Empress Theodora and their retinues from mosaics in the Basilica of San Vitale in Ravenna.

6

The Coming of Islam

AS WE HAVE SEEN, there were good reasons why Byzantine rule was unpopular in the Near East, and why there was so little resistance to the Moslem Arabs. But what was it that drove the Arabs north and west? Christianity had made some attempts to convert the desert tribes of Arabia. In the sixth century there was a Christian church as far south as Sana in the Yemen, but neither the Roman nor the Byzantine authorities ever relaxed their efforts to crush tribal raiding. They even imported lions from Africa and unloosed them in the desert to worry the Bedouins. Sometimes they found it cheaper to pay the tribes small subsidies as an alternative to thieving, but by the beginning of the seventh century they had returned to a policy of force. The Persians, too, were becoming more relentless in their suppression of tribal lawlessness.

Hence the Arabs were adopting a posture of increasing hostility towards both their imperial neighbours, when Mohammed began his mission. He came from one of the smaller trading clans of mercantile Mecca, which was then polytheistic. He visited the Syrian cities after the terrible Persian incursions during the second decade of the seventh century, and observed the fragility of Byzantine power. He admired both Christian and Jewish ethics, but he understood Judaism better – like the overwhelming majority of the desert peoples, he was repulsed by the Greek metaphysics of the Trinity. By contrast, Judaic monotheism was simple. But in 622 he moved from Mecca to Medina, where his claim to prophetic status was rejected by the Jewish community. Thereafter he took his own line.

The expansion of Islam began under Mohammed's successor, the first Caliph, Abu Bakr (632–4). Islam had no alternative to expansion, for the rapid spread of Mohammed's teachings among the Arabs had led to the end of inter-tribal warfare. The existence, in Syria and Mesopotamia, of huge minorities persecuted by imperial governments – for Persian Zoroastrianism, now in its dotage, like the Melkite and Orthodox authorities, had begun to persecute schismatics, heretics

The vaulted substructures of Crusader and Herodian masonry below the pavement at the south-east of the Temple Area, Jerusalem, sometimes associated with the Stables of Solomon.

and people of other religions – offered the Arabs their opportunity. They did not come with fire and sword, although they attacked the imperial armies with great ferocity. They offered toleration to all religions and respect for all houses devoted to God. Mohammed had deplored the Persian devastation of Christian churches, and had written: 'But who does greater wrong than he who bars the sanctuary of God from having his name mentioned in them and who busies himself to destroy them?' (Koran, Surah 3, 114). Abu Bakr issued orders to his general, Usama: 'Thou shalt mutilate none. Neither shalt thou kill child nor aged man, nor any women. Injure not the date-palm, neither burn it with fire; and cut not down any tree whereon is food for man or beast. Slay not of the flocks or herds or camels, saving for needful sustenance.' (He did add, it is true to say, 'And the monks with shaven heads smite thou thereupon'; as we saw, some of the Christian monks were violent persecutors.) Thus the Arabs did not inspire terror. In most of Syria-Palestine, their troops were seen as less murderous and destructive than the Persians or even the Byzantines.

It was under the second Caliph, Umar, 634–44, that the Holy Land passed into Moslem control. In Palestine, the decisive battle was fought at Ajnanain, on 30 July 634; this was probably in the Vale of Elah, where David killed Goliath. The Battle of the Yarmuk, following two years later, led to the rapid expulsion of the Byzantines from Syria. Caesarea, the last Byzantine stronghold, fell in 640. Next year, the Arabs destroyed the Persian army and went on to the conquest of Egypt, moving rapidly along the North African coast. Babylon fell in 641, Alexandria the next year, Cyrenaica in 643, Carthage in 698; Spain was invaded in 711, France in 717. The same year, the Emperor Leo the Isaurian halted Moslem progress towards Constantinople, and in 732 Charles Martel turned them back in France at the Battle of Poitiers. But in the preceding century of conquest, the Moslem Arabs had absorbed more than half the Christian world.

Jerusalem had remained formally loyal to Constantinople, for its patriarch was Chalcedonian. But it did not fight; it negotiated its surrender in 638. The *Annals of At-Tabari*, our Moslem source, says that the Christians of Jerusalem were given security 'for persons and property, for their churches and crosses.' The Caliph Umar took the surrender of the city in person and immediately demanded to be taken by the Patriarch Sophronius to see the Temple built by Solomon, whom Moslems treated, as they did most of the principal heroes of the Old Testament, as a prophet. But of course there was nothing to see, other than the great platform itself. The pagan Romans had set up several great statues there, but these had long been removed. According to the Moslem account, the Patriarch had to admit that the place had been turned into a dunghill and a midden. Umar was properly scandalized and in revenge the Moslems referred to the Holy Sepulchre Church as *al-Kumanah*, the dung church. Our Greek source, Theophanes, says that when Sophronius was showing Umar round, the Patriarch murmured to himself the phrase from Daniel 12:11, 'abomination of desolation', and he refers to the Caliph's interest in the Holy Places as 'diabolical hypocrisy.'

The arching interior of the Church of the Holy Sepulchre, with the Centre of the World in the foreground.

In fact it is difficult to find any early evidence of Moslem persecution of Christians in Palestine. There was still some doubt about where they stood in the religious spectrum. In Constantinople they were believed to be a weird Christian sect. In any case, there were very few of them to begin with. Even the invading armies were partly composed of polytheistic pagans. The Moslems moved in more as a military elite than as a settling horde, though gradually more and more of the land passed into the hands of big Moslem landowners. Originally Palestine was divided into two provinces or *Junds*, governed from Tabariyah (Tiberias) and Lydda (Ramleh). The authorities dealt with the various religious groups corporately, through their senior clergyman – in the case of the Christians their *catholicos* or patriarch. Christian communes were known as *millets*, and the millet system of government was maintained, without any essential change, until the arrival of the British in 1918. As the Arabs were a small, ruling elite, their taxation system was based on the assumption that non-Moslems enormously outnumbered 'the faithful.' Throughout the new Moslem territories, the followers of the prophet did not tax themselves, except for the relief of the poor, indeed, most of them received pensions. Taxes for the upkeep of the State were paid only by non-Moslems or *dhimmis*. These took the form of a poll-tax, but it was almost certainly less than the Byzantine poll-tax, and it was the only one all had to pay. The total weight of Moslem taxation was appreciably less than the multiple exactions of Constantinople. Moreover, cut off from Constantinople, there was no religious controversy among the Christians and no persecution. The *dhimmis* were left strictly alone, provided the practice of their religion was unobtrusive. The only restrictive provision was that they might not erect new churches.

But Christian stonemasons, artists and architects were not idle. They were promptly taken into Moslem employ (as were administrators, doctors, engineers and experts of all kinds). The new rulers did not take long to adapt themselves to the civilization of the Near East. The first caliphs had lived in Medina. In 656 the Caliph Uthman was murdered and his successor, Ali, moved the capital to al-Kufah on the Euphrates. Ali was the last of the orthodox caliphs, the companions of the prophet, and when he, too, was murdered in 661, Islam suffered the first of many splits. His followers refused to accept the doctrinal rulings of his successor, Muawiyah, Governor of Syria, and formed the *shiites*, or sectaries. Muawiyah, as head of the *sunni*, or traditionalists, made Damascus his capital. As he was descended from Umayyah, the nephew of Mohammed's great-grandfather, he introduced the principle of hereditary succession, and the Umayyad dynasty became the greatest of all the Moslem ruling houses, the only purely Arab one. It was under the Umayyad Walid, 705–15, that Islam reached its imperial apogee, stretching from Spain to India.

The Umayyads were Arabs, but lacking a deep-rooted historical culture of their own, their attitude was cosmopolitan and they availed themselves with freedom and delight of all that Byzantium (and Persia) had to offer. They wanted to rule peaceably over these former provinces of the Empire and absorb their civilization, which meant persuading the inhabitants to accept them. Hence, below the top level, the government was very largely staffed by Christians. But the Umayyads also showed their benevolence towards the Jews, and began the process whereby Islamic civilization absorbed a large element of Jewish culture. Indeed, at this stage it is worth noting that the change from Roman-Byzantine to Islamic rule marked the beginning of the Jewish recovery. In late Byzantine times the Jews of Palestine had been largely confined to a few villages in Galilee. Under the Umayyads Jews began to return to Jerusalem, to the quarter near the Wailing Wall between the Damascus and the St Stephen's Gates. They bought the slopes of the Mount of Olives facing the Temple, and Jewish pilgrimages were resumed, especially for the Feast of the Tabernacles. Jews also settled in Ramleh, Ascalon, Caesarea and Gaza. Under the *millet* system, the president of the rabbinical academy, or *gaon*, in Tiberias was recognized as head of the community. He also had international status among Jews, since the academy fixed the Jewish calendar every year for Jews all over the world. At Tiberias, in the seventh to eighth centuries, work continued steadily on collecting and editing the *Midrash* and on the textual studies of the *Massorah* – the present Massoretic text being based mainly on the work of the Tiberian scholars. And it was in seventh to eighth century Tiberias, possibly under the influence of Byzantine modes, that Eleazar Ben Kalir and other composers wrote many of the hymns still used in the synagogue.

But Jerusalem, too, became a centre of Jewish culture. The tendency to over-elaborate the Talmud produced the counter-tendency of periodic Messiahs. There were Serene of Syria and Abu-Isa of Isphahan; and Ahan ben David, though he did not exactly claim to be the Messiah, stressed the law of the Bible as opposed to the Talmud, and he and his followers, the Karaites, made Jerusalem their centre. Moreover, under Moslem rule, there was a tendency for scholars of the great Babylonian Talmudic schools of Sura and Pumbeditha to move to the newly tolerated Jewish communities further west, to Egypt, Kairouan, Spain and Jerusalem. So the Holy City became a centre of Talmudic studies also; these scholars were called Rabbanites, and their head was in charge of the Jerusalem Yeshivah, and called the *Gaon* of Jacob. The power of fixing the calendar now passed to Jerusalem, which in effect had two rival academies or miniature universities, which flourished in the city until the arrival of the Crusaders made their life impossible.

Under the Umayyads then, Jerusalem was both a Christian and a Jewish cynosure. The caliphs proceeded to make it into a Moslem one as well. This is not surprising since they venerated many Jewish patriarchal figures and saints, and regarded Jesus as a prophet. But they had a particular devotion to the Temple Mount, on account of Mohammed's famous dream, in which he took a night-journey on the winged horse El Buraq (lightning) in the company of the Angel Gabriel. The relevant passage in the seventeenth Surah of the Koran, 'The Night Journey', reads: 'from the sacred temple to the temple that is more remote, whose precinct we have blessed, that we might show him of our signs.' This was interpreted as a journey from Mecca to the Temple Mount at Jerusalem – a detailed account, attributed to Mohammed's contemporary, Al-Hassan, being given in the life of Mohammed written by Ibn Hisham, early in the ninth century.

This story was already believed by the Moslems when they took Jerusalem, and the Temple platform thus became the centre of their devotions in the city. Moslem architectural history, however, poses a number of problems, some of which may remain insoluble, and there are particular problems associated with their structures on the platform. Was Islamic architecture created essentially by borrowing from Christian forms? The prototype mosque, built at Medina by Mohammed himself in 622, was of wood, a square enclosure with walls of brick and stone, partly roofed; the roofs were of palm-branches covered in mud and resting on palm-trunks. The faithful knelt, facing north towards Jerusalem, in the Jewish fashion, but in 624 the orientation was changed towards Mecca. The next mosque of which we have knowledge was built at Kufah in Mesopotamia in 639, and this had marble columns, borrowed from another building. A mosque was built in old Cairo in 642, the first to possess the characteristic high Islamic pulpit or *mimbar*. The first minarets began to appear about the beginning of the eighth century, and the semi-circular *mihrab* or prayer-niche, shortly afterwards. The essential features of the Mosque had thus developed over less than eighty years; later came the colonnades or *liwanat*, surrounding the *sahn* or open court. The combination of these ecclesiastical devices provided for all the ritual requirements of the Islamic religion.

The first building the Moslems erected on the Temple platform, over the rock where Mohammed had alighted was a temporary wooden structure. In 691 this was replaced by the great Dome of the Rock, built by Byzantine workmen on orders from the Caliph Abdul Malik. This was not a mosque but a circular *mashhad* or 'place of witness', where pilgrims walked around Mohammed's rock under a dome and colonnade. It is the first real example of Moslem ecclesiastical architecture but it remained unique – for the next 400 years all mosques were built to the square pattern. Hence we must assume the inspiration for the Dome was Byzantine. It was, in fact, an aisled rotunda, developed from the Christian church. It was probably copied from the nearby Anastasis (Holy Sepulchre Church). Certainly there were domed churches in Syria long before the end of the seventh century, and rotunda churches within an octagon shape already existed in Palestine. But the Dome of the Rock is somehow more splendid than its prototypes, perhaps because of its magnificent position. It has been much battered and changed by time. The Dome itself was originally covered in gold leaf. All was stolen, revealing the lead beneath, and it is now covered in a gleaming but lifeless bronze alloy. In the sixteenth century, Suleiman the Magnificent put in about 50,000 ornamental tiles, and there have been other additions and modifications, not always prudent. The Dome of the Rock is not so much a work of the eighth century as a jigsaw puzzle of many periods.

Indeed, what ancient building in Jerusalem is not? The Caliph Malik's son and successor, Caliph Waleed, built on the platform a more regular mosque, the El Aqsa or 'distant place', with a silver dome. The Jews call it the Midrash Shlomo, 'Solomon's Study'. It seems to have been constructed originally from the nearby ruins of Justinian's great church, St Mary the New, which had been wrecked by the Persians and abandoned. But the present mosque dates in part from the eleventh century, not long before the First Crusade, and virtually all of it has been heavily reconstructed. Once it contained a pulpit donated by Saladin – a great pulpit-donor in his day – but this was destroyed when a madman set fire to the mosque in 1969. (The arsonist, a Christian fundamentalist from Australia, believed he was a descendant of King David, and had been commissioned by God to build a third Temple.) There is also a third major Moslem building on the platform, the eighth-century Dome of the Chain, which served as the treasury of the two other places of worship, and is supposed to be on the site where David held his court of judgment. For the Moslems, the platform as a whole forms what they call the Noble Sanctuary, the *Haram as-Sherif*, and it is the third in honour of their holy places, after Mecca and Medina.

The Moslems have controlled Jerusalem for most of the past millennium and a half, but it is a curious fact that their impression on the city has never been decisive. Certainly, the Old City of Jerusalem cannot be called a Jewish city in appearance or feeling, nor a Christian city, but it is not a Moslem city either. Jerusalem manifests a special religious fervour, but it is *sui generis* and not the

THE COMING OF ISLAM

The aisle around the Rock in the great Dome of the Rock, with the intricate and beautiful patterning of the Byzantine workmen.

property of a particular faith. Damascus, on the other hand, is quintessentially an Islamic city and it has a strong case to be considered the artistic capital of the Moslem world. Why is this? It had a large Jewish community in very early times; indeed, there were probably Jews in Damascus before they even set foot in Canaan, and it was later an important part of David's kingdom. One of the earliest Christian communities formed itself there and the Christians were in control for about 300 years.

The answer, probably, is that Damascus is the archetype desert-oasis town and so forms a natural setting for the Arab-Islamic genius. It is on the far side of the eastern slope of the Lebanese mountain barrier, which cuts it off from the Mediterranean, and it is the natural starting-point, or caravanserai, for both the great caravan routes of antiquity, which lead east to Mesopotamia and the Persian Gulf and south along the east side of the Jordan to the Gulf of Aqaba. Damascus is an immensely fertile oasis on the edge of the desert. Its river, the Barada, 'the River of Gold', rises in the Lebanon, and rushes down its slopes until it literally disappears in the Damascus oasis. But before it vanishes it not only waters the magnificent groves and orchards and gardens which surround the city, but gushes and trickles through endless channels and conduits, fountains and pools, in the city itself – sometimes being lifted up by wheels to the higher levels. Some of these arrangements go back to deep antiquity but they have been immensely improved and elaborated by the Arab water-engineers of the Middle Ages, for whom any opportunity to build a fountain-courtyard was an irresistible challenge. It was inevitable that the Umayyads should choose Damascus as their capital.

Early in the eighth century, after the Dome of the Rock had been completed, but probably while the Aqsa Mosque was still being built, the Caliph Waleed decided to build a great mosque in his capital and make it the fourth holiest place in Islam. Damascus was already a very large city. From 63 BC, in Pompey's time, it had been the capital of Roman Syria, and from about AD 200 a self-governing *colonia*. In the first half of the first century AD it had acquired an immense temple to Jupiter Damascenus (the Syrian god Hadad), whose precinct was about 400 yards long and 300 wide. The Emperor Theodosius had cleared away the temple but kept the precinct, and within it had built a fourth-century church to St John the Baptist, whose head he believed he possessed (and which is still, supposedly, enshrined in the present mosque). An enormous amount of Roman masonry and brickwork remains in the precinct wall.

In 705 Caliph Waleed cleared the church from the precinct, and in 714–15 he built his great mosque. It is the first indisputable masterpiece of Moslem architecture, if we allow that the Dome of the Rock is essentially Byzantine. What is new about the Damascus mosque is the main *liwan*, or arcade, which has three aisles, crossed by a central transept, with a dome over the top of it. The arches around the central court are horseshoe-shaped, very characteristic of the Moslem architecture of the west, mounted on pillars and columns. Above the main arcade

The arcaded interior of the Prayer Hall of the Umayyad Mosque in Damascus, which has three aisles, crossed by a central transept, with a dome overhead.

are semi-circular windows, two to each arch. Damascus was also the first mosque to be provided with a minaret. This was the tower, used by the *muhaddin* to call the faithful to prayer (as distinct from the Christians, who used clappers and later bells, and the Jews, who used the ram's horn). There were in fact four minarets, on each of four Roman towers at the corner. Only one of the towers remains, however, and the minaret above it dates only from 1488; another is from the twelfth, a third from the fourteenth century. (The earliest surviving minaret is on the Great Mosque of Kairouan near Tunis, and is about 724–43, a generation after the original Damascus minaret.)

The Great Mosque of Damascus is a kind of microcosm of the civilizations which have swept over the Holy Land. It is difficult to tell which part of the walls date from Roman-pagan or Christian-Byzantine or Moslem-Umayyad or later times, for there had been continual repairing and tinkering, and a major reconstruction after a fire in 1893. Across the North Precinct Wall is the tomb of Saladin, most of it actually built by the German Kaiser, Wilhelm II, in his fanatical Eastern phase. The glory of the mosque are the mosaics on the west side of the court, within the arcaded portico, of palaces and tree-covered hills and rivers – no humans, not even animals. But very little is left of the original

LEFT The Biblical 'Street called Straight' in Damascus.

BELOW The Umayyad palace at Khirbet Mefjer, near Jericho.

decorative scheme: the arcades were once lit by 600 lamps, and there was glitter and gold everywhere. In short, the sense of a particular time is absent. What is present, and very powerfully present, is a sense of timelessness, or rather (as Robin Fedden in *Syria and Lebanon* has well observed) of a continuing ageing process. The huge court stills and muffles sound as well as echoing it. Prayer seems to come effortlessly; the place is sanctified, and not by any particular creed or cult. It is said that there is not a single spider in the whole vast area.

It was in Damascus that the Moslems began to develop decorative skills of their own. Around 'the street called Straight' (Acts of the Apostles 9:11), which was of course a colonnaded street in the Oriental style of Graeco-Roman big cities, developed the great central *souk*. In its workshops the Syrian-Arab craftsmen developed the elaborate metalwork, the silks, and above all the glassware and painted tiles which took Islamic decorative motifs all over the world. It was the Umayyad princes who supplied the first impulse, for they were gay and discerning patrons, great spendthrifts, not too strict in their interpretation of the Koran. The great mosque at Damascus was built, in part at least, to attract the tourist-pilgrim trade and thus rebuff the challenge from the rival caliphs in Mecca. But the Umayyads also built palaces and hunting-lodges for themselves, particularly in and around the Jordan valley: at Quseir Amra, Amman, and Khirbet Mefjer, near Jericho, not far from Herod's winter-quarters. Some of the decorative schemes their artists employed took no heed of the Islamic rule which prohibited the depiction of the human person.

The Umayyads were also considerable military architects, building on a sound Byzantine foundation. They seem to have invented the important technique of machicolation, which allowed defenders to shoot or pour down missiles on assailants while remaining protected. Some so-called examples from Syria are in fact projecting stone latrines, but one of the earliest genuine examples, Umayyad work, is at Qasr al-Hair, near Rusafa, which is dated 729. The Moslems were equally ingenious in developing new architectural devices for strengthening the entrances to their forts and citadels. The right-angled or crooked gateway was used in the new walled city of Baghdad, built by the Abbasid caliphs after they had ousted the Umayyads. But the Moslem masterpiece of defensive gateways is to be found at Aleppo, in the great citadel. The *tell* on which it is built is of tremendous antiquity. Abraham is supposed to have grazed his flocks on the summit. It was a defensive post of the Hittites. Indeed, Aleppo was a Hittite town, their twin capital, along with Karchemish. Curiously enough, the twin-headed eagle which is found on some of their monuments was adapted in the twelfth century as the badge of the Seljuk sultans, and stolen from them by the Habsburgs, whose emblem it became. The Seleucids called Aleppo Beroea and made it into a great trading and mercantile city, on the route to their city of Seleucia on the Tigris. There is a Christian cathedral, said to have been built by St Helena, which was turned into a mosque in the twelfth century, though it still keeps its acanthus-leaf capitals from the fifth and sixth centuries. Aleppo is a

The impressive *tell* at Aleppo with the great citadel on top.

connoisseur's city, with its thirteenth-century souks, some of the oldest in existence, and its splendid khans or warehouses, built round courtyards on the college plan, and its fine private houses, richly adorned with decorative stonework. But rising out of all of this, on its *tell*, is the marvellous citadel built by Saladin's son, El-Malek-es-Zaher, whose gateway tells more about the Moslem military architects, their cunning and sense of elegance, than any book can. The series of angled obstacles are more complicated than anything Edward I of England, and his great military engineer, James of St George, contrived to design in the formidable castles they built together in Wales, and Zaher's designer added, for good measure, a gallery of wild beasts and fabulous dragons to terrify any intruder who had not yet been driven off by pitch, oil, arrow or rock.

When the Moslem Arabs first arrived in the Holy Land, they were barely literate. By the end of the seventh century they were learning to fashion a civilization of their own from the wreck of the Byzantine Near East, and by 750 or thereabouts, a formidable new civilization had been launched. If the Umayyads were great builders, their successors, the Abbasids, were vigorous and enthusiastic patrons of learning, too. The Bayt al-Hikmah, or House of Wisdom, founded by Al-Mamum at Baghdad in 830, was the first institute of higher learning of its kind since the disappearance of the old Museon at Alexandria. It specialized in acquiring rare manuscripts, especially from the Greek; it sent scribes to Constantinople to copy works in the libraries there, and sometimes it persuaded the caliphal government to make the handing over of manuscripts a condition of new peace-treaties signed between the Greek and Moslem rulers. But long before the founding of the House of Wisdom, the Arabs had been absorbing Greek learning through the Persians. Their first translations were in fact from the Persian adaptations of Greek technical manuals. By way of Persian astronomy, the Arabs discovered Claudius Ptolemy's *Syntaxis*, from which they took their basic astronomical information. Then they moved on to Euclid's *Elements*. By the first half of the ninth century they had combined ancient Babylonian, Greek and perhaps Indian sources to produce their system of algebra. In the introduction to the treatise on algebra which he published in 825, Al-Khuwarizmi explained that it would be found useful in measuring land, planning canals and other engineering works, as well as in trade and the division of property. He also produced a number of arithmetical works in which he made use of decimal notation, and it was the translations of these, by Gerard of Cremona in the twelfth century, which first introduced Arabic numerals and fractional notation to Europe.

Arabic learning tended to be practical rather than speculative, or, rather, it was both, but with the emphasis on the useful. During the reign of Harun al-Rashid (786–809), Arabs translated Hippocrates and Galen, and their medicine became superior to anything in the West until the Renaissance. From the Chinese they had acquired the art of making paper, which was manufactured in Baghdad and other centres from the end of the eighth century. This greatly reduced the cost of copying manuscripts and most Arab doctors acquired a library of the salient treatises. Not all their best doctors were Moslems. Yuhanna ibn-Masawayh, author of a famous treatise on the eye, was a Nestorian Christian, as were other doctors – and many of the scholars who translated the Greek classics. Christians with medical skill were more likely to find advancement among the Arabs. The first hospitals in England and France date from the twelfth century. In the Arab lands there were already thirty-four hospitals in existence by the mid-ninth century and these usually had regular incomes from State lands. There were special hospitals for women, and for certain categories of illness, and the medical profession was trained and organized under official supervision. In 978, the Bimaristan, the caliphal hospital in Damascus, had twenty-four doctors who lived in the establishment. Arab doctors produced valuable pharmacopeia and encyclopaedic works of medicine and treatment and a number of specialized studies. The famous Al-Razi, known in the West as Rhazes (he too was translated by Gerard of Cremona) first correctly diagnosed smallpox and measles, and he is credited with discovering the principle of vaccination.

The Arabs also did important work in geography, some of it highly original. The world geography of Al-Idrisi, commissioned by the Norman King Roger II at Palermo in the early twelfth century, was more comprehensive and accurate than anything produced since the decline of the Alexandrine school of late antiquity. The Arabs, unlike the Latin West, continued to use the spherical globe and they

A Syrian marble fountain-basin, dated AD 1277–8.

had better astronomical tables. On a wide range of subjects, such as physics, mechanical engineering, optics, hydraulics and hydrostatics, direction-finding and various forms of measurements, their theoretical knowledge and their technology was ahead of the West.

No doubt one reason why the Arabs progressed so fast in the eighth and ninth centuries, and were so creative for the next 200 years, was that they were in contact with a wide range of cultures, ranging as far as India and China (it was from the latter, for instance, that they got the mariner's compass). In the early Middle Ages, Arab Jerusalem was a much more cosmopolitan city than either London or Paris, or even Rome; and Baghdad, under the Abbasids, was the centre of the universe. The Arabs discovered the works of Aristotle in the eighth century; they began to develop their own systems of thought in the ninth and tenth centuries, and from the time of Ibn Sina (980–1037), known in the West as Avicenna, their methodical attempts to reconcile religious with philosophical knowledge, continued with Al-Ghazzali or Algazel (1058–1109) and Ibn-Rushd or Averroes (1126–98), dazzled those Western schoolmen who were fortunate enough to get hold of their writings in translation – usually through Spain or Sicily. There could have been no Thomist philosophy without them.

Alas, the intellectual achievements of the Arabs in the early Middle Ages were not matched by political or juridical skill. The Moslem rulers of the Holy Land could provide neither stability and continuity of government, nor any real security for property. Their intentions towards their non-Moslem subjects were good; they believed in toleration; but the practical results were increasingly oppressive. Under the Umayyads and the early Abbasids, there was some prosperity in the Holy Land. Jerusalem continued to hold its annual fair on 15 September, a relic of Byzantine times, and it was attended by merchants from the

THE COMING OF ISLAM

West who combined business with pilgrimage. But in the ninth century, the caliphs of Baghdad ceased to exercise authority except in religion. Individual governors seized power and carved out kingdoms for themselves. In 868, a Turk, Ahmad ibn-Tulun, became master of Egypt, and overran Palestine and Syria in the next decade. He created a military state with a fleet at Acre. There were times when the Holy Land was controlled by Baghdad, times when Egypt ruled, and times when no one's writ ran outside the gates of a strongly-fortified town. In 923 Moslem extremists destroyed Christian churches in Ramleh, Caesarea and Ascalon; and in 937 they attacked and damaged the Church of the Holy Sepulchre.

These outrages naturally provoked the Byzantine State, which had never formally relinquished its lost provinces. In the second half of the tenth century, Byzantine fleets and armies began raiding Syria. John Timisces penetrated as far as northern Palestine in 975. But these successes were, perhaps inevitably, followed by Moslem reprisals against their Christian subjects. The Patriarch of Jerusalem was burnt alive on the grounds that he was a Byzantine spy. At the end of the tenth century the Fatamid caliphs of Egypt, who were heretical *shiites*, got control of Palestine. Al-Aziz, the first to rule there, was a tolerant cosmopolitan. His vizir or chief minister was a Jew, at least by origin; his wife was the sister of the Patriarch of Jerusalem. Jews and Christians alike found favour at his court. But his son and successor, Al-Hakim (996–1021), was a Moslem fanatic, of an unusual variety, from which the present Druzes draw their salient tenets. He may, indeed, have been insane, and he certainly declared himself to be the incarnation of the deity. In 1009 he stopped all Christian pilgrimages and he ordered Christian churches, including the Holy Sepulchre, to be destroyed. After his death, it was possible for the Christians, with Byzantine money and help, to set about repairing the damage, but it was evident to all that the Christian community could look to no better security, for the practice of their religion and the protection of their Holy Places, than the whim of the individual Moslem potentate. It was in these circumstances that the notion of a crusade took hold.

7
Pilgrims, Crusaders and Saracens

THE BYZANTINE AUTHORITIES had retained an interest in the Holy Land because it was still, in their eyes, an integral part of their Empire, to be recovered as and when their military strength revived. The Christians of the Latin West had no such claim, but throughout the Dark Ages, their love for Palestine, as the supreme object of pilgrimage, never diminished; indeed, it constantly increased. At all times, and despite the length and hazards of the journey, a surprising number of men and women from France, Germany, England, Ireland, Spain and Italy found their way to the Holy Land.

The Latins, who had no territorial claim, were more welcome than the Greeks to the Moslem authorities. In fact, the Carolingians of the West and the Abbasid caliphs of Baghdad had much in common, notably common enemies in the Byzantine emperors and the Umayyad caliphs of Spain. In 797, three years before he was crowned Emperor of the West, Charlemagne sent an embassy to Baghdad to negotiate improvements in Christian access to the Holy Places. The three ambassadors were called Lantfrid, Sigismund and Isaac the Jew. Only Isaac survived the journey, but he successfully accomplished the mission. The Caliph, Harun al-Rashid, readily agreed to grant privileges to Latin pilgrims and, in effect, to the establishment of a Carolingian protectorate in Jerusalem. Isaac was back in Aachen in 802 with the terms, and shortly afterwards Charlemagne sent out funds to found a hospice for Latins in the Holy city. In 870, according to a traveller called Bernard the Monk, it had a chapel, a library and vineyard. No doubt its fortunes declined with those of the Carolingian empire; it may have ceased to exist in the tenth century. But other benefactors were found from among the kings and prelates of the West and there was always some kind of Latin-sponsored refuge for pilgrims to the city.

These pilgrims were of all classes and ages, and both sexes; sinners as well as saints. Many were criminals, on a 'judicial pilgrimage'. Murderers were often sent to the Holy Land wearing chains or an iron collar to live there for a

A bust of the Emperor Charlemagne in gold and enamel from the Treasury of Aix-la-Chapelle.

These Byzantine and Crusader pilgrims' crosses are cut into the wall leading down to the Chapel of the Finding of the Cross in the Church of the Holy Sepulchre.

penitential period in expiation of their crimes. Some of these Dark Age pilgrimage-penances were startlingly harsh. One Fortmund, a Breton aristocrat, left a description (c. 870–4) of the penance he seems to have imposed on himself: a voyage not only to Syria and Palestine, but to Egypt and Carthage, wearing chains round the waist and irons. He had committed a crime of blood and presumably escaped more formal punishment by swearing an oath to remain a pilgrim until released by ecclesiastical authority. After four years of travelling he reached Rome and asked the Pope to agree that his vow was now fulfilled. The Pope refused and Fortmund went back to the Holy Land, still chained, and visited Jerusalem again, Cana, the Red Sea, Mount Sinai and the site of Noah's Ark in the Armenian hills. In the end he got absolution at the monastery of Redon in Brittany, before the tomb of St Marcellinus. The Christian clergy of Jerusalem and other Holy Places were accustomed (for a fee) to provide evidence to such sinners that they had actually been to the sites contracted in their vow.

Pilgrims also brought back with them a palm-leaf, in the form of a cross, which they were given only after they had bathed in the waters of the Jordan. This was, in a way, the real climax of a pilgrimage because the Jordan itself was regarded as a relic, having been in contact with Jesus's own body. The pilgrims bathed *en masse* in a spirit of exaltation; they believed they would emerge from the water freed not only from the burden of guilt but the fear of purgatorial fires. During these centuries the prospect of Purgatory was only marginally less terrifying than Hell itself – the papacy had not yet worked out the mollifying institutions beloved of the later Middle Ages – and Christians were driven to Jerusalem by anxiety or

PILGRIMS, CRUSADERS AND SARACENS

One of the stations of the Cross which mark the route that Jesus took to Calvary.

even terror. There was an element of hysteria in the pilgrimages as well as gross superstition. Christianity was dominated by the concept of the relic, which if important enough was believed to radiate actual power, destructive if approached too closely, rather like a nuclear reactor. And of course the Holy Land had the most important relics in the world, including the actual footprints of Christ in many places. Moreover, the pilgrims did not see the Holy Land as a static place but as a theatre of impending events. Belief in an imminent millennium, or Second Coming, was real and universal. It was commonly held, for instance, that the Last Judgment would take place in the valley of Jehoshaphat on the east side of Jerusalem. When the millennium came, the Emperor, as a symbol of the dissolution of all earthly power, would come to Jerusalem at the head of his army and hand over his insignia to God. Then the Judgment would commence and pilgrims thought they might reserve seats at the tribunal, as in a vast amphitheatre.

During the tenth century, the excitement mounted, for many preachers assured their congregations in the West that the year 1000 would see the end of the world. Mass-pilgrimages began, often organized by powerful prelates or nobles. The passing of the year 1000 without any untoward event made no difference. Two years later, Count Fulk of Anjou, one of the most ferocious warriors of the West, took a great communal host from France to the Holy Land. Very likely the arrival of these growing numbers of Christians, some of them armed, helped to alarm the unstable Caliph Hakim: hence his anti-Christian pogrom of 1009. All pilgrimages were then stopped for a decade. Christian churches in the Holy Land, with the exception of the Holy Nativity in Bethlehem, were ransacked. But Hakim's reign of terror, which lasted until his death in 1020, did not deter Western pilgrims; on the contrary, it set up a wave of panic among sinners who feared that their path to the Holy Places, and so to expiation and relief from purging torments, would be permanently blocked. In fact the Arabs normally behaved well to the Christians, or at any rate tolerated them. The Moslem geographer al-Muqaddasi, writing just before 1000, says the Christians had complete freedom of worship and that native Christians flourished in Palestine, especially as lawyers and doctors. The reign of Hakim was an exception, and immediately he was dead toleration was restored and the Greek Emperor Michael the Paphlegonian made an agreement with the new Caliph for the rebuilding of the Holy Sepulchre church.

Indeed, Western pilgrims, fearful that the road might be blocked again, began to arrive in Palestine in ever-growing numbers. As soon as one possible date for the millennium was past, another would be conjured up by the clergy (for example, 1033, the thousandth anniversary of Calvary). In 1027, there were two mass-pilgrimages from France. In 1035, Duke Robert of Normandy led one formed by his subjects. In 1064 there was another huge pilgrimage from Normandy, and an even bigger one, of over 7,000 people, from Germany, led by Bishop Gunther of Bamburg. Both seem to have been prompted by a wave of millennarian excitement, and both had armed escorts. Gunther's fought a regular battle near Ramleh. Of course it made sense for pilgrims to be well-organized, for protection both against bandits and the exactions of local authorities. At Constantinople the Greeks charged each pilgrim a toll of a half gold piece; a mounted man, assumed to be rich, paid three. At Jerusalem, the Arabs forced every pilgrim who passed through the city gates to pay a whole gold piece. A poor man on an organized pilgrimage could always find someone to fork out on his behalf, but the individual pilgrim might have to beg. When Duke Robert got to Jerusalem in 1036, he found several hundred penniless pilgrims, without resources of any kind, starving outside the walls. Of course many pilgrims were sick – that was why they made the journey – and in fact died in Palestine in great numbers, being buried, if they were lucky, in the Field of Aceldama. As the century progressed, the problem of sick and dying pilgrims led to a group of merchants from Amalfi, around 1060, founding a new hospice or hospital, dedicated to St John, which provided beds for over 2,000 – about 50 patients, apparently, dying each day. A pilgrim travelling alone, without lay or ecclesiastical patrons, might find it difficult to get into a hospice, and might literally starve to death, die in the gutter and go to eternity without proper burial. Hence the popularity of the organized pilgrimage. But as they grew larger, and the armed escorts increased accordingly, they began to resemble, indeed became, armies. At what point did the mass-pilgrimage become a crusade? It is not easy to say.

The international circumstances which produced the First Crusade revolve

A twelfth-century map of Crusader Jerusalem.

around the irruption into the Near East of the Seljuk Turks. They came originally from the Chinese provinces and had become Moslems in the tenth century while serving as mercenaries under the Ghaznavid Dynasty which ruled Persia and North-West India. In 1055 their leader, Tughril Beg, seized Baghdad, and was made Sultan by the Caliph al-Kaim. In 1063 Alp Arslan became Sultan, and eight years later, at the battle of Manzikert (Malazgerd) in Armenia, he destroyed the Byzantine army. The Seljuks then turned south against the heretic Fatamid caliphs of Cairo who occupied Syria and Palestine, and captured Jerusalem, which they held until 1098 – when it returned, briefly, to the Fatamids, just before the crusaders arrived. The Seljuks saw themselves as the army of orthodoxy putting down heresy, also as a race superior in physical and military skills to the effeminate Arabs. The First Crusade, therefore, came at a time when the Moslem world was violently split on both doctrinal and racial lines.

The official pretext for the intervention of the West was a request for assistance by the Byzantine Emperor, following the rout at Manzikert. In fact the Byzantines had appealed to the Latin Christians as long ago as 838, and had been ignored. But by the closing decades of the eleventh century, the West was already engaged in successful, expansionary war with the Moslem world, especially in Spain and Sicily. The Italian maritime cities, led by Pisa, Genoa and Venice, were developing navies. All the evidence suggests that the population of the Latin West, led by France, was expanding very fast and this, combined with the laws governing the inheritance of land, tended to produce a huge surplus of younger sons of the knightly or would-be knightly class, available for overseas expeditions of conquest. That was how William the Conqueror had furnished his Hastings expedition of 1066, which the papacy had treated as a crusade against a usurping king who had been uncanonically crowned by a schismatic Archbishop of Canterbury. It was the force behind the Norman drive in south Italy and Sicily, and the systematic erosion of Moslem power in Spain. The assumption among the young men was that they would get land in reward. The assumption among the clergy was that the power of Rome would be extended. That ardent protagonist of papal power, Archdeacon Hildebrand, who had been behind William's invasion of England, was Pope by the time the Emperor Michael VII's appeal for help was received in Rome, and apparently he planned to lead a crusade himself in 1074. But his own bitter dispute with the 'other' Emperor, the German Henry IV, made this impossible. The idea was revived in 1098 by Urban II, a Frenchman and a Cluniac monk – the Cluniacs coordinated the mass-pilgrimages to Jerusalem – who had a passionate interest in the Holy Places and who preached a famous exhortatory sermon at a mass gathering of the French nobility. From the various texts of the Sermon which have survived, which may not be authentic, it is clear that the Pope's motives in promoting the crusade were mixed. The Church was engaged in the 'Peace of God' campaign to stamp out, or at least limit, private warfare in overcrowded France, and Urban thought that a regular, well-organized military expedition to the Holy Land would syphon off the land-hungry, bellicose elements from the West. He put forward territorial claims to the Holy Land on behalf of Christianity and evidently thought in terms of colonization. That, of course, was not what the Byzantines had in mind. When they asked for help from the Latins, what they meant was mercenaries, who would serve for pay in the regular army of Byzantium, and so reconquer the lost territories for the Greek Crown (and Orthodox Church).

In fact the crusade got out of hand before it even started for Palestine. The common people, unable to distinguish between a mass-pilgrimage and a military crusade, were unwilling to leave the rich spiritual rewards Pope Urban had promised entirely to the men of war. They formed huge, excited mobs, which the local authorities were unable to control and therefore pointed hastily in the vague direction of Palestine. Worse: there was a widespread belief in the West that Hakim's attack on the Christians in 1009 had been instigated by the Jews, and among the Latins anti-Semitism and sporadic pogroms, had become a growing feature of religious enthusiasm. Once the crusading mobs were assembled and on the move, they turned ferociously on the Jewish communities, especially in the Rhineland; there were horrific massacres and long, dismal additions to the martyrology of Judaism. The mobs then coalesced, and struggled across the Balkans to Byzantium, where they were regarded by the Greeks in horror as a reincarnation of the 'barbarian hordes' who had destroyed the Western Empire in the fifth century. The Emperor was glad to ferry them across the Bosphorus and unleash them in the wilds of Anatolia, where most of them perished. Very

PILGRIMS, CRUSADERS AND SARACENS

few of these crusaders ever reached the Holy Land.

The regular crusade, led by the Count of Toulouse (southern France), the Duke of Lorraine, Robert of Flanders and Stephen of Blois (northern France), Bohemond of Taranto (Sicily), and accompanied by the Pope's representative Adhemar, Bishop of Le Puys, was more successful. It began on a footing of legality at Constantinople, where the chief nobles acknowledged Byzantine suzerainty over any territories they might occupy, in return for financial help. Adhemar, moreover, was prepared to act under the authority of the Patriarch of Jerusalem who was in exile in Cyprus. Hence in both Church and State the crusade began as an attempt to restore the Greek Empire and Christian Orthodoxy. But Bishop Adhemar died early in the campaign and thereafter the interests of the Orthodox Church, let alone the local Christians – who had never asked for Western help at all, and much preferred to make their own deals with the Moslem authorities – were almost totally ignored by the Latins. Once the breakout into Syria and Palestine was achieved, the greatest of the barons quickly carved out counties and principalities for themselves and, in fact if not always in name, disregarded their obligations to the Greek Emperor. Jerusalem fell in 1099. The Fatamids and the Seljuks hated each other more than either feared the Latins – at this stage – and the Christians might have taken the city by negotiation. But they preferred to use force and stormed it, like the Romans before them, on the weak north side. They then massacred all the Moslems and the Jews they found within – only the governor and his staff being allowed to bribe their way to safety.

William, Archbishop of Tyre, the historian of the Latin states, makes it clear in his account that the slaughter was premeditated:

Each marauder claimed as his own in perpetuity the particular house that he had entered, together with all that it contained. For before the capture of the city the pilgrims had agreed that after it had been taken by force whatever each man might win for himself should be his forever by right of possession, without molestation. Consequently the pilgrims reached the city most carefully and boldly killed the citizens. They penetrated into the most retired and out-of-the-way places and broke open the most private apartments of the foe. At the entrance of each house, as it was taken up, the victor hung up his shield and his arms, as a sign to all who approached not to pause there by that place as already in possession of another.

One eye-witness, Fulcher of Chartres, chaplain to Baldwin of Boulogne, says that 10,000 Moslems were killed in the Aqsa Mosque.

Having feasted on blood, visited the Holy Places, and bathed in the Jordan, most of the Latin knights returned to Europe with their troops, leaving behind a mere 300 knights and 2,000 foot-soldiers. This was the military 'colony'. It was small, but it must be remembered that a generation before, William I had conquered England with a mere 10,000 men. The permanent crusaders created two 'counties', Edessa and Tripoli, the Principality of Antioch, and in Palestine proper they set up a feudal Kingdom of Jerusalem, of which Baldwin was elected the first King. The enterprise was hazardous from the start, for the Europeans

Baldwin I, here depicted on a coin, was the first king of the Crusader Kingdom of Jerusalem.

were so few, too few, in fact, to undertake a regular campaign of consolidation. Each spring, fleets from the Italian cities arrived with thousands of pilgrims for the festival of Easter, and the Latins took advantage of the presence of these forces to take towns on the coast. In 1103 they captured Acre, the best harbour on the coast; in 1109 Tripoli; Beirut and Sidon fell to them in 1110 and Tyre in 1124. But Ascalon, further to the south, did not fall until 1153, by which time the Latins had already lost Edessa to the resurgent Moslem forces.

Indeed, the strategic weakness of the Latins throughout was that they never really dominated the territory beyond the line of the Lebanese mountains and the Jordan. They took neither Damascus nor Aleppo, and the inland desert route was always open to the Moslems. Communications between Egypt and Mesopotamia were thus maintained and it was only a matter of time before the warring Moslem factions came together in their hatred of the Christian invaders. In 1154, the orthodox Kurdish warrior Nur ad-Din occupied Damascus on behalf of the Caliph, and when he took over Egypt in 1163 the restoration of Moslem military unity was a *fait accompli*. Nur ad-Din died in 1174, bequeathing the completion of his work to his nephew Saladin, who was recognized by the Caliph as Sultan of Egypt and Syria. In 1187 Saladin destroyed the Latin army in a great battle at the Horns of Hattin to the west of the Sea of Galilee. The fall of Jerusalem followed inevitably. A crusade to retake it for Christianity was launched by the three most powerful monarchs of the West: Frederick Barbarossa, Emperor of Germany, Richard the Lionheart of England, and Philip Augustus of France. But Frederick was drowned *en route* and his huge army melted away; Philip performed his duties as a Christian warrior only in the most perfunctory manner, and returned to Europe; and Richard calculated that he just lacked the strength to mount a siege of the Holy City. In fact Saladin, too, was by this stage (1192) near the end of his resources (and his life) as we know from an account written by his secretary, Beha ad-Din. He was a great warrior, a courteous and civilized opponent and a pious Moslem, but no administrator or financier; he would almost certainly have evacuated Jerusalem if Richard had persisted a little longer. But Richard did not know this and withdrew from the crusade in sadness.

In the thirteenth century, Jerusalem was briefly regained by the Christians as a

RIGHT The triumphal entry of Richard the Lionheart and Philip Augustus into Acre. Richard lacked the necessary strength to attempt a siege of Jerusalem and had to abandon the crusade to win the city back from Saladin.

BELOW Saladin's mimbar or pulpit which was built for Aleppo is now in the El Aqsa mosque, Jerusalem.

result of the diplomatic skills of their excommunicate Emperor, Frederick II. Indeed, it is likely that the Latins, had they not been so belligerent and hasty, could all along have obtained their reasonable objectives by bargaining. What they scorned to hold by treaty they lost by force. In 1244 Jerusalem fell to a Turkish army and was never again retaken. The Latin states to the north crumbled at the same time. The beleaguered Christians might, indeed, have recovered all by forming an alliance with the invading Tartars of Hulagu Khan, who swept into the Near East in the mid-thirteenth century. Many of them, including their general, Kitbugha, were Nestorian Christians. An alliance between the Latin West and the Nestorian East could have extinguished the very heart of Moslem power and restored the vision of a Christian Asia which had faded early in the seventh century. But in fact the Latins, by now reduced to their great fortress of Acre, sided with the Moslems against men they regarded as savages. Kitbugha was defeated in 1260 by the Mamluk mercenaries of Egypt, at the foot of Mount Gilboa where Gideon had once beaten the Midianites – one of the most decisive, and least known, battles of history. Acre, in turn, fell to the Moslems in 1291, and with it the attempt to create a Latin colony in the Holy Land came to an end.

This bald account does scant justice to one of the most fascinating of all historical experiments. The Greek West had ruled the Near East for 900 years, from 300 BC to after AD 600. Palestine and Syria had then been Orientalized again. The Latin states of the twelfth and thirteenth centuries can be seen as an attempt to reassert the traditional primacy of Western civilization, or they can be seen as the first, faltering wave of European colonization which, in the fifteenth century, was to turn dramatically towards the west and the great oceans, and thereafter to encompass the whole earth. In a sense it was both. We might sum up the crusades as a tiny but distinct hyphenation between the colonizing spirit of the ancient Greeks, going back to before the days of Homer, and the enormous migratory armies which continued to populate the world from Europe almost until our own time. But, though distinct, it *was* tiny: it lacked the sheer human resources to accomplish its objective.

Just as the Greeks of Hellenistic times had sought to encourage settlers from the mainland to consolidate their new territories in the East, so opinion-formers in the Latin West begged Christians to join the crusades. Preaching the Second Crusade, St Bernard of Clairvaux wrote to the nation of the English:

Are you a shrewd businessman, a man quick to see the profits of this world? If you are, I can offer you a splendid bargain. Do not miss this opportunity. Take the sign of the cross. At once you will have indulgence for all the sins which you confess with a contrite heart. It does not cost you much to buy and if you wear it with humility you will find that it is worth the Kingdom of Heaven.

The Saint hinted at the sort of people most likely to respond to this 'bargain' when he asked, in his innocence: 'Is is not an extraordinary and special act of divine charity that God Almighty should treat murderers, robbers, adulterers, the

The Knights of the Holy Ghost embarking for the Crusades. This voyage was never, in fact, undertaken, but the chevaliers were ordered to hold themselves ready in the event of their departure being resolved upon in Rome. The crusaders, on horseback have the bow on their breastplates, and the vessels are decked with banners on which are blazoned the arms of the Pope, of England and France among others.

foresworn and criminals of every kind as though they were righteous men, and worthy to be enlisted in his service?' Indeed, those who were actually out in the colony and knew what it had to offer, stressed the material, rather than the spiritual blessings in their propaganda. Fulcher of Chartres wrote from Jerusalem in 1127:

We who had been Occidentals have become Orientals; the man who had been a Roman or a Frank has turned into a Galilean or a Palestinian; and the man who once lived in Rheims or Chartres is now a citizen of Tyre or Acre. ... Some of us already possess over here houses and followers which they have received as of hereditary right. We have married as wives Syrians or Armenians or even Saracens who have been baptized*... Our dependents and relatives join us here day by day, leaving behind (perhaps reluctantly) their possessions. Those who were poor over there have been made rich by God over here. Instead of empty coffers they have piles of besants. Those who had not even a village there, have towns here – the gift of God. Why should any therefore return to the West, when they have found an East like this?

This was the philosophy not only of a colony, but of a ruling-class colony. Yet curiously enough very few crusaders of whom we have record came from the European ruling or knightly class. Late twelfth-century lists of English crusaders show that most of them were poor or average folk. A list from Lincolnshire shows a clerk, a smith, a skinner, a potter, a butcher and a vintner. A group of forty-three crusaders from Cornwall included a tailor, a smith, two chaplains, a merchant, a shoemaker, a miller, two tanners and two women. One ship, the *St Victor*, travelling to the Holy Land in 1250, carried 453 crusaders. Of these, fourteen were knights, with ninety retainers; and there were seven clergymen.

*Any crusader who married an unbaptized Moslem woman could be sentenced to be castrated; and, for good measure, his nose cut off.

CIVILIZATIONS OF THE HOLY LAND

But there were 342 commoners, including forty-two women – fifteen of them accompanying their husbands. How many of these were temporary crusaders, how many settlers? We do not know. We do know that many middle-class people organized themselves into crusading confraternities to maintain settler outposts in the East. This vicarious crusading was typical and in the end fatal. The Church was altogether too astute in inventing ways in which the pious could win the eternal rewards of crusading by merely handing over cash and never setting foot outside their native provinces. The crusades, in fact, detonated the great inflation of indulgences which in the end wrecked the crusading venture by starving it of the one commodity it could not do without – live crusaders.

There were never enough of them. Their life-expectancy in the East was lower even than in Europe. They were killed off by dysentery and malaria. Leprosy was another scourge – its noblest and most pathetic victim being King Baldwin IV, who came to the throne in 1173 at the age of sixteen, and magnificently held Saladin at bay until the disease finally burnt him up ten painful years later. Compared to the Moslems, the Latins had virtually no medical science. Arab literature of the time is full of horrific anecdotes about the ignorance and brutality of crusader surgeons. A more sinister feature of the colony was a low birthrate; or, at least, a low rate of survival of European infants. Kings, princes and great feudatories of the states persistently failed to produce heirs. So disputed successions, quarrels over heiresses, a failure of leadership at crucial moments added to the troubles of the beleaguered Latins, who never numbered more than 100,000.

The Latin states were on the whole well governed, or rather better governed that neighbouring Moslem territories. The law-code, the Assize of Jerusalem, which was later written down when the Latins retreated to Cyprus, is infinitely superior to the prevailing justice administered in Cairo or Damascus. There was, for one thing, much greater security for property, and therefore some encouragement for agricultural investment. In fact the decay of the land in Syria and Palestine, an accelerating feature of Moslem rule, was for a time arrested or even reversed. Moslem peasants were better off under Latin lords than under their own masters. It is significant that the Jews, who had good cause to fear the Latin Christians, nonetheless preferred on the whole to live in villages administered by them.

But if the Latins were better at ruling, their culture in almost every other respect was inferior to Islam. Indeed, they did not really possess a culture of their own. They learnt very little from the Arabs, who had so much to teach. They were fairly quick to adopt the more superficial and obvious refinements of the East: loose clothes, cushions, hot baths, food and spices. But hardly any learnt to speak Arabic, let alone master its enormous and valuable literature. The cross-fertilization of cultures, which was so fruitful in the Spain and Sicily of the same period, simply did not take place in the Latin Holy Land. The crusaders were brave (if also foolish and quarrelsome) men, but they were philistines down to the

RIGHT The great entrance to the Citadel at Aleppo, in Syria.
OVERLEAF LEFT The Crusaders 'Castle of the Sea' at Sidon.
OVERLEAF RIGHT Krak des Chevaliers is the largest and most impressive of the Crusader defensive castles.

PILGRIMS, CRUSADERS AND SARACENS

tips of their halberds – more philistine, indeed, than the gifted race of warriors who gave the land its name. The only Latin colonist to achieve any kind of intellectual distinction was William of Tyre, and he was completely educated in Europe. Indeed, the crusaders were poor representatives of their own culture, which in the twelfth century was undergoing a dazzling renaissance. No echo of the schools of Paris or Bologna was to be heard in Jerusalem or Antioch, which had only the most rudimentary educational establishments; and this at a time when Damascus could number seventy-three colleges and Cairo seventy-four – including the great al-Azhar, founded as a public university as early as 970.

The fault chiefly lay with the Latin Christian clergy, who were, almost without exception, immensely rich, corrupt, lazy and ignorant. They did not trouble to establish contact with the centres of Arabic culture in Damascus, Aleppo or Baghdad, or to translate the Koran into Latin (that was done in Toledo). The Latins did not take seriously any missionary activities, which necessarily would have involved the study of Oriental languages, until it was all over. In 1303 the Council of Vienne commanded that chairs of Greek, Arabic, Hebrew and Syriac be established in Christian universities, but by that date the last crusader had left the Holy Land.

Nevertheless, the crusader states did to some extent act as carriers of Islamic culture to the West. It was through the Holy Land that the craft of paper-making reached Europe at last. Paper had been manufactured in the Middle East as early as 712 and there was an actual paper-factory at Baghdad in 794. Yet it was not until the establishment of the Latin states that Westerners began to make paper, at any rate in quantity. They also learnt a good deal about the production of fine glass, which had been a speciality of the Syrians (especially in Damascus) since at least the eighth century. The British Museum has a rare enamelled glass beaker, dating from the thirteenth century, which was evidently made in Syria but mounted and used as a mass-chalice when it eventually reached Europe. The fact that it has figures on it suggests it may have been made to European taste. Certainly in the fourteenth to fifteenth centuries some Damascus craftsmen made beautiful objects specially for Latins: the British Museum possesses another Syrian beaker which has figures of the Virgin and Child, St Peter and Paul, and is inscribed in Latin. There were undoubtedly Christian craftsmen in the Syrian glass industry. In 1277, Bohemond IV, Prince of Antioch, signed an agreement with the Doge of Venice to bring the secrets of the craft to Europe. Thereafter, an *atelier* in Venice was set up with the help of Syrian experts and with raw materials shipped from the Holy Land, and the Venetian glass-industry was born. The crusaders also brought back many fine *objets d'art*; few survive, but we get tantalizing glimpses in the records. In Joinville's *Life of St Louis*, we hear that the head of the Assassins sect, the famous 'Old Man of the Mountains', sent the French King:

a very well made figure of an elephant, another of an animal called a giraffe, and apples of different kinds, all of which were of crystal. With these he sent gaming boards and sets of

One of the crusading kings of Europe, Louis XI of France, later St Louis.

chessmen. All these objects were profusely decorated with little flowers made of amber, which were attached to the crystal by delicately fashioned clips of good fine gold.

Of the enormous literature which the crusades inspired in Europe, only one poem of any importance was actually written in the Holy Land, or Outremer as the Latins called it: the so-called *Chanson des Chetifs*, produced at Antioch a little before 1149. Archbishop William of Tyre wrote the only great prose-work, his *History of Outremer* in twenty-three books, carrying the story up to 1183. He at least took the trouble to learn something about the Arabs and Islam, and apparently produced a *History of the Mohammedan Princes from the Appearance of the Prophet*; it has not survived but fragments of it are reproduced in William of Tripoli's *Tractatus de Statu Saracenormum* of 1273. The crusaders were not notable writers or readers either. They destroyed what had apparently been a fine Arab library in Tripoli, and they sold off the Jewish library they found in Jerusalem. But they did create a new library in Tyre, in the cathedral, and they re-established the Christian library in Jerusalem, which had links – albeit tenuous – with the original library of very early Christian times used by Eusebius and Jerome. There was a *scriptorium*, or book-producing studio, in the Holy Sepulchre Church, which created some elaborate illuminated volumes, and its artists seem to have transferred themselves to Acre after the Holy City fell. At any rate it was at Acre, in the time of St Louis, that the marvellous *Paris Arsenal* Bible, based on Byzantine manuscripts, was produced. The painters of Outremer worked in a medley of Western and Eastern styles and some of their ikons are still to be found in the remote monastery of St Catherine in Sinai.

PILGRIMS, CRUSADERS AND SARACENS

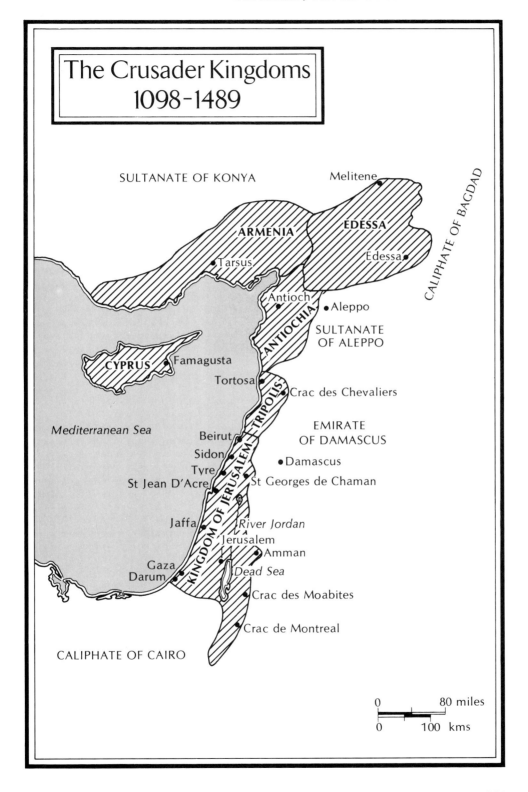

The elaborate interior of the Armenian Church of St James within the walls of Jerusalem, dating from the Crusader period.

But of course it must be remembered that the crusaders were in the Holy Land for a practical purpose; they were not there to create a culture, or for that matter to defend one, but to keep the flag of Christianity flying. One of their more friendly Moslem critics, Usama, Amir of Shaizar, wrote: 'Among the Franks – God damn them! – no quality is more highly esteemed in a man than military prowess. The knights have a monopoly of the positions of honour and importance among them, and no one else has prestige in their eyes.' In the end, then, they must be judged by their churches and their castles. They built many churches, but few survive at least in anything approaching their original form. Some are now mosques, as at Sebastiyeh, Hebron, Ramleh, Gaza and Nabi Samwill. At Lydda there is the Church of St George. There is a splendid fortress-church at Qiryat al-Inab, the site of Emmaus; and crusader churches can be found at Ain Karim, reputedly the birthplace of John the Baptist, and Saffuriyeh. The style is mainly French, of the twelfth century. It seems likely that most of the builders and sculptors who worked for the Latins were, in fact, born in Europe. For instance, the five Romanesque capitals at Nazareth, dating immediately before the 1187 catastrophe, can clearly be classified as Burgundian work.

In Jerusalem the crusaders rebuilt and adorned the Holy Sepulchre church – they seem to have set to work on the rotunda as soon as they took Jerusalem in 1099. Dimly, through the ravages of centuries and the neglect engendered by the savage faction-fighting of the Christian sects charged with its custody, can be seen

PILGRIMS, CRUSADERS AND SARACENS

the noble basic structure of this twelfth-century church. But by its nature it is a place to arouse piety and a sense of historical awe rather than aesthetic satisfaction. Much more appealing architecturally is the small and austere Church of St Anne at Jerusalem, also much changed, which was once offered to the Anglican community as their base in the Holy City, and which they foolishly declined to accept. Other churches within the walls which date from crusader times are the Armenian Church of St James, the Church of St Elias (or Elijah) and the Syrian-Jacobite St Mark's. It is a meagre total. Many crusader churches and chapels in and around the city were pulled down by the Moslems (the Church of the Tomb of the Virgin in the valley of Jehoshaphat, is a survivor), and even in recent decades there has been further vandalism, for the Monastery of St Cross, outside modern Jerusalem, was smashed in the 1967 fighting. Some indication of what we have lost is provided by the magnificent Cathedral of Tortosa, with its triple apse, rib-vaulted aisles and barrel-vaulted nave – immensely strong and purposeful without, for it was built for defence as well as worship, but within a museum of stone carving of the most delicate and confident workmanship. Tortosa, of course, was a pilgrimage-centre because St Peter was supposed to have said mass there, and it was the site of the earliest altar dedicated to the Virgin Mary, but ecclesiastical buildings of similar quality were put up at many other centres of crusader settlement, only to vanish. Tyre, for instance, had eighteen churches, of which William's cathedral was one; all have now gone.

Tortosa also possesses the well preserved ruins of a great crusader castle, with

The soaring barrel-vaulted nave of the Cathedral of Tortosa, built for both worship and defence.

two concentric systems of defence based on sea-filled moats. Crusader castles have, on the whole, fared much better than crusader churches; with few exceptions they are ruinous, but at least none has vanished utterly without trace. Thanks to the work of the late Robin Fedden, we know a good deal about their history and how they functioned. There are many more than the modern traveller expects – at least sixty of considerable importance and size, scattered over the coastline of south-east Turkey, Syria, the Lebanon and Palestine, and in the various mountain spines which guard the approaches to the Holy Land. There were even two on the Gulf of Aqaba, and others on the old desert trail past the eastern side of the Dead Sea. Many were so grouped as to be within signalling distance of each other, on the old Roman system.

The expenditure on these huge and numerous castles must have been formidable; they absorbed perhaps half the revenues of the Latin states. But there was no alternative. Really strong and well designed castles saved manpower, and manpower was the commodity the Latins always lacked. It was never very difficult to raise money in Europe, or indeed to persuade trained fighting men to come out for a season. But with permanent, regular troops the states were ill-provided, and there were too few knights enfeoffed for feudal service. Indeed, the states could not have survived at all without the monk-knights of the two great orders, the Templars and the Hospitalers. Both, as it happened, came to their frontier role by accident. Hugh of Payens, a knight from Champagne, founded the Templars under Baldwin II to protect pilgrims on the road from the port of entry (originally Jaffa) to Jerusalem, which was then infested with bandits. The knights were not priests: they merely swore to be obedient, poor and chaste – and in due course became arrogant, rich and (so it was claimed) addicted to unnatural vice. They came from well born families and wore full armour, to which was added a white tunic, eventually adorned with a red cross. St Bernard, who drew up their rule, approved at the Council of Troyes in 1128, contrasted their valour and piety with the selfish worldliness of secular knights, but in fact the Templars, because of their peculiar role in protecting the pilgrims and their money, rapidly became involved in banking. Baldwin II gave them their first quarters in the ruined Dome of the Rock, believed to be the site of Solomon's Temple – hence their name. This circular building, which the Templars reconstructed, and which was known as the Templum Domini, featured on many coins and other examples of Christian iconography until the end of the Middle Ages, gradually becoming confused with the original Temple of Solomon. It inspired the circular Templar churches which the knights built all over Europe – in London, Aix, Metz, Laon to name only four.

The Hospitalers arose from the foundation originally financed by the Amalfi merchants in the eleventh century. As the need for first-rate fighting men became apparent they ceased to be mere stretcher-bearers and hospital attendants and became fully-fledged knights like the Templars; and like them, from 1137, took on frontier duties. With their rise in the social scale they abandoned their original, humble patron, John the Almsgiver, and from 1155 claimed to be knights of the more glamorous St John the Baptist. Both the Templars and the Hospitalers quickly freed themselves from local ecclesiastical shackles and reported directly to the Pope – the high road to wealth and power for any medieval ecclesiastical institution. During the twelfth century they were heavily endowed with land, commercial property, fees, feudal dues and hard cash – the wealthy of the West, as always, preferring to pay up rather than serve overseas – and had estates all over Europe as well as massive properties in the Holy Land. With these they financed the construction and maintenance of many of the castles.

The crusader castles were originally built largely to secure communications and trading routes. One of the earliest was Safed, set up in 1102 to cover the Acre–Damascus road and Jacob's Ford over the Jordan. Toron (about 1105) and Subeibe (about 1130) protected the Tyre–Damascus road. Beaufort (1139) controlled the routes to Sidon and the Leontes valley. Belvoir (1140) protected the road down the Jordan valley and the Pont de Judaire. In 1142 two enormous new castles were first completed: the Kerak at Moab, called the Petra Deserti, Stone of the Desert, which dominated the trade route between the Red Sea and the Dead Sea (and the valuable Dead Sea salt-pans); and Crac des Chevaliers, taken over and remodelled by the Hospitalers to guard the approaches to Tripoli from the north-east. No route of any importance was outside a few miles' range of a big castle which could quickly send out a squadron of cavalry. But castles were also set up to protect agricultural settlements. As in Europe, they were nearly always the centre of a 'lordship', the headquarters of an economic unit, a supply and storage-centre, estate office, bank, prison, courtroom and residence as well as a fortress; in private hands they were expected to show a profit, not least from the services they provided in baking bread and milling grain (most had windmills or water-mills).

Thus the castle at Safed, held by the Templars, was the centre of a lordship of 260 villages, with a population of 10,000. It was built, like all of them, to economize on manpower. Its full peace-time garrison consisted of 50 knights, a 50-strong squadron of light Turcopole cavalry; 30 lay brothers and 300 archers, who provided the main defensive force for the walls. There were 820 men for general duties and 400 slaves. If necessary, over 2,000 people from the surrounding districts could be accommodated, in some discomfort, within the outer walls, and this is evidently what happened in the end, for 2,000 is the number given of those slaughtered by the Saracens when the castle finally fell.

Crac des Chevaliers was even bigger. When the French took over the League of Nations mandate for Syria and the Lebanon after the First World War, they found that Crac had become a village. They persuaded, or forced, the inhabitants to leave, and faithfully restored the immense structure, which is incomparably the finest of the crusader castles – some, including T.E.Lawrence, rate it the finest in the world. It is in essence a late-twelfth-century concentric castle, with a

OVERLEAF Sahyun Castle seen from the east. The Crusader castles were built originally to protect trading routes.

particularly clear, steep and striking approach, which makes the defences of the gateway, both natural and artificial, very formidable indeed. The defences which form the inner wall have, on the exposed south and west, a steep battered glacis leading up to three enormous towers which are the core of the stronghold. The central one was the residence of the fortress commander, or of the Grand Master of the Order when on a visit. There is also a Great Hall of splendid proportions, with three vaulted bays and a sunlit loggia, and a powerful-looking chapel. Strictly speaking, Crac never fell. It was besieged several times but the Moslems did not get into the circuit of the walls until the last siege of 1271, and even then the knights fought on in the three towers of the inner ward, laying down their arms only on the receipt of orders to surrender – which turned out to be forged.

The knights were certainly brave. They fought to the last at the blundered battle in the Horns of Hattin and they never yielded a castle easily. They made use of mercenaries on a large scale, and trained their vassals to fight; but even so their armies were tiny. They never possessed enough men both to garrison their castles and to field an army. Hence when the army was lost at Hattin the castles were denuded and many fell in consequence. In the thirteenth century the knights tended, by contrast, to try to defend all their castles and thus could not put a big army in the field, leaving the Moslems to take the castles one by one. The knights tried, as we know from surviving documents, to recruit agricultural colonies which were also defensive units, like Israeli kibbutzim. At Beth Gibelin, near Ascalon in 1168, there is a record of thirty-two Latin families being brought from Europe and given 150 acres each, in bourgeois tenure, in return for payment in kind and military service. There were too few such colonies, partly because recruits from Europe were lacking, partly because the knightly orders lacked the money to set them up. This is paradoxical because both, and especially the Templars, were very rich – indeed, that was the main complaint against them, which led to the plundering and suppression of the Templars by the French Crown early in the fourteenth century. But the truth is that though the estates of the knights in Europe were highly profitable, their establishments in Outremer were often starved of money, for the knights lacked the administrative competence to overcome the formidable institutional obstacles which prevented the transfer of funds from one to the other. Such was the nature of medieval man.

It was also of the nature of medieval man to distribute, rather than to concentrate, authority. The Latin Kingdom was feudal, as were the other crusader states. In theory the king had wide powers, especially in military affairs, but these were slowly eroded in the twelfth century and transferred to the barons. The two knightly orders were also independent powers in the state. So, most disastrously, were the maritime towns of Italy, especially Genoa and Venice. They played only a small role in the First Crusade, but immediately they realized that the Christian colony was a going concern they moved into the picture, hiring out their fleets in return for immensely valuable trading and financial concessions in the ports. The King had no fleet – no one seems to have thought of setting up a naval order of fighting monks, which might have saved the entire enterprise – and therefore had no alternative but to meet the conditions laid down by the Italians. With their self-policed trading quarters and their own courts, they, like the military orders, were virtually outside the power of the throne. Their privileges were written on the walls of the Holy Sepulchre Church in glittering letters of gold. They put their commercial interests before any other consideration. It was the Venetians who diverted the Fourth Crusade into a disastrous attack on Constantinople – their trading enemy – from which the Byzantine Empire never really recovered, and the subsequent bitterness between Latins and Greeks bedevilled all efforts to relaunch the crusader movement on an effective scale. The maritime cities were not even loyal to the basic principles of Christian militancy, since they always traded with the Moslems if they found it profitable, and sold them arms. They were interested in making money. But then, what were the crusaders themselves interested in? There were few idealists among the settlers of Outremer.

The Latin settlements would not have lasted even as long as they did without the divisions in the Moslem world. These divisions made the successful invasion possible and their perpetuation eroded the overwhelming advantages which demography gave the Moslems throughout. The Moslems were not merely split; like the Greek and Latin Christians, they actually fought each other. The orthodox Sunnis had to contend not only with the Shia Fatamids of Egypt but with various militant sects, such as the Ismailians, who plundered caravans and in many ways corresponded to the Albigensian heretics of France. They expected the coming of a Mahdi or Messiah, like the Jews, and thus were in a recurrent state of violent excitement, easily stirred up by a charismatic 'reformer'.

In 1090 an Ismailian extremist, Hassan i-Sabbah, set up a terrorist centre in the Elburtz mountains in Persia. His disciples formed the sect of the Assassins and spread over large areas of the Moslem Near East. Much nonsense has been written about this sect, which had nothing to do with hashish. Their surviving literature shows that the Assassins were gnostics, who believed in a secret corpus of knowledge revealed only to a few (them), containing the keys to the universe. They thus went back to the neo-Platonists and beyond that to the numbers mysticism of ancient Egypt and the Chaldeans. But unlike most gnostics they were fanatical puritans who believed it was right to kill the wicked in God's name. Their dagger-men, or *fida'is*, swore solemn oaths and were, in fact, not unlike the Jewish *sicarii* who flourished in the time of Jesus. In 1140 they captured the castle of Masyaf near Aleppo and this became their headquarters, from which, among others, the 'Old Man of the Mountains', Rashid al-Din Sinan, operated. They were particularly anti-Turk, for the orthodox Caliphs used the Turks as a kind of savage Sunni police-force, to stamp out heresy. Hence on the whole they sided with the crusaders, though they also attacked (and assassinated) them from time to time. They twice wounded the orthodox leader Saladin, who never managed to stamp them out. He gave to smashing the Assassins the same priority as the destruction of the crusader states, though other orthodox Moslems saw the Latins as a useful barrier against the spread of Ismailian heresies of various kinds and therefore wished them to survive. The success of Saladin's policy lay in the fact that he was briefly able to unite the Moslem world by damping down the Shia heretics in Cairo (burning all their literature in the process), and thus was able to throw the weight of Egypt into the balance against the crusader states. That meant the end of the Latin Kingdom of Jerusalem.

The process of expelling the Latin Christians was completed after the Mamluks seized power in Egypt. The policy of buying Turkish slaves or recruiting Turkish tribesmen as mercenaries, led to the Turkish element outnumbering the original Kurds in the Sultan's army – and the Turks eventually ousted the ruling Kurdish dynasty. All the Mamluks were former slaves. Power rested in their hands, for they elected a Mamluk as Sultan when the old one died, and fresh Mamluk slaves were continually recruited. They not only extinguished Latin civilization in the Holy Land, but ruled in the Near East for nearly 300 years. They were the best troops in the region until the Ottomans came with their janissaries, but they never fully adapted themselves to the use of firearms. Hence the Ottoman triumph in 1517 and the centuries of their empire in Palestine.

It is possible, of course, that the Latins might have created a more viable state in Syria-Palestine if they had been able to merge the various Christian Churches, including their own, into a homogenous community; or even if they had shared their rule and their privileges with the Eastern Churches. In fact they treated the native Christians little better than Moslems or Jews. Of course it is true that these Christians did little to help the crusaders' cause, though some groups, such as the Syrian Jacobites, moved into crusader territory. The Christian Copts of Egypt flatly declined to rise against their Moslem masters. There were exceptions among the native Churches. The Christians of the Lebanese Mountain threw in their lot with the crusaders, united themselves with Rome, and thus formed what became known as the Maronite Church. The Armenians, who hated the Seljuks and the Byzantines, and who had migrated to Cicilia in the 1060s, were even more useful allies of the crusaders. They fought on, side by side with the knights, until the bitter end, and even after the crusaders retreated to Cyprus the Armenians remained in alliance with them and the papacy until the Mamluks virtually destroyed their homeland. Their last king, Leo VI, died an exile in Paris in 1393.

In the long run, the crusades did immense harm to Christianity in the Near East, and so to the basic civilization of the region, which was both economic and cultural. After all, the Fertile Crescent, as we have seen, had been productive of advanced cultures since the beginning of recorded history, even before it. Christianity in this area was the legatee not only of the Byzantine and, before that, the Hellenic culture but also, and equally important, of the Aramaic culture of the ancient East. When the crusaders arrived, the Christians were very probably in a majority. Their various Churches, though divided by regions and within them, all flourished after a fashion, and had organized hierarchies of bishops. By the time the crusaders left, all these Churches were in decline, some fatally so; some now survive merely as tiny, colourful specimens, like rare plants or animals. Syriac (like Coptic in Egypt) disappeared as a language except in the liturgy. Aramaic, in its Syriac form, is said to be still spoken in a few Lebanese mountain villages. But the Christian decline still continues, as in the case of the Maronite Church, the last major survival of these exotic religious establishments. The near-eclipse of Oriental Christianity and the cultural impoverishment of the Islamic world which the crusading violence helped to promote, accelerated the economic decay of the region. The Mamluks deliberately destroyed the coastal towns of Palestine and Syria to prevent a crusader re-entry and the interior, cut off from its Mediterranean links, forged in the days of Byblos in 3000 BC, became poor and depopulated. During the long, weary centuries of Mamluk and Ottoman governance, or mis-governance, civilization barely survived at all in the Holy Land.

Hence it is not without sadness that one examines today the great cavernous underground ruins of the crusader fortress of Acre, whose mysteries the archaeologists are still slowly exploring, or visits the fishing village of Ruad, an island off the coast, which is built into the shell of an enormous crusader redoubt, and where the Christian knights hung on for eleven years after they had been forced to evacuate Acre in 1291. They were certainly tenacious, those bachelor Christian soldiers! But the longer and more desperately the crusaders fought, the more certainly they sealed the doom of the native Christian Churches and the Christian civilization, that they had come east to defend. In the relations between Christianity and Islam, force always failed while negotiations usually succeeded. The entire crusading movement was an exercise in counter-productivity.

Of course the end of the active phase of the crusades did not bring the pilgrimages to an end. On the contrary. The fourteenth and fifteenth centuries were the classic era of the Holy Land pilgrimage. Under the auspices and protection of Venice, the one power which really profited from the crusading movement, pilgrims, at least if they were wealthy, could travel into and around the Holy Places in reasonable security and comfort. Many diaries and accounts of these voyages survive, as do travel-contracts and phrase books – one, found in the library of Charles V of France, was entitled *How to Ask in Arabic for the Necessities of Life*.

The main trouble the pilgrims had to contend with was that the Mamluks and the local governors and officials constantly and arbitrarily increased the tolls and changed the methods of payment. They tolerated the pilgrims but milked them. The master of the ship had to provide the Moslem officials with a list of his pilgrims and a fee of six Venetian *gros* a head: Venetian currency was generally accepted in the East and sometimes minted with Koranic inscriptions, a practice which much incensed the Pope. Then the pilgrims waited until permission to proceed inland was received, being kept in the meantime in malodorous underground dungeons, a privation which merited a seven-year indulgence. The

tours inland were often managed by the Franciscans, who took over the running of the Holy Places on behalf of the Latins and even collected the tolls for the government. The Venetians were the outstanding carriers, running fleets every year in March and September. They offered all-in tours, which included transport, both by sea and land, food and lodging, and all tolls and taxes. A pilgrim who arrived in Venice had to wait only a few days to get a berth. He was reasonably safe since the Venetian navy patrolled the Mediterranean routes and the Moslem governor was well bribed to ensure the protection inland of Venetian-sponsored pilgrims. Tourist-shipowners had to be licensed by the Venetian government in order to work the pilgrim trade, and a copy of their actual contract with the pilgrim had to be posted with a magistrate in Venice to prevent exploitation. The Venetian maritime laws laid down the maximum numbers to be carried in ships and the precise sailing dates.

The pilgrimages to the Holy Land petered out at the end of the fifteenth century, partly because the rapacity of local officials made the Venetian package-tour operations uneconomic, partly because men were ceasing to believe in the mechanistic form of Christian devotion represented by indulgences and the cult of relics, but chiefly because the West was turning outwards to the great oceans – the transatlantic route to America, the Cape route to India and the Far East. The cult of the Holy Places disappeared, to be resumed, in due course, in the form of secular tourism. From the sixteenth to the nineteenth century the Holy Land slumbered: neglected, poor, under-populated, eroded, overtaxed, ill-governed and very largely illiterate. Its dreams were then disturbed both by Messrs Thomas Cook and the early Zionists; but it was not reawakened until General Allenby, in an unusual gesture of Christian humility, entered Jerusalem on foot in 1918.

From that moment began a new cycle in the history and the civilization of the Holy Land. The story of that cycle, which began to gather momentum when the State of Israel was founded in 1948, has only just begun, and is in any case outside the scope of this book. But it fits dramatically into the general perspective of the Holy Land's history, which has always revolved around the clash of rival cultures and the struggle for mastery of these frontier territories between East and West. At the same time, the creation of the Jewish state adds a note of novelty, even of paradox. For Israel, while the heir to the political, economic and indeed military traditions and technology of the West, is also – is it not? – one of the residual legatees of that great Semitic culture which flourished in the Fertile Crescent during remote antiquity. It can thus claim credentials from both East and West, and the further merit of acting as a bridge between them; something the Near East, as we have seen, always needs. But the Arabs, too, are legitimate heirs to the ancient civilizations of the Near East; and they too have helped to make the land holy. The inheritance is wide enough to accommodate both. The history of the Holy Land and its civilizations suggests that such an accommodation will not be reached easily. But it indicates, even more strongly, that truth is many-sided, and that the path of wisdom is best followed in the light of the ecumenical spirit.

The magnificent refectory of the Hospitaller Palace in Acre. Below the column is a secret passage connecting the refectory with the compound.

Bibliography

The best and most authoritative guide to the early history of the Holy Land is the *Cambridge Ancient History*, especially Volumes I–IV in the new Third Edition (1970–5). Other works which I have found particularly illuminating in preparing this book, or which I suggest for further reading, include:

W. F. Albright *Archaeology of Palestine*, London 1960.
P. Alphandery *La Chrétienté et l'idée de la croisade*, 2 vols., Paris 1954–9.
A. Alt *Essays on Old Testament History and Religion*, New York 1968.
E. Anati *Palestine Before the Hebrews*, London 1963.
Thomas Arnold and Alfred Guillaume (eds.) *The Legacy of Islam*, Oxford 1965 ed.
Aziz S. Atiya *Crusade, Commerce and Culture*, Indiana 1962.
D. Baly and A. D. Tushingham *Atlas of the Bible World*, New York 1971.
H. H. Ben-Sasson (ed.) *A History of the Jewish People*, Harvard 1976.
E. R. Bevan *The House of Seleucus*, 2 vols., London 1902.
L. E. Browne *The Eclipse of Christianity in Asia*, Cambridge 1933.
Ivan Engnell *Critical Essays on the Old Testament*, trans., London 1970.
Robin Fedden *Crusader Castles*, London 1950;
— *Syria and Lebanon*, London 1965.
Francesco Gabrieli *Arab Historians and the Crusades*, trans., London 1969.
M. Gaster *The Samaritans*, London 1923.
John Gray *Archaeology and the Old Testament World*, London 1962;
— *A History of Jerusalem*, London 1969.
R. W. Hamilton *The Church of the Nativity, Bethlehem*, Jerusalem 1947.
R. K. Harrison *Introduction to the Old Testament*, London 1970.
H. St-J. Hart *Foreword to the Old Testament*, London 1951.
A. H. Hourani *Minorities in the Arab World*, Oxford 1947.
J. Jeremias *Jerusalem in the Time of Jesus*, London 1969.
A. H. M. Jones *The Cities of the Eastern Roman Provinces*, Oxford 1937;
— *The Herods of Judaea*, Oxford 1938;
— *The Later Roman Empire*, 3 vols., Oxford 1964.
H. J. Katzenstein *The History of Tyre*, Jerusalem 1973.
J. N. D. Kelley *Jerome*, London 1975.
Kathleen Kenyon *Digging Up Jericho*, London 1957;
— *Royal Cities of the Old Testament*, London 1971;
— *Digging Up Jerusalem*, London 1974.
J. Lindblom *Prophecy in Ancient Israel*, Oxford 1962.
Hans Eberhard Mayer *The Crusades*, trans., Oxford 1972.

Enid B. Mellor (ed.) *The Making of the Old Testament*, Cambridge 1972.
J. Maxwell Miller *The Old Testament and the Historian*, London 1976.
M. Noth *The History of Israel*, trans., 2nd ed., London 1972.
James Parkes *A History of Palestine from 135 AD to Modern Times*, London 1959.
Anne Perkins *The Art of Dura Europos*, Oxford 1973.
James B. Pritchard *Gibeon: Where the Sun Stood Still*, Princeton 1962.
Gerhard von Rad *The Problem of the Hexateuch and Other Essays*, trans., Edinburgh 1966.
Jonathan Riley-Smith *What Were the Crusades?* London 1977
M. Rostovtsev *Caravan Cities*, Oxford 1932.
H. H. Rowley *From Moses to Qumran: studies in the Old Testament*, London 1963.
Alfred Rubens *A History of Jewish Costume*, London 2nd ed., 1973.
J. J. Saunders *Aspects of the Crusades*, Canterbury 1962.
R. C. Smail *Crusading Warfare*, London 1956.
M. Smith *Palestine Parties and Politics that Shaped the Old Testament*, New York 1971.
M. Stern *Greek and Latin Authors on Jews and Judaism*, Jerusalem 1974.
W. W. Tarn and G. T. Griffith *Hellenistic Civilisation*, London 1952.
V. Tcherikover *Hellenistic Civilisation and the Jews*, Philadelphia 1959.
H. St-J. Thackeray *Josephus, the Man and the Historian*, New York 1929.
R. de Vaux *The Bible and the Ancient Near East*, trans., London 1972.
M. Weinfeld *Deuteronomy and the Deuteronomic School*, Oxford 1972.
G. E. Wright *Shechem: the Biography of a Biblical City*, London 1965.
Yigael Yadin *Hazor: the Rediscovery of a Great Citadel of the Bible*, London 1975.
S. Zeitlin *The Rise and Fall of the Judean State*, 2 vols., Philadelphia 1962–7.

Acknowledgments

The illustrations in this book appear by kind permission of the following museums, agencies and individuals. References to colour illustrations are printed in *italics*.

Ashmolean Museum 11, 15;
Biblioteca Ambrosiana, Milan 80 above;
Bibliotheque Royale, Brussels 189;
Y. Braun 165;
British Library 80 below (Add. Ms. 43725 ff 72v–73);
British Museum 8, 14, 21, 30, 33, 50, 56, 63, 65, 69, 71 above, 71 below, 75, 84, 115;
Françoise Foliot/Bibliothèque Nationale 53;
John Freeman 59;
David Harris 2, 20, 28, 46, 57, 67, 113, 116, 129, 133, 134 below, 187;
Israel Government Tourist Office 158, 171;
Joint Expedition to Shechem 39;
A. F. Kersting 26–7, 91, 93 *below*, 94–5, 98, 101, 132 left, 137, 139, 142–3, 144, 154, 155, 159, 175, 177, 180, *197*, 199, *203*, 205, 208–9, 214;
Mansell Collection 86, 92 above, 108, 110 above, 110 below, 125, 147 left, 147 right, 150, 153 left, 157, 166 above, 166 below, 184, 192 left, 192 right, 193 above, 195, *200 above*, 202;
Middle East Archive 6, *25 below*, 44, 68, 93 above, *161*, *164*, 178 above, 178 below, 186, 193 below, *200 above*;
National Museum, Damascus 134 above;
Réunion des musées nationaux *25 above*;
Ronald Sheridan *1*, 18, 23, 36, 48, 61, 73, 77, 92 below, 96, 104, 123, 127, 132 right, 152, 153 right, *162–3*, 168, *198*, *200 below*, 204;
State Antiquities Organization, Baghdad 35;
Victoria and Albert Museum 182;
Weidenfeld and Nicolson Archive 130 left, 130 right, 140;
Yigael Yadin 128.

Index

Page numbers in *italic* refer to the illustrations and their captions.

Aaron, 60, 78
Abbasids, 179, 180, 182, 185
Abdul Malik, Caliph, 174
Abimelech, 35
Abishua, 78
Abraham, 32–3, 34–7, 38, 51, 140, 179; *36*
Absalom, 36, 52, 55
Abu Bakr, Caliph, 169–70
Achish of Gath, 52
Acre, 183, 192, 194, 202, 213; *193, 214*
Actium, 109
Adhemar, Bishop of Le Puys, 191
Adonijah, 55
Adonis, 155
Adora, 105
Aegean, 9
Aelia Capitolina, 131, 145, 146, 148
Agrippa I, 112
Agrippa II, 124, 125–6
Ahab, King of Israel, 68–70
Ahasuerus, 135
Ahiran, King of Byblos, 16
Ain Karim, 204
Ain Qadais, 40, 43
Ain Qudeirat, 43
Ajnanain, 170
Akhenaten, 21, 42
Akkadian, 16
Al-Aziz, 183
Alalakh, 35
Albright, W.F., 21, 37, 45, 64
Aleppo, 179–80, 192, 201; *180, 193, 197*
Alexander, Bishop, 148
Alexander Jannaeus, High Priest, 103, 105, 107, 112

Alexander the Great, King of Macedonia, 22, 23, 87–9, 90, 97; *86*
Alexandreion, 104
Alexandria, 79, 89, 98, 99, 100, 112, 170
Ali, Caliph, 172
Allenby, General, 9, 215
Alp Arslan, Sultan, 189
Amalfi, 206
Amarna, 47
Amarna Letters, 20–1, 38, 39, 45; *21*
Amelekites, 40
Amman, 97, 179; *98*
Ammon, 40, 44, 54, 55, 76
Amorites, 8, 15, 24, 47
Amos, 63, 82
Amrit, 29
Amun, 62
Ananias, 125
Anatolia, 9, 10, 14, 190
Annals of At-Tabari, 170
Anthony, Mark, 109
Anthony, St, 156, 157
Antigonus, 109, 112
Antioch, 90–1, 99, 106, 112, 113, 118, 152–3, 201; *91, 154*
Antiochus II, King of Syria, 90
Antiochus III, King of Syria, 97, 98–9
Antiochus IV Epiphanes, King of Syria, 90, 102, 106
Antipater, 107, 109
Antipater of Sidon, 99
Apamea, 91
Apochrypha, 79, 85
Apollo, 90, 97
Aqaba, Gulf of, 54, 66, 176, 206

Aqiba, Rabbi, 79, 83, 130, 131, 132
Aquila, 146
Aquinas, St Thomas, 147
Arabia Petraea, 136, 145
Arabs, 76; architecture, 173–80; invade Holy Land, 167, 169–80, 182–3; learning, 180–2; monophysites, 167
Arameans, 9
Araunah the Jebusite, 58
Archelaus, 118, 124
Aristeas, 117
Aristobulus III, 112
Aristotle, 182; *Politics*, 89
Ark of the Covenant, 50, 55, 59, 60–2, 70, 135; *61*
Armageddon, 9
Armenia, 189, 212
Armenian Church, 167
Arsinoe, 90
Artaxerxes I, King of Persia, 76
Arvad, 23
Ascalon, 50, 113, 146, 172, 183, 192
Asclepios, 97
Ashdod, 50, 53
Asher, 65
Asheret, 29
Ashtaroth-Karnaim, 71
Ashur, 33
Ashurbanipal I, 34, 65
Ashurnasipal II, 56
Assassins, 201–2, 211–12
Assize of Jerusalem, 196
Assumption of Moses, 112
Assyria, 8, 61, 135; *65*; battle of Karkar, 70; campaigns against Holy Land, 72–3;

218

INDEX

clothing, 65; eponym lists, 53; law-codes, 33; rise of, 22; sacks Samaria, 69
Astarte, 29
Atargatis, 138
Athens, 48, 113
Attalus, 89
Augustine, St, 147, 152; *The City of God*, 105
Augustus, Emperor, 109–11, 113–14
Aurelian, Emperor, 141
Avaris, 39
Avicenna, 182
Avtina family, 117
Azequah, 74

Baal, 15, 29, 69
Baal Hamman, 45
Baal Malkart, 100
Baalbeck, 138; *139*
Babel, Tower of, 32
Babylon, 8, 120; besieges Tyre, 23; chronology, 37; Jewish exile in, 70, 73–5, 100; law–codes, 32; Moslem conquest, 170
Babylonian Chronicle, 73–4, 75
Baghazkoy, 24
Baghdad, 24, 112, 179–83, 185, 201
Baibars, 91
Baldwin I, King of Jerusalem, 191; *192*
Baldwin II, King of Jerusalem, 206
Baldwin IV, King of Jerusalem, 196
Banias, 97, 114
Bar-Kochba revolt, 130–1, 145; *130*
Barada, river, 176
Baradi, Jacob, 167
Barsauma, 158
Basil, St, 157
Battir, 131
Beatitudes, Mount of, *165*
Beaufort, 207
Beersheba, 35–6, 72
Beha ad-Din, 192
Beit Jibrin, 146
Beit Mersim, 56
Belvoir, 207
Ben Asher family, 79–81
Ben David, Ahan, 173
Ben Gioras, Simon, 126
Ben Jacob, Samuel, 79-81
Ben Kalir, Eleazar, 172
Ben Naphtali family, 79–81
Ben Sirach, 85, 98, 120
Ben Torta, Rabbi Johanan, 130
Ben Yair, Eleazar, 126

Ben Zakkai, Johanan, 130
Ben Zakki, 132
Benaiah, 83
Beni-Hassan, 63–4
Benjamin, 65
Berenice, 124, 125
Berlin Papyrus, 42
Bernard of Clairvaux, St, 194–5, 206
Bernard the Monk, 185
Beroea, 179
Berossus of Babylon, 37, 99
Beth Alpha, 135
Beth Gibelin, 210
Beth She'an, 19–20, 21, 44, 47, 56, 58; *20*
Beth She'arim, 135
Bethar, 131
Bethel, 45, 64, 74
Bethlehem, 131, 155, 156, 188; *155*
Bethshemesh, 56
Bethsur, 104
Bilga family, 103
Black Obelisk of Shalmaneser III, 64, 72
Boethus family, 112
Bohemond IV, Prince of Antioch, 201
Bohemond of Taranto, 191
Bona da Ferrara, *157*
Bordeaux Pilgrim, 151, 153
Bosra, 140, 145
British Museum, 13, 64, 65, 73, 75, 81, 201
Brook Cherith, 160
Brook of Kedron, 54
Burchhardt, 136
Byblos, 16–18, 21, 22, 35, 58, 97, 113; *18*
Byzantium, 36; Christianity, 165–7; and the Crusades, 189–91, 211; leaves Holy Land, 167, 169, 170; monks, 157–9; rules Holy Land, 145–6; tries to regain Holy Land, 183, 185

Cabbala, 100
Caesarea, 114, 118, 123, 125, 126, 146–7, 170, 172, 183; *113, 123, 130*
Cairo, 173, 201, 212
Callirrhoe, 118
Canaan, 13–20, 29, 43–9
Capernaum, 133; *61, 134*
Carolingians, 185
Carthage, 23, 29, 88, 170
Cassius, 109
Catal Huyuk, 10
Celts, 111
Cestius Gallus, 126
Chahba, 140
Chalcis, 153

Chansons des Chetifs, 202
Charlemagne, Emperor, 185; *184*
Charles Martel, 170
Chemosh, 40, 66–8
China, 181, 182, 189
Christianity, 106; Crusades, 183, 188–213; divergence from Judaism, 135, 146; early churches, 148–52, 159–65; and fall of Jerusalem, 129; monks, 156–9; in Moslem Palestine, 171–2; origins of, 123, 146–7; pilgrimages to Holy Land, 185–8
Chronicles, First Book of, 55
churches, Crusaders, 204–5; Jerusalem, 148–52; mosques, 173–4; synagogues, 131, 133–5
Cicero, 153
Citium, 106
Cleopatra, 111, 141
Cluniacs, 190
Code of Gortyn, 81
Codex Alexandrinus, 81
Codex Sinaiticus, 81; *80*
Codex Vaticanus, 81; *80*
Coelesyria, 98
Constantine, Emperor, 147, 150, 151, 155; *150, 155*
Constantinople, 145, 170, 171, 180, 188, 211
Cook, Thomas, 215
Coptic Church, 165, 212
Cornelius, 146
Council of Chalcedon (451), 158, 165–7
Council of Ephesus (449), 147, 158
Council of Jamnia, 83
Council of Nicea (325), 150
Council of Troyes, (1128), 206
Council of Vienne (1303), 201
Crac des Chevaliers, 207–10, *199*
Crete, 9, 49, 65, 81
Crusades, 36, 152, 173, 183, 188–213; *195*
Cuspius Fadus, 124
Cypros, 114
Cyprus, 59, 88, 196
Cyrenaica, 170
Cyril of Scythopolis, 157
Cyrus the Great, King of Persia, 75–6, 121; *75*

Dagon, 15
Damascus, 9, 54, 55, 69, 113, 135, 172, 176–9, 181, 192, 201; *177, 178, 200*
Damasus, Pope, 153
Dan, 65

219

INDEX

Daniel, 79, 102
Daniel, Book of, 105–6, 120
Darius I, King of Persia, 75, 76
David, King of Israel, 52–5, 58, 64, 82, 121, 135; *53*
Dead Sea, 153, 206, 207
Dead Sea Scrolls, 81, 102, 120–1; *101*
Deborah, Song of, 38
Decapolis, 20, 105, 138, 145
Decian, Emperor, 147
Delos, 90, 100
Demetrius, 90
Deuteronomy, Book of, 42
Didyma, 106
Dio Cassius, 131
Diocletian, Emperor, 145
Dionysias, 140
Dionysos, 97
Diospolis, 146
Dog River, 92
Dophkah, 43
Dothan, 44
Druses, 138–40, 183
Dura Europos, 135, 141; *134*
Dushara, 136

Ebal, Mount, 38
Ebers Papyrus, 42
Ebionites, 146
Ecclesiastes, 84
Ecclesiasticus, 85, 120
Edessa, 191, 192
Edom, 44, 54, 74, 84–5, 136; *137*
Edward I, King of England, 180
Edwin Smith Medical Papyrus, 42
Egypt, 8, 49; chronology, 53; civilization in, 13; dietary regulations, 42–3; *Execration Texts*, 19, 20, 45; exodus from, 39–40, 43; Greek colonies, 89; Hebrew migration to, 37, 39; Jewish community, 77, 173; king-lists, 37; literature, 81, 82; Mamluks, 212; medical papyri, 42; monasticism, 157; Moslem conquest, 170; prophets, 51; religion, 62; rules Holy Land, 9, 14, 18–21; trade with Byblos, 16–19; writing, 16
Eilat, 67
Ekron, 50
El Elyon, 40
Elah, Vale of, 170
Elburtz mountains, 211
Eleazar, 127–9
Eleazir, 102

Eliezer, 34
Elijah, 60, 70, 157
Elim, 43
Elisha, 51, 70
Elutheropolis, 146
Emmaus, 204
Engedi oasis, 131
Enki, 33
Ephesus, 100
Ephraim, 74
Esau, 34
Esdraelon, Plain of, 9, 52, 135
Eshmoun, Temple of, 58
Eshnunna, 24
Essenes, 121
Esther, Queen, 135
Euclid, *Elements*, 181
Eudoxia, Empress, 151, 152
Eunapius, 141
Eusebius, 148–50, 202; *147*; *Ecclesiastical History*, 147, 148; *Onomasticon*, 38, 148
Eutychides, 90, 106
Evagrius, 153
Even Shetiyah, 59–60
Exodus, 39–40, 43, 135
Exodus, Book of, 40
Ezekiel, 51, 74, 75
Ezekiel, Book of, 23
Ezion-Geber, 66
Ezra, 75, 78, 79, 118

Fatamids, 183, 189, 191, 211
Fedden, Robin, 179, 206
Felix, 124
Fertile Crescent, 8, 212, 215
Flavian, 158
flood myths, 33–4; *8, 33*
Florus, 124
Fortmund, 186
France, Crusades, 189–90, 191–5, 204; hospitals, 181; Moors in, 170; pilgrimages, 187, 188
Franciscans, 215
Frederick II, Emperor, 194
Frederick Barbarossa, Emperor, 192
Fulcher of Chartres, 191, 195
Fulk, Count of Anjou, 187

Gad, 65
Galen, 181
Galilee, 109, 118, 131, 145, 172
Galilee, Sea of, 10
Gamala, 97
Gath, 50
Gaulanitis, 105
Gaza, 21, 44, 50, 97, 136; 157, 172, 204
Geba, 72

Genesis, Book of, 31, 33, 34, 37, 40
Genoa, 210
Gerard of Cremona, 181
Gerasa, 138
Gerasimus, St, 156
Gerizim, Mount, 38, 49, 60, 72, 78, 97, 104, 136
Gezer, 21, 48–9, 54, 55, 56–7, 103; *57*
Al-Ghazzali (Algazel), 182
Gibbon, Edward, 141
Gibeon, 47–8, 54
Gibeon, pool of, 48; *48*
Gideon, 194
Gihon spring, 54, 73; *73*
Gilboa, Mount, 52, 194
Gilead, 10; *71*
Gilgal, 44, 47, 50, 82
Gilgamesh, 21, 33, 34; *33*
Glueck, Nelson, 66
Greece, 16, 49, 51, 87–106, 113, 194
Gunther, Bishop of Bamburg, 188
Gush-Halav, 135

Habiru, 37–8, 40
Hacilar, 12
Haddu, 29
Hadrian, Emperor, 130, 131, 135, 138, 148, 155
Haggai, 76
Al-Hakim, Caliph, 151, 183, 187–8, 190
Hakotz family, 103
Halebiyah, 167
Halicarnassus, 81, 88
Hammurabi, 24, 37, 45; *25*
Hammurabi, Code of, 24, 32, 42, 78, 81
Hananel, 112
Hanassi, Rabbi Yehuda, 135
Harding, G.L., 121
Harun al-Rashid, Caliph, 181, 185
Hasidim, 102, 103
Hasmon, Mattathias, 102
Hasmoneans, 100, 102–5, 107, 109, 111, 121
Al-Hassan, 173
Hassan i-Sabbah, 211
Hatti, 29
Hattin, Battle of, 192, 210
Hauran, 85, 111, 138
Hazor, 44, 45–7, 56, 66, 72; *46*
Hearst Papyrus, 42
Hebrews, conquest of Canaan, 43–9; early history of, 31–43; law codes, 32–3; campaigns against Philistines, 49–50, 52, 53; *see also* Jews
Hebron, 36–7, 53, 54, 204; *36*

INDEX

Helena, St, 150, 151, 155, 179; *153, 155*
Heliopolis, 138
Helios, 97
Henry IV, Emperor, 190
Heracles, 100
Hermes, 97
Herod Agrippa, 118
Herod Antipas, 118, 132
Herod the Great, 36, 45, 58, 68, 104, 107, 109–18, 122, 124, 126–7, 160; *108, 133, 129*
Herodium, 114
Herodotus, 81
Hezekiah, 72–3; *71, 73*
Hezekias, 109
Hijaz Oasis, 85
Hilarion, 157
Hildebrand, Archdeacon, 190
Hippocrates, 181
Hiram, King of Tyre, 22, 55, 58; *56*
Hirbet el-kheleifeh, 66
Hisham, 45
Hittites, 9, 14, 21, 24, 43, 49, 81, 179
Homer, 22, 141
Horonites, 76
Hospitalers, 206–7
Hugh of Payens, 206
Hulagu Khan, 194
Hyksos, 9, 13, 14, 18, 19, 39, 44
Hyrcanus, 107, 109

Ibleam, 48
Ibn Hisham, 173
ibn-Masawayh, Yuhanna, 181
Ibn-Rushd (Averroes), 182
ibn-Tulun, Ahmad, 183
Ibni-Adad, 45
Idi, 24
Al-Idrisi, 181
Idumaea, 105, 107, 126, 136, 145
Imhotep, 42, 97
Isaac, 32, 36, 135
Isaac the Jew, 185
Isaiah, 23
Ishtar, 61–2
Isidore of Milo, 167
Isis, 97
Islam, 167; controls Holy Land, 167, 169–80, 182–3; Crusades against, 188–213; divisions in, 211–12; tombs of the patriarchs venerated by, 36–7
Ismailians, 211–12
Issus, battle of, 97
Italy, 49, 87, 189–90, 192, 210–11

Ituraea, 188

Jabin, 45
Jacob, 32, 34, 36, 37, 38, 60, 65
Jacobite Church, 167, 212
Jaffa, 103, 206
James of St George, 180
Jamnia, 130, 131
Jarmo, 10
Jason High Priest, 100
Jebel Druze, 111, 138–40
Jebel Usdum, 35
Jebusites, 54, 62
Jehoiakim, King of Judah, 74
Jehoshaphat, valley of, 187, 205; *153*
Jehu, King of Israel, 64, 69–70, 72; *65*
Jemdet Nasr, 34
Jerash, 138; *140*
Jeremiah, 74, 79
Jericho, 9–13, 44–5; *6, 11, 14*
Jerome, St, 131, 151, 152–6, 158, 202; *154, 157*
Jerusalem, 21; *77, 93–6, 104, 110, 132, 161–4, 168, 175, 189, 193, 203, 204*; Babylon captures, 73–4; Church of the Holy Sepulchre, 148–51, 183, 202; *144, 171, 186*; Crusades, 191, 192–4, 201, 202, 205; David conquers, 54–5; early Christians in, 146, 147–56; Hebrew capture, 44; Herod the Great and, 109, 114–18, 124; importance of, 9; under Greek rule, 100–2; under Islamic rule, 170, 172–4, 182–3; Jewish War, 125–31; under Maccabees, 102–4; pilgrimages to, 148–9, 151, 185–8; Seljuks capture, 189; under Solomon, 57–61, 64; water supply, 73
Jerusalem, Temple, *116*; Solomon builds, 45, 57–61, 64–5; *59*; law-books discovered in, 72; Romans destroy, 74, 114; restored under Persians, 76; and Seleucids, 98–9; Herod rebuilds, 115–18
Jerusalem National Museum, 69
Jerusalem Talmud, 132
Jesus Christ, 65, 122–4, 133, 148, 151, 173, 187; *132, 134, 187*
Jewish War, 125–31, 135

Jews, anti-Semitism, 190; early synagogues, 131, 133–5; exile in Babylon, 70, 73–5, 100; in Moslem Palestine, 172–3; relationship with Romans, 118–21; under Greek rule, 97–103; under Persian rule, 75–7; *see also* Hebrews
Jezebel, 68–9, 100
Jezreel, 9, 70
El-Ji, 136
El-Jib, 47–8
Joab, 54
Job, 106
Job, Book of, 84–5
John the Almsgiver, 207
John the Baptist, St, 157, 176, 204, 207
John of Byzantium, 167
John of Gischala, 126
John Hyrcanus, High Priest, 103, 104–5, 107, 113, 136
Joinville, Jean, Sire de, *Life of St Louis*, 201–2
Jonathan, 52
Jonathan, High Priest, 103, 104, 126; *104*
Joppa, 21
Joram, King of Israel, 70
Jordan, River, 121, 186
Joseph, 38, 65
Joseph (son of Tobias), 98
Josephus, 38, 54, 55, 65, 83, 103–5, 111–12, 114–18, 121, 123; *125*; *Antiquities of the Jews*, 98, 99; *Contra Apionem*, 78; *Jewish War*, 125, 126–8, 148
Joshua, 10, 38, 39, 40, 44–9, 56, 58, 105; *46*
Joshua, Book of, 44
Josiah, 72
Jotapata, 125
Judaea, 98; under Greek rule, 97, 100–3; under Herod the Great, 109–18; under Roman rule, 118–21, 145; Jewish War, 124–31
Judah, 65, 66, 70–3
Judaism, development of, 74; divergence from Christianity, 135, 146; Greek view of, 100; Messiah, 121–4; Pharisees and, 107
Judas of Gamala, 124
Judas Maccabeus, 102, 104
Julian, Emperor, 106, 158
Jupiter, 29
Justin Martyr, 130
Justinian, Emperor, 81, 90, 151, 155–6, 165–7; *155, 166*

221

INDEX

Juvenal, Bishop of Jerusalem, 158

Kadesh, battle of, 14
al-Kaim, Caliph, 189
Kairouan, 173, 177
Kalat Seman, 158–60
Kanawat, 140
Karaites, 173
Karchemish, 73, 179
Karkar, battle of, 70
Karnak, 47, 50
Kaufman, Yehezkel, 42
Kedron Brook, 73
Kefar-Biram, 135
Kenyon, Dame Kathleen, 10, 12, 13, 45, 54, 58, 103, 148; *6*
Kerak, 207
Khattusas, 24
Khirbet al-Yahud, 131
Khirbet Bel'ameh, 48
Khirbet et-Tubeiqeh, 104
Khirbet Mefjer, 45, 179; *178, 200*
Khirokitia, 12
Khorsabad Annals, 72
Al-Khuwarizmi, 181
Kings, Book of, 57, 58
Kish, 34, 64
Kitbugha, 194
Knights of the Holy Ghost, *195*
Koran, 170, 173, 201
al-Kufah, 172, 173
Kuyunjik, 8
Kurds, 212

Laban, 34
Labaya, 38–9; *21*
Lachish, 45, 47, 56, 58; *30*
Lachish Letters, 74
Lantfrid, 185
Laodicea, 113
law-codes, 24, 42, 81
Lawrence, T.E., 207
League of Nations, 207
Lebanon, 16–18, 53, 138–40, 206, 212
Leo VI, King of Armenia, 212
Leo the Isaurian, Emperor, 170
Letter of Aristeas, 117
Levi, 65
Levites, 64, 116–17
Leviticus, Book of, 42, 43, 118
Libya, 65
Lipir–Ishtar Lawcode, 81
Longinus, Cassius, 141
Lorraine, Duke of, 191
Louis XI, King of France, 202

Lycia, 113
Lydda, 64, 146, 171, 204

Maccabean revolt, 100, 102–3, 120
Macedonia, 68, 87
Machaerus, 114
Madaba, 131
Madaba Map, 152; *93*
Magharians, 121
Magnesia, battle of, 88
Maimonides, 133
Makarius, Bishop, 150
Malichus, 109
Mallowan, Sir Max, 69
Maltaia, 61
Mamluks, 91, 194, 212, 213
Al-Mamum, 180
Manetho, 37, 99
Manizkert, battle of, 189
Mar Saba, 157; *158*
Marduk, 75
Mari, 24, 34, 35, 45
Mariamne, 111–12, 115
Marissa, 97, 105
Maronite Church, 212, 213
Masada, 114, 125, 126–9; *128, 129*
Massorah, 172
massoretes, 79–81
Masyaf, 211
Mecca, 169, 174, 179
Medes, 75
Medina, 169, 172, 173, 174
Mediterranean, 87, 141
Megiddo, 9, 21, 44, 48, 54, 56, 57, 59, 64, 66, 70, 72
Meir, Rabbi, 133
Melito, Bishop of Sardis, 148
Melkart, 22
Melkites, 167, 169
Menachem, 125
Menahem, King of Israel, 72
Menelaus, High Priest, 102
Meneptah stele, 39–40
Merom, waters of, 45–7; *46*
Meskalamdug, 32
Mesopotamia, 7–8: civilization in, 13; flood epics, 33; human sacrifices, 29; Jews in, 77; law-codes, 32; origins of Hebrews in, 37–8; persecuted minorities, 169–70; trade routes, 176
Messiah, 60, 121–4, 173
Michael VII, Emperor, 190
Michael the Paphlegonian, Emperor, 188
Midianites, 194
Midrash, 172
Milkom, 49
Millo, 55, 57, 74, 76
Mishnah, 132, 135

Mitanni, 21, 49
Mizpah, 74
Moab, 10, 40, 44, 54, 66–8, 207
Modein, 102
Mohammed, 169–70, 173
Monastery of St George, 160; *25*
Monophysites, 165–7
Moscus, John, 157
Moses, 40–3, 51, 60, 100, 135, 153; *28*
Muawiyah, 172
al-Muqaddasi, 188
Mycenae, 48, 49

Nabateans, 74, 105, 136, 167
Nabi Samwill, 204
Nablus, 38, 130, 136
Nabonidus, 75
Nahr el Kelb, 92
Nanna, 24
Naphtali, 65
Nazareth, 160, 204
Neapolis, 38, 130, 146
Nebuchadnezzar, 73, 76
Negev, 153
Nehemiah, 75, 76, 79
Nehemiah, Book of, 76
Nero, Emperor, 124
Nestorian Church, 165, 194
Nicolas of Damascus, 111
Nicopolis, 113
Nile, River, 8
Nimrud, 69; *69, 71*
Nineveh, 34; *30*
Nippur, 33
Nova Trajan, 140
Numbers, Book of, 36
Nur ad-Din, 192
Nuzu, 34; *35*

Oak of Mamre, 36
Odenath family, 141
Odenathus, 141
Old Testament, 31–2, 34, 52, 78–85; *see also individual books*
Olives, Mount of, 124, 151, 152, 172; *152, 153*
Olympic Games, 112–13
Omri, King of Israel, 66–8, 70
Onias, House of, 100
Origen, 146–7, 148, 155; *147*; *Hexapla,* 146
Orontes valley, 70
Ottoman empire, 212, 213

Pachomius, St, 157
Palestine, Christian churches, 160–5; Crusader castles, 206; decline, 212–13; under Fatimids,

INDEX

183; Greek communities, 105; under Greek rule, 97–8; monks, 157; Moslem conquest of, 170–4, 182–3; pilgrimages to, 185–8; religious sects, 121; Semites migrate to, 8
Palmyra, 141; 26–7, 142–3
Pamphilus, St, 147
Panion, 97
Parthians, 109, 135
Paul, St, 85, 100, 105, 107, 122, 123, 146
Paul of Samosata, Bishop of Antioch, 141
Paula, 153–6
Pelagius, 156
Pella, 146, 148
Pentateuch, 31, 42, 60, 70, 72, 78–9
Pergamum, 113
Persia, 58, 75–7, 87, 97, 165, 167, 181
Persian Gulf, 141, 176
Peter, St, 146, 205
Petra, 105, 111, 136, 145, 167; *137*
Petrie, Sir Flinders, 16, 39
Pharisees, 103, 105–7, 118, 122
Phasael, 109, 115
Philadelphia, 97; *98*
Philip (son of Herod the Great), 118
Philip the Arab, Emperor, 140
Philip of Macedon, 88
Philip Augustus, King of France, 54, 192; *193*
Philipopolis, 140
Philistines, 9, 19–20, 49–50, 52, 53; *50*
Philo, 16, 121, 132
Philodemus of Gadara, 99
Phoenicia, 9, 14, 51, 54, 56, 70, 98; *56, 69*
Phoenicia, Alexander the Great conquers, 88; craftsmen, 58–9, 76; influence on Israelites, 64; rise of, 22–3; writing, 16
Pilate, Pontius, 123, 124, 136, 151; *123*
pilgrimages, 148–9, 151, 185–8, 213–15
Pithom, 39
Plato, 141; *Phaedrus*, 51
Poitiers, battle of, 170
Polybius, 98
Polyeidus the Thessalian, 88
Pompey, 107, 109, 136; *110*
Pritchard, James, 47
prophets, 51–2, 79, 120–1
Ptolemais, 113

Ptolemies, 90, 91, 97–8, 100; *92*
Ptolemy I, 89, 90
Ptolemy II, 90
Ptolemy IV Philopater, 98–9
Ptolemy, Claudius, 53; *Syntaxis*, 181
Pumbeditha, 173
Pyramid Texts, 42

Qadesh, 40–2, 43
Qalat Shergat, 32–3
Qarn Sartaba, 104
Qasr al-Hair, 179
Qinnesrun, 153
Qiryat al-Inab, 204
Qumran, 102, 106, 107, 120–1; *101*
Quseir Amra, 179

Rachel, 34
Rad, Gerhard von, 83
Rahab, 45
Ramah, 51
Ramesses II, 14, 39
Ramesses III, 50
Ramleh, 171, 172, 183, 188, 204
Raphia, 98
Ras Shamra, 12, 13–14, 43, 84
Rashid al-Din Sinan, 211
Rav Ammi, 132
Rav Assis, 132
Ravenna, 166
Al-Razi (Rhazes), 181
Red Sea, 54, 207
Rehoboam, King of Judah, 66
religion, Canaanite, 29; early Hebrews, 40–3; Greek, 89–90, 106; Solomon and, 62; *see also* Islam; Judaism
Resafa, 167
Reuben, 65
Reuchlin Codex, 81
Rhodes, 88, 109, 113
Richard I, King of England, 192; *193*
Rizpah, 47
Robert, Duke of Normandy, 188
Robert of Flanders, 191
Roger II, King of Sicily, 181
Roman Catholicism, 62
Roman empire, battle of Magnesia, 88; and Christ, 122–3; destruction of the Temple, 114; Jews revolt against, 124–31, 135; and Maccabean revolt, 103; and the Nabateans, 136; rules Holy Land, 107, 109, 118–40, 145; treatment of Samaritans, 136

Rome, 112; *127*
Ruad, 213

Saachar, 65
Sabas, St, 157; *158*
Sadducees, 100, 103, 105–6, 122
Safed, 207; *133*
Saffuriyeh, 204
Sahyun Castle, 208–9
Sakkara, 42
Saladin, 37, 174, 177, 192, 196, 211–12; *193*
Salome, 107
Samaria, 68–9, 70, 72, 97, 98, 104, 111, 113–14, 145; *68*
Samaritans, 76, 78–9, 136, 158
Samuel, 50–1, 66, 79
Samuel, Books of, 50, 52–3, 61
Sana, 169
Sanhedrin, 109, 111, 122, 130, 135; *110*
Saracens, 207
Sarah, 34
Saragon I, 37
Saragon II, 69, 72
Sasa, 135
Sassanids, 135
Saul, King of Israel, 19–20, 47, 50–1, 52, 53, 121
Scythopolis, 20, 104
Sea Peoples, 14, 21–2, 49, 50
Sebaste, 68, 113–14, 146
Sebastiyeh, 204
Seleucia, 179
Seleucids, 89, 90, 91, 97–103, 121, 179; *92*
Seljuks, 179, 189, 191, 212
Semites, 8–9
Sennacherib, 23; *30, 71*
Septuagint, 79, 146
Serabit el-Khadem, 43
Sergius, St, 167
Sesostris III, 38
Setj-Anu, 19
Severus, Julius, 130
Shaddai, 40
Shalmaneser III, 64, 70
Sheba, Queen of, 66
Shebna, 63
Shechem, 38–9, 44, 49, 53, 54, 64, 66, 72, 97; *39*
Sheshbazzar, 76
Shia, 172, 211, 212
Shiloh, 50
Shuruppak, 33–4, 37
sicarii, 124, 127, 211
Sicily, 87, 182, 189–90, 196
Sidon, 22, 23, 97, 113, 192, 207; *198*
Sigismund, 185

223

INDEX

Siloam, 63; *63*
Siloam Tunnel, 73; *73*
Silva, Flavius, 126–7; *128*
Silwan, 63
Simeon, 65
Simon, High Priest, 103
Simon Maccabee, 120
Simon Stylites, St, 158–60; *159*
Sinai, 16, 146
Sinai, Mount, 43, 60; *28, 44*
Sodom, 35, 154
Solomon, King of Israel, 48, 53, 55–63, 66, 68, 70, 82, 83–4, 111, 140, 170; *53, 57*
Song of Solomon, 83
Sophronius, Patriarch, 170
Soueida, 140
Spain, 170, 172, 173, 182, 185, 189–90, 196
Sparta, 113
Starkey, J.L., 45, 74
Stephen St, 151
Stephen of Blois, 191
Stoic philosophy, 106–7
Strabo, 141; *Geography*, 100
Straton's Tower, 97, 114
Sueibe, 207
Suleiman the Magnificent, 148, 151, 174
Sumer, 16, 24
Sunnis, 172, 211
Surah, 170, 173
Susa, 24
Symmachus, 146
Syracuse, 88
Syria, 9, 49, 65; Christian churches, 160; Crusader castles, 206; decline, 212–13; early churches, 174; glass industry, 201–2; Greek communities, 105; under Greek rule, 97; under Islam, 176–80; migrations to, 44; monks, 157; persecuted minorities, 169–70; religious sects, 121; under Roman rule, 145; Semites migrate to, 8

Taanach, 44
Tabariyah, 171
Tabgha, 160
Tabor, Mount, 160
Tacitus, 62–3
Talmud, 60, 118, 120, 131, 173
Tanis, 36, 39
Tanit, 45

Tartars, 194
Tel Tainet, 58
Tell El-Amarna, 14
Tell el-Retabeh, 39
Tell es-Sandahannah, 105
Templars, 206, 207, 210
Temptation, Mount of, 160
Thebes, 19, 39–40; *50*
Theodora, Empress, 167; *166*
Theodore of Mopsuestia, 85
Theodoret of Tyre, 159
Theodosius, 146, 152
Theodosius I, Emperor, 91
Theodotion, 146
Theophanes, 170
Therapeuta, 121
Theudas, 124
Tiberias, 132–3, 171, 172; *133*
Tiglath-Pileser III, 57, 72; *71*
Timisces, John, 183
Tirzah, 72
Tishpack, 24
Titus, Emperor, 126; *127*
Tobiad family, 103
Tobias, 97
Tobias, House of, 100
Torah, 78, 79, 105, 122, 131
Toron, 207
Tortosa, 205–6; *205*
Toulouse, Count of, 191
Trajan, Emperor, 135, 136, 138, 140
Tripoli, 97, 113, 191, 192, 202, 207
Tughril Bey, Sultan, 189
Turks, 189, 191, 194, 206, 211, 212
Tuthmosis III, 20
Tyre, 29, 53, 66, 100; rise of, 22–3; fall of, 23, 70, 88–9; *86*; Hellenization, 97; Herod the Great and, 113; Crusaders in, 192, 202, 205; temple, 23

Ugarit, 13–16, 22, 29, 44
Umar, Caliph, 170
Umayyads, 172–9, 180, 182, 185
Umayyah, 172
Ur, 32, 34, 37, 64, 75
Ur Nammu, 24, 32, 37, 81
Urban II, Pope, 190
Usama, 170
Usama, Amir of Shaizar, 204
Usha, 131
Uthman, Caliph, 172

Vaux, Père de, 10, 121
Venice, 201, 210–11, 213, 215
Venus, 148, 155
Veronese, Paolo, *153*
Vespasian, Emperor, 38, 126, 129–30, 132, 135
Vester family, 73

Wadi Gharandel, 43
Wadi Kelt, 160; *25*
Wadi Qilt, 114
Waleed, Caliph, 174, 176
Walid, Caliph, 172
'Weld-Blundell Prism', 15
Wellhausen, Julius, 31, 32, 34, 42, 60
Wenamun, 51
Wilhelm II, Kaiser, 177
William, Archbishop of Tyre, 191, 201, 205; *History of Outremer*, 202
William I, King of England, 190, 191
William of Tripoli, *Tractatus de Statu Sarecenormum*, 202, *The Wisdom of Amenope*, 84; *84*
Woolley, Sir Leonard, 32, 33–4
Wright, G.E., 38–9

Xerxes, King of Persia, 76

Yadin, Yigael, 45, 47, 56
Yahweh, 40–3, 49, 50–1, 55, 60–2, 70, 72, 74–5
Yaqar-Ammu, 19
Yarmuk, battle of the, 167, 170
Yebneel, 130
Yemen, 169
Yenoam, 44
Yorghan Tepe, 34

Zadokites, 100
Zaher, El-Malek-es, 180
zealots, 124, 125, 126–9; *128*
Zebulun, 65
Zechariah, 76
Zedekiah, 74, 121
Zeno, 106–7
Zenobia, Queen, 141
Zerubbabel, 58, 76
Zeus, 29, 90, 97
Zeus Olympus, 102
Zionism, 215
Ziusudra, 33
Zoan, 36
Zoroastrianism, 75, 169–70